Island Studies Series

Series Editor
Lino Briguglio

Small Worlds, Global Lives

Island Studies Series

Series Editor
Lino Briguglio
Islands and Small States Institute, Foundation for International Studies, University of Malta

The *Island Studies* series focuses on issues which particularly affect small inhabited islands, with special reference to politically independent ones.

The study of islands, as distinct from other geographical entities, is developing as a special area of interest because islands tend to face special problems associated with smallness, insularity, fragile eco-systems and proneness to natural disasters, which render them very vulnerable in the face of forces outside their control. This condition sometimes threatens their very economic viability.

Small island states also offer unique perspectives for study, especially in the areas of tourism and leisure, geography, anthropology, sociology, economics, the environment and sustainable development in general.

The volumes in this series include studies of a general nature and case studies, aimed at scholars and practitioners engaged in the study and in the management of small islands.

Also published in the same series:

Sustainable Tourism in Islands and Small States: Issues and Policies
Edited by Lino Briguglio, Brian Archer, Jafar Jafari and Geoffrey Wall

Sustainable Tourism in Islands and Small States: Case Studies
Edited by Lino Briguglio, Richard Butler, David Harrison and Walter Leal Filho

Banking and Finance in Islands and Small States
Edited by Michael Bowe, Lino Briguglio and James W. Dean

Insularity and Development: International Perspectives on Islands
Edited by Emilio Biagini and Brian Hoyle

Small Worlds, Global Lives

Islands and Migration

Edited by
Russell King
and
John Connell

PINTER

London and New York

First published 1999 by
Pinter, *A Cassell Imprint*
Wellington House, 125 Strand, London WC2R 0BB
370 Lexington Avenue, New York, NY 10017-6550

British Library Cataloguing in Publication Data

A catalogue record for this book is available from the British Library
ISBN 1–85567–548–X

Library of Congress Cataloging-in-Publication Data

Small worlds, global lives: islands and migration/edited by Russell King and John Connell.
   p.   cm. — (Island studies series)
  Includes bibliographical references and index.
  ISBN 1–85567–548–X (hc.)
  1. Emigration and immigration. 2. Population geography.
3. Islands. I. King, Russell, 1945– . II. Connell, John.
III. Series: Island studies series (London, England)
JV6035.S53 1999
304.8—dc21
                                             99–17802
                                                 CIP

Typeset by York House Typographic Ltd, London

Printed and bound in Great Britain by The Cromwell Press, Trowbridge, Wilts.

Contents

Contributors   vii

List of Figures   ix

List of Tables   x

1   Island Migration in a Changing World   1
John Connell and Russell King

2   From the Periphery of the Periphery:
Historical, Cultural and Literary Perspectives
on Emigration from the Minor Islands of Ireland   27
Stephen A. Royle

3   The Azores: Between Europe and North America   55
Allan M. Williams and Maria Lucinda Fonseca

4   Emigration and Demography in Cape Verde:
Escaping the Malthusian Trap   77
Annababette Wils

5   Insiders and Outsiders: The Role of Insularity,
Migration and Modernity on Grand Manan, New Brunswick   95
Joan Marshall

6   Migration as a Way of Life: Nevis and the Post-war Labour
    Movement to Britain   115
    Margaret Byron

7   The Breath of 'The Beast': Migration, Volcanic Disaster, Place
    and Identity in Montserrat   137
    Stuart B. Philpott

8   Caribbean Migration: Motivation and Choice of Destination
    in West Indian Literature   161
    Brian Hudson

9   Between the Devil and a Warm Blue Sea: Islands and the
    Migration Experience in the Fiction of Jamaica Kincaid   177
    Rachel Hughes

10 'My Island Home': The Politics and Poetics of the Torres Strait   195
    John Connell

11 Speaking of Norfolk Island: From Dystopia to Utopia?   213
    Peter Mühlhäusler and Elaine Stratford

12 Islanders in Space: Tongans Online   235
    Helen Morton

13 Spatial Population Mobility as Social Interaction:
    A Fijian Island's Multi-local Village Community   255
    Carsten Felgentreff

14 The Changing Contours of Migrant Samoan Kinship   277
    Cluny and La'avasa Macpherson

15 Islands as Havens for Retirement Migration: Finding a
    Place in Sunny Corfu   297
    Gabriella Lazaridis, Joanna Poyago-Theotoky and Russell King

Index   321

Contributors

Margaret Byron is Lecturer in Human Geography at King's College London.

John Connell is Associate Professor of Geography at the University of Sydney.

Carsten Felgentreff is Lecturer in Human Geography at the University of Potsdam, Germany.

Maria Lucinda Fonseca is Associate Professor of Geography at the University of Lisbon.

Brian Hudson is Senior Lecturer in the School of Planning, Landscape Architecture and Surveying, Queensland University of Technology, Brisbane.

Rachel Hughes is Tutor in the Department of Geography and Environmental Studies, University of Melbourne.

Russell King is Professor of Geography and Dean of the School of European Studies, University of Sussex.

Gabriella Lazaridis is Lecturer in Gender Relations and European Studies at the University of Dundee.

Cluny Macpherson is Associate Professor of Sociology at the University of Auckland.

La'avasa Macpherson is Research Associate in the Department of Sociology, University of Auckland.

Joan Marshall is Adjunct Professor in Geography at Carleton University, Ottawa, and in Agricultural Economics at McGill University, Montreal.

Helen Morton is Postdoctoral Fellow in the School of Sociology and Anthropology, La Trobe University, Bundoora, Victoria, Australia.

Peter Mühlhäusler is Professor of Linguistics at the University of Adelaide.

Stuart B. Philpott is Professor of Anthropology at the University of Toronto.

Joanna Poyago-Theotoky is Lecturer in Economics at the University of Nottingham.

Stephen A. Royle is Lecturer in Geography at Queen's University Belfast.

Elaine Stratford is Lecturer in the Department of Geography and Environmental Studies, University of Tasmania, Hobart.

Allan M. Williams is Professor of Geography and European Studies at the University of Exeter.

Annababette Wils is Research Associate in the Department of Geography and Geology, Vassar College, Poughkeepsie, New York.

Figures

2.1	Irish islands mentioned in the text	35
2.2	The pattern of settlement in Clare Island in 1956	37
2.3	Tory Island in 1997	47
2.4	Dursey Island in 1997	50
3.1	The Azores	58
3.2	Emigration from the Azores, 1960–88	63
3.3	Age–sex pyramids for all of mainland Portugal, the Alentejo, Central Portugal and the Azores, 1991	71
4.1	Map of Cape Verde	78
4.2	Population of Cape Verde, 1582–1990	80
4.3	Emigration and immigration trends for Cape Verde, 1900–88	82
4.4	Remittances per migrant and GDP per capita, Cape Verde, 1973–88	83
4.5	Births and net migration for Cape Verde, 1960–88	85
6.1	The Caribbean region including migrant destinations prior to World War II	116
6.2	Migration from the Leeward Islands, 1839–45	119
6.3	Age and sex structure of the Nevisian population at the 1891, 1911, 1921 and 1946 censuses	124
10.1	The Torres Strait region	196
11.1	Norfolk Island: location and topography	215
13.1	Map of Fiji showing location of Kabara	256
13.2	The spatial distribution of the registered members of Naikeleyaga's *yavusa* (October–November 1993)	267
15.1	Corfu: location of British retirement migrants	300

Tables

2.1	Population of Ireland and its islands, 1841 and 1991	28
2.2	Population development of islands which were inhabited in 1971, Republic of Ireland, 1841–1991	30
2.3	Buildings and services, selected Irish islands, *circa* 1900 and 1997	34
3.1	The Azores: population change, 1981–91	70
4.1	Estimate of age-specific male and female migration, Cape Verde, 1970–90	88
4.2	Estimated compared to registered net migration, Cape Verde, 1970–89	89
4.3	Estimated births and registered births, and three estimates of total fertility rates of Cape Verdean women, 1970–90	90
5.1	Grand Manan, New Brunswick province and Canada: mobility statistics according to the 1991 census	99
6.1	Main migration destinations of returnees present in Nevis, 1991	130
13.1	Naikeleyaga's present population: *yavusa* membership, 1993	263
13.2	Spatial distribution of the registered members of the *yavusa* of Naikeleyaga village, 1993	265

1

Island Migration in a Changing World

John Connell and Russell King

'Islands' and 'migration': two notions rarely juxtaposed but which are invariably closely related. One, a geographical entity, is easily defined: a piece of land surrounded by water, usually the sea. The other is a geographical process – human relocation across space – which is less easy to pin down since it shades, often imperceptibly, into other types of mobility (tourism, visiting, commuting) which are not normally regarded as 'migratory types'. Nevertheless, the key interaction which is the subject of this book – namely that islands have an unusually intense engagement with migratory phenomena – is clear. What this book seeks to do is to examine how this relationship expresses itself in different islands, and how it has changed and diversified during the manifold processes of globalization.

Whilst islands may be easily defined, they vary enormously and sooner or later one comes up against the hoary question: how big is an island? Is Britain one? Greenland? Australia? Whilst, in a sense, Australia has been 'made' by immigration, the British case is more complex: the long history of emigration is an under-researched part of British colonialism and of Britain's role in the creation of a specific epoch of globalization, whilst immigration from the colonies and former colonies formed an important element of Britain's economic growth in the post-war decades. At the same time there is much rhetoric in Britain which invokes the 'island race' and the notion of islanders being distinct from the populations of continental Europe.

Ignoring the 'big islands', this book takes the pragmatic approach of focusing on much smaller islands, where recent migration has been of substantial social and economic importance. It is on such small islands and island groups, some with populations of only a few thousand (or even a few dozen) people, that islanders are constantly reminded that their way of life and their identity have much to do with insularity and isolation on the one hand, and with migration and mobility on the other.

The book also explores the wide range of types and meanings of migration involving islands. Many islands have a long history of emigration, and the haemorrhage of population has taken many forms: permanent emigration, temporary emigration with subsequent return, circular migration, migration of one particular segment of the population such as young men. Destinations, too, have varied: neighbouring islands, a nearby land mass, or countries and continents on the other side of the globe. Only retrospectively is migration permanent; hence the changing relationships between migrants, their kin and their home islands are of great significance in tracing the wider meaning of migration. Islands have also been a focus for immigration. Historical logic tells us this, at least as far as the original settlement of islands is concerned. But there are other forms of inward movement such as the settlement of new populations as a result of warfare and conquest and, more recently, the attraction of islands for rich people seeking havens of peace and quiet – a new form of 'conquest' which reflects certain islands' new role in the global system, and which may be as resented as the earlier form of colonialism. In various forms, then, islands and island realms have increasingly become arenas of migration, mobility and multiple meanings, as individuals and households move – whether physically and/or metaphorically – between different worlds and different spaces (Chapman, 1991).

This opening chapter presents an overview of the migration experiences of islands and islanders: a survey that is preliminary and far from exhaustive, yet seeks to reflect on the wide range of migratory types, island locations and historical periods. This triple agenda – thematic, geographical and historical – poses obvious problems for the organization of material. Accordingly we privilege a chronological approach in seeking to examine the diverse ways in which migration has been participated in and perceived. The main focus will be on the last fifty years, but we start with a more historiographical note on the early appreciation of the heuristic role of islands as models of geographic space and of island populations historically prone to fluctuate through migration.

Early Observations on Islands and Migration

Early geographers were fascinated by islands as 'little geographical worlds', arguing that 'only by a careful study of a small unit can one discern and evaluate the relations between physical facts and human destinies' (Brunhes, 1920, p. 52). One obvious human destiny is that of migration. Ellen Churchill Semple wrote copiously about the migration proneness of islanders in a long chapter entitled 'Island peoples' in her influential (but subsequently largely discredited) book *Influences of the Geographic Environment* (Semple, 1911). Here is a not untypical quote concerning Oceania in which her determinist philosophy – inherent in the title of her volume – is given full rein yet in which her sensitivity to insular migration processes can also be appreciated. Writing of this island-rich part of the Pacific, she says:

> ... where every landscape is a seascape, where every diplomatic visit or war campaign, every trading journey or search for new coco-palm plantation means a voyage beyond the narrow confines of the home island, there dwells a race whose splendid chest and arm muscles were developed in the gymnasium of the sea; who living on a paltry 515,000 square miles of scattered fragments of land but roaming over an ocean area of twenty-five million square miles, are not more at home in their palm-wreathed islets than on the encompassing deep. Migrations, voluntary and involuntary, make up their history. (Semple, 1911, p. 299)

And again, speaking more directly of emigration:

> A small cup soon overflows. Islands may not keep; they are forced to give, live by giving. Herein lies their historical significance. (Semple, 1911, p. 416)

But, as centres of human settlement and commercial and cultural activity, islands have taken on other historical roles too. Semple described these in detail, albeit with a resolutely turn-of-the-century colonial tone. Generalizing, two historical 'ideal types' can be recognized. The first is the familiar role of the island as an 'isolated world', with a 'precarious, restricted and threatened life'. Standing aside from the march of civilization elsewhere, in the modern era such islands 'survived by emigration', as Braudel's account of Corsica, the Mediterranean island of emigration *par excellence*, demonstrates (Braudel, 1972, pp. 158–60). This historical role of islands as fountain-heads of emigration is analysed more systematically below and is recurrent throughout several of the chapters of this book.

Other islands, *per contra*, have a nodal location, so that 'sailors and traders, colonists and conquerors flock in from every side' (Semple, 1911, p. 424). As a result of these immigrations from near and far, such islands develop populations which are cosmopolitan and polyglot. Examples are numerous from various periods in the past. Semple quotes Pliny's description of Delos as the crossroads of the Aegean Sea, drawing to itself the trade and people of the entire Mediterranean Basin. Likewise ancient Dioscoridis (Socotra) attracted settlers from both Arabia in one direction and East Africa in another (Semple, 1911, p. 425). In the European colonial era Mauritius, with the best harbour at the western entrance to the Indian Ocean, quickly acquired a dense population which reflected its colonial history as a Dutch, French and English possession, in addition to its majority populations of Hindu Indians and African slaves. Malta is another excellent example of a 'cross-roads island': at the navel of the Mediterranean, its population reflects ethnic and linguistic mixtures of Phoenician, Arab, Sicilian and British colonial influences. The Hawaiian islands, at the centre of the Pacific, have had a comparable, if more belated, role.

The case of Malta also shows how these historical roles can vary through time, with obvious consequences for migration patterns. Malta is central to the Mediterranean but peripheral to Europe. Once its strategic colonial role as a key element of British maritime control of the Mediterranean was lost and the navel/naval function started to be run down, strong emigration currents developed to the 'mother country' and to other parts of the developed anglophone world such as North America and Australia. In fact, during the 1950s and 1960s, Malta had one of the highest rates of emigration in the world (Jones, 1973). Return migration was the dominant trend of the 1970s and 1980s as the Maltese economy started to recover and broaden from its earlier narrow reliance on a military function to a more balanced structure based on tourism, services and light industry. Contemporaneously foreign settlers – mainly British and including many people who had served or holidayed in the island in previous years – created an expatriate population of relatively wealthy retired and semi-retired immigrants (Warnes and Patterson, 1998).

Early studies also emphasised the diversity of islands – and the range of migration experiences that resulted. Semple (1911), quite rightly, saw size as an important variable. Smaller islands were, and are, more vulnerable to natural hazards (especially in the tropics), and were more prone to colonial transformation – except where they were too tiny and remote to be of great significance. Shifting perspectives on natural resources – especially minerals – have also influenced the fate of islands, attracting and repelling population. A handful of islands have had brief lives, in the nineteenth and twentieth centuries, as sources of raw materials such as phosphate. Clipperton Island had a short but exotic population history in the nineteenth century whilst Banaba (Ocean Island), in Kiribati, experienced the dismal fate of having its entire population resettled to distant Fiji to make way for mining operations; the end of mining in 1979, the year that Kiribati gained independence, left a ruined island, to which return migration has been trivial (Binder, 1977). Nearby Nauru, where phosphate mining was mainly undertaken by migrants from Kiribati and Tuvalu, now faces the imminent end of mining, and has contemplated the resettlement of part of its population in Australia and the Philippines, given the absence of economic alternatives.

A number of islands, such as Spitzbergen and Bahrain, still owe either all or a significant proportion of their population to migration into the resource sectors. In New Caledonia, the 'nickel boom' that accompanied the Vietnam War brought massive, politically engineered immigration from France and its other Pacific territories, turning the indigenous Melanesian population into a minority in its own land, and influencing the whole political evolution of the island (Connell, 1988). On Lihir, in Papua New Guinea, where the largest gold mine in the southern hemisphere has just opened, substantial immigration from elsewhere in the country has brought rural real estate speculation and considerable tension between islanders and migrant workers, as previously occurred in other Papua New Guinean islands – Bougainville and Misima (Connell, 1997, pp. 160–1). Oil has revalued and revitalized certain islands. In the Caribbean, oil brought relative prosperity to Trinidad, as it did to Bahrain in the Persian Gulf, whilst, even at second hand, oil refining on Aruba led to substantial immigration in the inter-war years. Since the 1970s, development in the Orkneys and Shetlands has been stimulated by the North Sea oilfields, whilst various island groups – the Paracels and the Spratlys in the China Sea, Timor and recently even tiny Rockall in the North Atlantic – have gained such value, because of their presence close to oil deposits, that they have acquired a new strategic significance.

Other resources, usually of less value than minerals, have rarely been of equivalent economic importance, yet, without fisheries, Iceland, the Faroes, Newfoundland, and St Pierre and Miquelon would have tiny populations. The timber industry has contributed to immigration into Vancouver Island, and several of the islands of eastern New Guinea and the western Solomon Islands. Resources, and resources diplomacy, have increasingly played a more substantial role in migration to and from islands than the older agendas of colonial and military strategy.

In the twentieth century, post-Semple, islands have become particularly vulnerable to global economic events. The movement of peoples within and between islands has intensified in volume, increased in distance and become more complex in pattern and purpose. International population flows are now the major regulators of demographic change in many of the islands and island states discussed here. The development of modern modes of transport, and a relative decline in its cost, have influenced the structure of migration. Whereas, in the past, movements tended to be circular, repetitive and usually over short distances, permanent and relatively long-distance migration has become more general. These changes have incorporated islands, and islanders, into transnational contexts in a way that could not have been conceived by Semple.

Islands and Migration in History

The above remarks clearly demonstrate that out-migration from islands is not merely a contemporary phenomenon. In this section of the chapter we essay a more ordered chronological account. Although this book is primarily concerned with the twentieth century, the period when islands and emigration have been closely linked, the history of emigration does not begin there, and nor is that the end of the story.

In several European islands concern over emigration dates back many centuries. As early as 1422 legislation existed in the Isle of Man to prevent emigration, whilst in the Azores and Canary Islands emigration to Spanish America had started before the end of the seventeenth century (Aldrich and Connell, 1998, p. 99). At much the same time that the Isle of Man was first experiencing depopulation, the last phases of colonization were finally setting people on some of the more remote Polynesian islands in the South Pacific. Even more recently, the migration streams of colonial expansionism populated such lonely Atlantic islands as Bermuda and St Helena, and the Indian Ocean islands of Mauritius and Réunion.

Though the Isle of Man experienced emigration centuries ago, it was not generally until the nineteenth century that global social and economic changes – notably the demise of slavery and the declining productivity of tropical sugar plantations – resulted in more substantial emigration from islands. In the twentieth century, a loose combination of rising social and economic expectations, continued upward demographic pressure, the declining significance of subsistence economies and the rise of global communications, accelerated this process, once again initially from islands closest to the European mainlands. Ireland, the Canaries, Madeira and many Mediterranean islands experienced two main waves of emigration, one during the decades spanning the late nineteenth and early twentieth centuries, and the second during the 1950s and 1960s.

After emancipation from slavery in the Caribbean, there was extensive migration from some of the smaller islands where sugar production had been barely economic, even with slavery. Chapters 6 and 7 provide details on this period for Nevis and Montserrat, but several other islands were affected too. St Eustatius, with a population of 8000 at the end of the eighteenth century, had just 1300 people at the start of the twentieth. St Barthélémy went through a similar demographic decimation, which continued into the twentieth century. In the 1920s, emigration from Saba, St Maarten and St Eustatius was substantial enough to provoke fears of almost complete depopulation of these islands. Then tourism, in the second half of the twentieth century, allayed such fears. A number of island groups, initially the US Virgin Islands, St Maarten, the British Virgin Islands, and subsequently Anguilla, Montserrat, Bonaire and the Turks and Caicos, made the transition from being islands of emigration to places of immigration (McElroy and de Albuquerque, 1988). Tourism, and subsequently retirement migration, brought similar population turnarounds outside the Caribbean, primarily in the Mediterranean (see the account of Corfu, Chapter 15) and some of Britain's offshore islands such as the tax havens of Jersey, Guernsey and the Isle of Man. As we shall see in more detail later on, these were elite not mass migrations.

On a very different scale, the 1950s, 1960s and 1970s marked a new era of emigration in many island realms. The 'long boom' created a demand for labour in metropolitan states, and accelerated incipient processes of migration from distant, mainly colonial territories. In Britain, migrant workers were actively recruited from Caribbean islands for the health and transport industries. In the Pacific, temporary work schemes took Polynesians to New Zealand. Similar recruitment and employment schemes in France, Denmark, The Netherlands and the United States took migrants from many islands ranging in location from the tropical Caribbean to the cool islands of Greenland and the Faroes. Most colonial islands saw a wave of post-war migration to the metropole. More than three quarters of a million migrants left Puerto Rico between 1945 and 1964. Extensive recruitment from France's Caribbean islands of Martinique and Guadeloupe, and the Indian Ocean island of Réunion, saw a quarter of a million emigrants leave each of these islands in two decades, though their populations remained more or less the same thanks to a high rate of natural increase (Aldrich and Connell, 1998, pp. 101–2; Condon and Ogden, 1996).

It was also during these post-war decades that global population structures changed, with the establishment of migrants from distant territories and former colonies, many of them islands, as more or less permanent residents of the metropolitan states. It was no longer possible to think of many islanders as living primarily on their home islands, but rather as cosmopolitans inhabiting diverse worlds and with identities constructed by their experiences of living in more than one geographical, cultural and socio-economic environment. As several chapters in this book show, home countries, and islands, became 'constructed and reconstructed' by overseas migrants, through changing or disrupted ties with islands of origin (see especially Chapters 10 and 12). Island identities were reshaped by migrants' activities at the distant metropolitan shores, but rarely were migrants any less islanders after migration. Indeed alien environments, even racism, fostered new experiences of identity, whilst unexpected events – such as volcanic eruption – established island identities where only vague conceptions hitherto existed, as Philpott demonstrates for Montserrat (Chapter 7).

By the 1980s the most rapid phase of emigration from many islands and island groups had ended: the economies of metropolitan states faced recession, their governments actively discouraged immigration and a conservative backlash challenged migrants. By contrast, in some islands, especially where colonial administration had supported bureaucracies or tourism had developed, relative economic prosperity resulted in some return migration. This was particularly evident in Caribbean islands, but similar processes can be observed in many other island realms such as the Faroes, some islands off the west coast of Ireland (Chapter 2), and in several Mediterranean islands. Where island populations once seemed in freefall, with the 'end ever nigh' (i.e. ultimate depopulation), the rate of emigration declined and a phase of relative population stability ensued. Changing island demographics also contributed to this new phase of stability. By the 1980s, many island populations had passed through the demographic transition, so birth-rates were much lower than in the past. Moreover, the removal of reproductively active young adults by migration played a major contributing role in holding down fertility, as Wils shows for Cape Verde (Chapter 4).

Return migration was not the only component of the 'migration transition' experienced by many islands in the period since the 1980s (McElroy and de Albuquerque, 1988). In several island territories, such as St Maarten, the Turks and Caicos and the Cayman Islands, there was illegal immigration from neighbouring independent islands, and even the arrival of refugees from more impoverished islands such as Haiti and Cuba. The other side of this migration coin was evident in some dependent territories, such as the Cook Islands, where islanders rejected independence if it were to mean the loss of freedom to migrate to the metropole. Elsewhere, the Marshall Islands and the Federated States of Micronesia built migration provisions into their compacts of Free Association with the United States (Connell, 1990). Hence, once again, as global economic growth foundered, migration became more complex within island realms.

Retirement migration has been the last distinctive population change to affect many islands, initially those close to Europe and North America. The phenomenon had already begun in the Channel Islands in the nineteenth century; by the 1970s it was evident in Madeira, the Canary Islands and parts of the Mediterranean, before subsequently spreading through the eastern Caribbean. Retirement migration is itself a diverse phenomenon. Particularly in the Caribbean, many retirement migrants are the island's own returnees, retiring home after a lifetime's work abroad. In the Mediterranean and the Canary Islands the retirees are former tourists who often become permanent or seasonal residents through the purchase of island properties in anticipation of a 'retirement to the sun' (see Chapter 15 on Corfu). Elsewhere, the tax haven incentive has given islands a new niche role in the global economy, stimulating the settlement of wealthy retired or semi-retired foreigners.

Given the small scale of islands, pressures from retirement migrants, tax exiles and property developers may cause local tensions. In the Isle of Man and the Channel Islands, the impact of tax and retirement migration has proved highly significant. In the 1950s the Isle of Man government abolished death duties and reduced income tax, attracting more than 6000 'new residents' in the following decade. Further immigration was opposed because of the rise in property prices, concern over the impact on Manx culture, and feelings that the island was becoming an 'offshore retirement home'. In 1989 the Manx parliament introduced legislation to control immigration more carefully (Boyle *et al.*, 1998, pp. 168–9; Prentice, 1990). The Channel Islands experienced similar problems with what some perceived as too much immigration resulting in high property costs and pressure on space and amenities. Nevertheless such immigrations reflect new perceptions of the attractiveness of island life, and changing evaluations of the social and economic value of islands.

Migration as a 'Way of Life' on Islands

Several chapters in this book describe the various processes by which emigration has come to dominate island societies. Key themes which echo through many accounts are the restricted resource base due to such factors as smallness, rugged terrain and climatic vagaries; high rates of demographic increase; and the lure of more wealthy and modern economies elsewhere. The physical factors associated with a harsh environment are well exemplified in Royle's account, in the next chapter, of emigration from the small islands off the coast of Ireland, especially the Blasket Islands which feature prominently in his survey. Here was a tough life of primitive simplicity; rocks, cliffs, wind and rain were the essential elements of the physical setting of this remote Atlantic outpost of Europe. Fish, sheep and potatoes were almost the only products of the micro-insular economy: poor competition with the richer rewards of the United States where most Blasket Islanders eventually emigrated.

As we saw earlier, Semple's determinist philosophy laid great stress on the role of environmental factors in human destiny; and islands, with their limited ecological diversity and vulnerability to hazard, exemplified places that were particularly prone to the vagaries of physical geography. Yet the environmental conditions in most islands have rarely been as harsh as those of the Blasket Islands where, before emigration led to the complete depopulation of the islands in 1953, people's lives were 'shot through with the purity of isolation' (Cunningham, 1998, p. 43). In any case, environmental factors are highly dependent on technology as well as mediated through culture and the 'island way of life'.

Thus the reasons for migration from islands are inevitably more complex. In fact, the variety of reasons put forward to explain island migrations sometimes seems interminable and the problems of generalization considerable. Yet apart from migration as a direct result of natural hazard (see Chapter 7), the major influences are basically economic, although they are often accompanied by significant socio-cultural changes. As countless studies have shown, migration is primarily a response to real and perceived inequalities in socio-economic opportunities and standards of living. These inequalities result from uneven development, in turn a function of the penetration and consolidation of capitalism in what have often been global peripheries; and under traditional regimes of mercantilism, colonialism and industrialization, small islands were amongst the most peripheral spaces in the world. At the behavioural level, migration remains a time-honoured strategy of moving from a poor area of few opportunities to a rich one in the search for socio-economic mobility at home or abroad. A major influence on migration decision-making has been radical changes in expectations over what constitutes a satisfactory standard of living, a desirable occupation and a suitable mix of accessible services and amenities. Aspirations are constantly being revised upwards as small islands are increasingly less able to meet new needs and demands. Thus, in Tonga, parents express the wish that their children 'work at something better than agriculture' even though they themselves are farmers; this 'something better' invariably means scarce white-collar jobs which carry a lot of prestige. Likewise, in Fiji, young people 'are taught to value white-collar occupations and to detest farming' (Naidu, 1981, p. 18). With or without migration, a growing disdain for agriculture has accompanied the change of island economies away from primary production towards the tertiary sector.

There are many islands and island-regions of the world where migration has undoubtedly become a 'way of life'. Few can be in any doubt that this applies throughout the fragmented, microcosmic domain of the minor Irish islands (Chapter 2); indeed migration has been a constant element in more than a century and a half of Irish history. Caribbean islands such as Nevis (Chapter 6) and Montserrat (Chapter 7) give very much the same picture, and it is no accident that St Kitts-Nevis and Montserrat were more deeply affected by emigration to Britain during the 1950s and 1960s than any other part of the Caribbean (McElroy and de Albuquerque, 1988). The collapse of sugar cane and other plantation crops removed the traditional livelihood of these islands under colonial rule and, quite simply, during the early post-war decades, around 30–40 per cent of the total population, including most of the able-bodied young adults, left the islands for the 'mother country'. Both the analyses of Byron, on Nevis, and Philpott, on Montserrat, are historically grounded, and make the important point that post-war migration to Britain at that time was merely the latest 'opportunity' for poverty-stricken and dispossessed labourers to find work within initially the regional and subsequently the global labour markets (cf. Richardson, 1986). Whilst return migration and the growth of tourism in the 1970s and 1980s provided a new migration phase for many Caribbean (and other) islands, the case of Montserrat was given a tragically dramatic new twist with, first, Hurricane Hugo in 1989, and then the eruption of the Soufrière volcano in the late 1990s; both were powerful influences on migration.

Island Demography and Migration

The demographic context of migration from islands represents a rather different angle of analysis, although often there is a close connection with some of the economic factors mentioned above. Demography and migration are interrelated in a two-way dialectic: population pressure may be a powerful 'push factor' for emigration; but out-migration in turn changes the demographic structure of the residual population because of the selectivity of the migration stream. As Cleland and Singh (1980) and Cruz *et al.* (1987) have shown, the demography of small islands is distinct, marked by cycles of wide population swings, with increases or decreases well beyond the parameters of natural change (birth and death rate balances). This volatility is explained by a combination of factors – the opening up of new opportunities, external market fluctuations, natural hazards, epidemics, forced or voluntary migrations – which impact at a more intense scale on small islands. Beller (1987, p. 17) lists some examples of these processes: the rapid population growth of many Caribbean islands during the seventeenth and eighteenth centuries when slaves were imported to expand the plantation system; major population declines, even decimation, experienced by some Pacific islands during the nineteenth century due to successive epidemics; the rapid expansion of population driven by a high birth-rate combined with the elimination of disease, such as the eradication of malaria in Mauritius in the mid-twentieth century; and the sudden population losses through massive emigration following crop failure or unfavourable market shifts – the experience of many small Caribbean and Mediterranean islands.

Whilst there are thousands of sparsely populated islands around the world (not to mention uninhabited ones), generally islands carry dense populations when compared to adjacent areas of mainland or to the larger political units of which they are a part (Semple, 1911, pp. 447–9). Mention has already been made of densely settled Malta. Reliance on trade, shipping or a history of plantation agriculture often explain this fact, but the ultimate consequence is usually a high propensity to migrate, stimulated by the obviously restricted land area, narrow base of the economy, and often high natural increase rates.

As Wils shows for Cape Verde (Chapter 3), the relationships between migration, demographic trends and 'development' are far from straightforward and by no means always follow the simple model of emigration being a response to rapid population growth, poverty and overcrowding on islands, leading then to population decline as departures snowball. The Cape Verde data from the late nineteenth century seem to suggest that emigration actually accelerated population growth as remittances gave the population the means to buy food and improve health and hence lower mortality. At a later stage, however, as emigration increased in scale over the period from 1950 to 1990 and involved more women, population growth did slacken off due to the reduction, by emigration, of reproductively active cohorts of the population and to their lowered fertility through exposure to the small-family norms of the destination countries in Europe and North America.

Further interesting demographic questions concern the notion of a minimum threshold for the survival of an island population, and the relationships between migration and population stability. On the Irish Blaskets emigration emerged as a way of life but then took away all human life from

the islands. A number of estimates have been made, on the basis of historical and archaeological data, computer simulations and intuition, of the minimum population required to enable an isolated island community to
remain viable. Estimates seem to range from around 30–50 on the one hand
to around 300–500 on the other.[1] These are essentially theoretical calculations with limited recourse to rigorous notions of land carrying capacity and
hazard assessment. In practice those who have predicted the demise of
island populations have often had their forecasts confounded by reality.

Planners, too, have had their depopulation schemes frustrated. For the
Polynesian Tokelau Islands, the New Zealand authorities devised a policy to
relocate the entire population to New Zealand. Despite a 'rush to New
Zealand' (Hooper and Huntsman, 1973, p. 392; Wessen *et al.*, 1992), the population of the three atolls stabilized at around 1500 (with about three times
that number in New Zealand). In the nearby Polynesian island of Niue,
depopulation loomed after a spectacular decline took its population from
5200 in 1966 to less than 2000 in 1994. On the other side of the world the
United Kingdom discouraged the repopulation of Tristan da Cunha in 1963,
and at the same time questioned the continued population of the British
Virgin Islands and Montserrat. Lowenthal and Comitas (1962, p. 207) asked:
'Should Britain and the West Indies ... try to revitalize Montserrat's economy or conserve scarce funds and skills, encourage Montserratians to move
to Trinidad, and abandon the island to the goats and beachcombers?'
Although Montserrat underwent a demographic revival in the 1970s and
1980s, Hurricane Hugo (1989) and the recent cycle of volcanic eruptions
have led to this question being re-posed. Similar sentiments about 'non-
viable populations' were expressed over the Dutch Caribbean island of
Saba, and the Australian government sought to depopulate Christmas and
the Cocos (Keeling) Islands, its tiny Indian Ocean territories. None of this
occurred and, despite emigration and some resettlement (as in the case of
Tokelau), the islands retained their populations and, in some cases, eventually experienced renewed growth.

Elsewhere there are many cases of migration creating, or contributing to,
more of an equilibrium – albeit one where the majority of islanders, or
island families, are involved in migration, often repeated in a constant cycle
of departure, return and re-emigration. Further up the Irish west coast from
the Blaskets, Achill Island exemplifies the way in which different types of
migration create different levels of demographic equilibrium (McGrath,
1991). All the censuses between 1881 and 1951 record a population of
around 5000 for the island: this equilibrium was maintained by a long-estab-
lished regime of seasonal migration which took mainly young male family
members to work in the harvests in Scotland and Northern England. This
short-term work in Britain provided a cash revenue which was integrated
with ongoing subsistence farming in Achill. A demographic stability was
maintained, the high birth-rate being cancelled out by a smaller-scale semi-

permanent emigration overseas to the USA. During the 1950s, emigration to Britain became more permanent as seasonal workers stayed on and moved into the construction and other manual sectors of employment. This led to a new demographic equilibrium at around 3000 to 3200 inhabitants (all censuses since 1971), based on larger-scale settlement abroad, significant return migration, and a lower birth-rate. This demographic stability does not necessarily imply economic stability since Achill's livelihood is based on an uneasy balance between declining agriculture, some highly seasonal tourism, and an important inflow of remittances, savings and pensions.

On a number of other islands there has been a similar balance between sustained emigration and population stability. For a century, on the Greek island of Mykonos, high population growth was counterbalanced by high emigration which kept the island population almost constant (Hionidou, 1995). Almost exactly the same situation has occurred for more than 25 years in the Pacific island micro-states of Tonga, Samoa, Tokelau, and Wallis and Futuna, where emigration is perceived as a 'safety-valve' in economies of limited diversity. Consequently migration may be sponsored or planned for, with households or extended kinship units encouraging the migration of particular individuals, often to a specific destination, where islanders' own kin are already established. Studies of Tonga have pointed to the emergence of a new institution, the 'transnational corporation of kin' (Marcus, 1981; see also James, 1991), whereby kinship groups colonize and exploit economic opportunities across a range of locations.

At the household level parents may actively hope to produce remittance earners and hence try to have several children. Carrier's study of the remote island of Ponam (Papua New Guinea) shows how population growth and migration are regarded as economically rational: childrearing for migration is seen as an economic investment generating an eventual economic return (remittances). Hence 'Ponam is not a society passively allowing or suffering the migration of its members', and Ponam people claim that 'children are our garden and we survive by eating the fruit' (Carrier, 1984, p. 49). Islanders aim to maximize the number of their children in order to ensure continuity of migration and with the intention of producing a child who might secure well-paid, white-collar employment beyond the island, and in so doing sustain island life.

Many chapters in this book amply demonstrate the continued viability of island communities, and the manner in which what are now multi-local island populations have, to a very substantial extent, 'gone away without leaving' (Nietschmann, 1979, p. 20). As the diaspora is bound into village life, questions of who is, and who is not, an island resident become increasingly difficult to answer, as Felgentreff demonstrates in Eastern Fiji (Chapter 13).

Migration and Island Economies

The impacts of migration on island economies are dependent both on the nature of the migration and on the structure of the economy. Migration has generally resulted in static or declining agricultural production, relative to population trends; agricultural output is most likely to fall where family migration is more common, as in large parts of Polynesia such as Tonga, the Cook Islands and Niue. The cultivation of food crops has been a particular victim of migration, which also makes land reform and rationalization more difficult because of the absence of title-holders. Migration also seems inevitably to transform consumption patterns: earlier wants become needs and their purchase must be financed by further migration to waged employment. The economic history of islands cannot, however, merely be reduced to a transition from subsistence to cash cropping and consumer goods. More broadly there has usually been a dramatic expansion in the service sector, especially in public services, a development of tourism, and even the emergence of manufacturing and offshore financial centres in favoured sites. According to Bertram and Watters (1985), many islands have evolved MIRAB economies, dependent on migration, remittances, aid and bureaucracy, although the heavy urban bias of the new bureaucratic and commercial sectors suggests that MURAB would perhaps be a better acronym (Munro, 1990). While such acronyms are appropriate as descriptive economic labels, they are detested by national governments because of their implications of mendicancy.

The departure and 'loss' of migrants from island societies is usually compensated, to some degree, by resource transfers from migrants to kin who remain on the island. Remittances, which maintain social ties and act as insurance premiums for migrants, thereby fuelling an ideology of return migration (Rubenstein, 1979), are principally used to repay debts, finance migration moves for kin, purchase consumer goods and improve housing. These processes have been documented by studies on many migration-dependent island societies. In some island states, such as Cape Verde, Tonga and Samoa, remittances are the single biggest source of foreign exchange – greater than exports or aid – and this situation is even more true of outlying islands in many countries. In Cape Verde, where remittances account for about half the national income, 'it is no exaggeration to say that this ensures the survival of the country' (Lesourd and Réaud-Thomas, 1987, p. 117); there and elsewhere, migrants have become 'the most valuable export' (Shankman, 1976, p. 28).

Remittances have usually contributed to creating an aura of greater prosperity, but their use – for housing and consumption – tends to reinforce the traditional set of values and hence supports, rather than challenges, the established social hierarchy. In the Polynesian states especially, remittances have become crucial to welfare, have significantly raised and maintained living standards, and eased national balance of payment problems, despite their contribution to inflation. They can also help to generate new employment, especially in services and the building sector. However, increased demand for consumer goods sees the re-export of remittances since these goods are unlikely to be manufactured locally. Acquisition of consumer goods fuels demand for further purchases, which can only be met by further migration. In general, only a small fraction of remittances has been invested in economic growth since in many islands (and other emigrant areas) a history of failed projects has discouraged such investments; marketing and managerial skills may also be in short supply. There has, however, been a transition in remittance use from consumption to investment, at least where real opportunities exist and where the immediate needs of improved welfare and consumption have already been met (Connell and Brown, 1995). On the other hand, the social construction of the meaning of 'productive investment' must also be borne in mind: in Tonga, for instance, it is less about a narrow economic definition of increased production and more about 'the well-being of kin and advancing to a sound social status' (Fuka, 1985, p. 93).

Migration, Colonialism, Globalization

Several chapters in this book demonstrate that migration from islands is not merely a response to overpopulation, environmental stability, poverty and economic need, but can also be seen within the important historical contexts of colonialism and globalization. Ever since the migrations of slavery and indenture, migration was one early component of globalization (King, 1995); these forced migrations were integral to the establishment of the world capitalist system, and were the first stages in the creation of a global labour market which is, in the Marxist analysis, the key historical explanation for international migration (Potts, 1990).

Under colonial regimes, as we have already noted, many islands' rural economies were fundamentally transformed by plantation agriculture. Monocrops formed the vital wealth of many islands but it was a fragile dependency, cruelly exposed when the market for the plantation crop collapsed or soil fertility became exhausted. Chapters 6 and 7 describe some of this background to the various phases of emigration from Nevis and Montserrat in the Caribbean, and the pernicious historical links between plantation agriculture and migration are analysed in a more general sense by King (1998): plantations require huge inputs of labour, brought in by forced and planned immigrations of slaves and indentured workers, who are then left in destitute unemployment when the plantations fail, leaving emigration as the only strategy for survival.

As an interesting historical antecedent to the well-known history of plantation agriculture and emigration from the Caribbean, Braudel (1972, p. 155) considers the case of the 'sugar islands' of the 'Mediterranean Atlantic'. Madeira, the Canaries, the Azores, Cape Verde and São Tomé were all ravaged by the monoculture of cane sugar which prevents the growing of other crops in rotation and restricted the land available for food crops. Of course, sugar was not necessarily the only, nor the most direct, cause of subsequent emigration; as both Wils (Chapter 4, Cape Verde) and Williams and Fonseca (Chapter 3, the Azores) show, whaling-boats played an important trigger role, but the economy and the ecology had already been destabilized by the relatively short-lived sugar cane boom.

Globalization is the main analytical framework employed by Williams and Fonseca to examine the development of emigration from the Azores, islands which were mid-Atlantic staging-posts for Portuguese colonialism and whose subsequent emigration trends reflected this west-facing orientation. Those chapters on the Caribbean and Pacific emphasize the impact of globalization both in terms of its pervasive ideology and in the way in which the most critical changes have been within living memory rather than in the days of plantation economies, significant though these may have been in historical terms.

Implicitly underlying the structure of migration is the constant dialectical interplay between individual human actions and institutions: between everyday life and structure. Although the notion of structuring is rather rarely made explicit in discussions of migration, most of the following chapters tend to highlight human agency rather than a procrustean structuralist bed. The emphasis is on the active role of islanders in shaping their own history, rather than on the mechanistic influence of external forces or the pervasive presence of a hostile environment. This is particularly evident in the Torres Strait (Chapter 10) where a single migrant was able to transform the whole nature of the relationship between land and people, not only in his home island, but indirectly throughout Australia.

A different perspective on the structure/agency debate is given by the case of Grand Manan, the small Canadian island described by Joan Marshall in Chapter 5. Here the existence of a relatively rich and diverse resource base has created a flexible and robust economy and, in turn, a sense of independence. Lacking an economic imperative, migration levels have been low. Limited migration has contributed to a greater wariness of outsiders and a powerful insider–outsider cleavage in local society.

Moving Back and Moving In

Because of the limited opportunities for local investment, and the often poorly developed social facilities on islands, return migration has all too often been of the old or the 'failures' without skills or capital, who therefore make little contribution to economic growth, and may instead introduce new discontents and demands. The contribution of return migrants to island development is likely to be greater in larger islands where growth potential is more apparent and where newly acquired skills have some utility. In Barbados, for example, those who had acquired skills overseas were in much greater demand than those who had merely taken up unskilled employment, though, on balance, return migration to this island appears to have been favourable to national development (Gmelch, 1987; Thomas-Hope, 1985). In Cape Verde the clothing industry has succeeded almost entirely because of the capital and skills of return migrants, and an Institute of Emigration has been established to collect data on emigrants and assess their financial and technical potential, while the state organizes visits home for second, third or fourth generation migrants (Connell, 1988, p. 30). On the other hand many migrants seem to have conservative attitudes to development in their home islands, as was noted earlier. Migrants from the Cook Islands, for example, have specifically remitted money to their home islands not for business development but for the construction of churches and village halls so that village life would remain 'traditional'. On the Maltese island of Gozo huge churches, financed largely by emigrants' donations, loom over tiny villages (King and Strachan, 1980). Consequently there is ambivalence in several island states towards return migration. In Samoa and Niue there is even local opposition to returnees: the view is that migrants have 'voted with their feet' to leave and that the scarce existing jobs and opportunities should be reserved for those who remain. In Malta in the 1970s this same view found expression in legislation which prevented returnees from acquiring public sector jobs for a period of three years after their return.

Return migrants themselves often suffer ambivalence over questions of their own identity. The national boundedness of islands and the small scale of their societies create a distinct sense of belonging based on face-to-face relationships, often reinforced by strong moral codes and rituals, or even distinct ethnicities, cultures and languages. On the Shetland island of Whalsay islanders can be defined by what amounts to an ideology of 'being Whalsa', which involves 'versality of skill, linguistic distinctiveness and inventiveness, controlled behaviour and emotion, equalitarianism, and the powerful sense of community on which Whalsay people explicitly pride themselves' (Cohen, 1982, p. 32). These are positive attributes which it would be difficult for others to emulate or, more obviously, be seen to emulate.

If return migration is rarely easy, insularity – as much a state of mind as a geographical concept – protects and defends local cultures, leading to a suspicion of outsiders who are seen as a potential threat to the 'moral community' of islanders. A variety of local terms are used to describe the in-migrants in various islands: 'from-aways' (Grand Manan), 'come-overs' (Isle of Man), 'blow-ins' (Irish islands) and 'down-islanders' (Virgin Islands) are just some examples. In the Grand Manan case (Chapter 5), insular identity expresses itself in strong social and community structures linked historically to local culture and family lineage. Now it is being challenged by the 'disembedding' effects of immigration, tourism and the off-island capital interests of the fishing industry.

On Grand Manan, as was briefly noted earlier, migration levels are rather low. Hence the insider–outsider dynamic which is played out in the relationship between native islanders, returnees and 'from-aways' is seen in rather sharp relief, all the more so since those who have left the island (e.g. for higher education) are reabsorbed as if they had never been away, a relatively unusual occurrence. In some other islands, the main dimension of the insider/outsider dialectic is expressed in the tensions between non-migrants and returnees. It is well-known that returnees (and migrants who do not return) often evolve complex and ambiguous identities which may swing between the potentially enriching experience of being members of two worlds (cf. Galtung, 1971) and the damaging and demoralizing state of being suspended in between. On the other hand, where mobility is at a very much higher level, affecting a majority of the population, migration itself becomes part of island identity, sometimes to the extent that those who have not been abroad are regarded with pity or contempt, as individuals who are not fully developed.

Most of the islands discussed in this book are characterized by out-migration rather than in-migration (though all experience some elements of return migration). An exception is Norfolk Island (Chapter 11), where immigration to take advantage of low taxes has occurred. Here there has been acute resentment at so many newcomers who appear to have superior access to certain services. Similarly, in Greenland, the presence of Danes, paid higher wages than the locally born though doing the same job, has emphasized income inequalities and placed indigenous Greenlanders in subordinate social and economic positions, particularly because few Danes learn Greenlandic or interact with the local people (Nuttall, 1992, pp. 102–4). More frequently, immigrants are in disadvantaged social and economic positions. In both the British and United States Virgin Islands during the 1960s and 1970s tensions between Virgin Islanders and 'down-islanders' mounted; migrants experienced insecure legal status, persistent job and wage discrimination, grudging social acceptance and political disenfranchisement. Where islands experienced economic growth, as in the Cayman Islands, immigration has transformed the demographic structure; after the successful establishment of financial services, barely half the population was island-born by the end of the 1980s – migrants had come from 117 different states! (Aldrich and Connell, 1998, pp. 106-7). These are exceptional islands, where immigration has changed, or threatened to change, the nature of island life. More commonly it is return migration that slowly changes islands but, wherever and however it occurs, and especially on the smallest islands, migration and change incite resentment, envy, tension and new perceptions of identity.

Writing of Passage

Individuals' testimonies are a vital part of understanding migration as a social process, especially at the relatively micro-scale of small islands. Social anthropologists, oral historians and ethnographers have long appreciated this, and a biographical approach is now becoming firmly entrenched as a key methodology within the wider interdisciplinary field of migration studies, as a number of recent studies have shown (Boyle *et al.*, 1998, pp. 71–3, 207–34; Chamberlain, 1995; Findlay and Li, 1997; Halfacree and Boyle, 1993; King *et al.*, 1998). However, a biographical approach can also embrace autobiography, as the migrants 'give voice to themselves' via a wide range of literature which involves not only conventionally published autobiographical accounts but also letters, diaries, short stories, poems, songs and orally recounted folklore (King *et al.*, 1995, pp. x-xiii). The simple power of the testimony of individual migrants is often stunning, capturing in a few words a whole world of meaning. Take for example the woman from the tiny Canadian island of Grand Manan who is so acutely aware of her island identity when she visits the mainland that she feels she has 'Grand Manan' written across her forehead (Marshall, Chapter 5). Maurice O'Sullivan's emigration from Great Blasket Island in the 1920s reveals another series of poignant, even painful encounters with the island's 'other' – in this case the railway stations at Dingle and Tralee, the street urchins of Cork and the traffic of Dublin. Although already in his twenties, this migration was O'Sullivan's first journey off the island. At Dingle he saw his first train:

> I had only gone a few steps when an echo came back from the whole town of Dingle with the whistle the train threw out, and as for myself I was lifted clean from the ground. I looked around to see if anyone had noticed the start it took out of me, but nobody had ... I had four hours to spend in the city (of Tralee), but ... I had no intention of leaving the station, for I had no trust in the train but that it might go at any time. I put my bag against the wall and kept my eye on it always ... (O'Sullivan, 1933, pp. 251–8)

In *Writing Across Worlds* (King *et al.*, 1995), several chapters are based on island migrations and island authors, amongst them the emigration novels of the Barbadian writer George Lamming (Alexander, 1995), the works of francophone writers from Martinique and Guadeloupe (Aldrich, 1995), the theme of migration in Maltese fiction (Cassola, 1995) and the 'Samoan worlds' of Albert Wendt (Connell, 1995). In Chapter 8 of the present book Brian Hudson offers a broader analysis of the creative literature from the anglophone Caribbean, structuring his account around a series of key migration themes such as intra-Caribbean migration, migration to England and North America, return migration and the migration-shaping role of the education system under colonialism. Rachel Hughes (Chapter 9) takes up the particular role of Antiguan émigré novelist Jamaica Kincaid, to examine the subjectivities of 'home', the significance of migration for women, and the relationships between women and the essentially masculine venture of colonialism: themes that are redolent of other islands in other regions.

Music, too, offers a focus on islands and migration. Consumers in developed countries have long had a fascination with islands, particularly those of the Mediterranean, Caribbean and Pacific. Mediterranean islands became some of the earliest tourist destinations and Aldrich (1993) gives some illuminating insights into the libertarian exploits of northern European writers and artists who congregated on Capri in the early decades of the twentieth century. In the Caribbean and Pacific, islands were deliberately exoticized for the tourist gaze in musical form. Bob Hope and Bing Crosby took the Road to the Islands and Bali Hai (transplanted from Indonesia to the Pacific!) became the *South Pacific* of global dreaming. Calypso music from the Caribbean etched powerful images of island paradise for people living under the grey skies of the northern hemisphere, typified by Harry Belafonte singing of yellow birds in banana trees. These themes were avidly taken up in the West. Billy Ocean sung of a 'Caribbean Queen', imaging a world where island women merely awaited Western men's attentions and affections, whilst the Beach Boys, in *Kokomo*, intoned 'Bahama, come on pretty mama/Martinique, that Montserrat mystique/Afternoon delight cocktails and moonlit nights'.

Other popular music provided relative realism; Elton John's *Island Girl* spoke of a Jamaican migrant in New York 'down where Lexington cross Forty-Seventh Street ... turning tricks for the dudes in the big city' to a chorus which asked 'what you wantin' wid de white man's world .... Black boy want you in his island world'. Reggae music pointed to a far from idyllic Caribbean. Bob Marley, in *Buffalo Soldier*, traced the problems of slavery from its African roots to the role of black people in the American Civil War, whilst *No Woman No Cry* emphasized the problems of slum life in the Kingston ghetto of Trenchtown. A new, more problematic, Caribbean identity was evident, linked outwards in two time dimensions: to the Rastafarian homeland in Ethiopia, and to the diaspora where the Fugees, migrants from Haiti, sung of difficult times and Boney M queried 'How can we sing the Lord's song in a strange land?' Migration created the contemporary Caribbean, found expression in its music, and contributed to the diffusion of reggae throughout and beyond the Caribbean diaspora.

In Ireland, where migration has become something of a way of life, themes of exile and migration have almost dominated the folk and popular musical canon. Boyle *et al.* (1998) draw attention to the music of The Pogues, an Anglo-Irish group formed in London in 1982, and of their principal singer–songwriter Shane McGowan, most of whose albums contain reflections on the experience of emigration, alienation and transformation of identity. Indeed, 'a more telling expression of the ambiguity of the Irish culture of migration is hard to find' (Boyle *et al.*, 1998, p. 226). That ambiguity is evident in the music of most migrants, as in the Torres Strait, where the music of Christine Anu (Chapter 10) not only reflected these uncertainties of identity but also evoked a powerful image of an island home that was hitherto largely unknown in the Australian consciousness. Music transformed island identity, both in the islands and in the wider nation.

At their simplest, literature and song attest to the significance of migration as a contemplative theme. More substantively, literature emphasizes the fundamental ambiguity of the migration experience, the uncertainties of arrivals and departures, the abiding significance of birthplace and childhood, all of which cast doubt on simple notions of economic rationale. The complexity of roots and routes, the meanings and metaphors of shifting identities and aspirations, are more evident in these creative forms than in most social surveys. They ensure that, at one level, migration can never be viewed as unidimensional and that, at another, it is always an unfinished process.

The End Game: The Romance of Islands

Islands have often been constructed as idyllic places; in literature, as in More's *Utopia* (1516), they may appear simply as utopian spaces, alternatives to harsher realities. In a more contemporary era such books as *Treasure Island* (Stevenson, 1883), *The Coral Island* (Ballantyne, 1857), *Robinson Crusoe* (Defoe, 1719) and *The Swiss Family Robinson* (Wyss, 1812) have largely portrayed pleasant realms, despite difficulties in establishing new regimes, and in the face of a few accounts, such as William Golding's *Lord of the Flies* (1954), which trace the social tensions and introspections of island life. Such European literary fascination for places seemingly on the edge of the world has ensured that islands are both imaginary and real.

Isolation has ensured some resistance to globalization. Though islands were sometimes important nodes in otherwise empty seas, remoteness protected many from colonization and colonialism: resources were sometimes slight or easily exhausted, goods were too few to trade, land areas insufficient for plantations and even landing sites inadequate. Marginality was enhanced by exclusion from shipping routes and the construction of dystopia: from Alcatraz to Norfolk Island (Chapter 11). Yet in the present century, marginality has been re-evaluated. Islands have become promising pieces of real estate, and specialized agents retail islands from the Hebrides to the Caribbean as sites for individual utopias. Others have found islands to be places of escapism, retreat and meditation. Tom Neale's solitary decade on a tiny Cook Island, recounted in *An Island to Oneself* (Neale, 1966), is one recent version of the more formal contemplative life that attracted monks to islands such as Lindisfarne in the North Sea and Skellig Michael off the west coast of Ireland (Moorhouse, 1997). Islands, in both negative and positive ways, have been exclusive sites for willing and unwilling migrants.

Isolation remains problematic, despite air transport and telecommunications. In favourable locations, including several Caribbean islands, this combination ushered in mass tourism, electronic industries, financial havens and the reversal of population decline. For Anguilla, Bermuda and the Cayman Islands it was a new dawn. For isolated coral atolls such as Woleai in Micronesia, telecommunications brought the ability to plan for imminent cyclones, ship arrivals or health problems, but no change to economic life. Even in much larger Guam, the advent of television 'makes us homesick for places we have never seen' (Underwood, 1985) and migration accompanied technological modernization.

Technology continues to transform island lives, whether in the recent arrival of television in St Helena (which simultaneously brought *The Simpsons* and the O.J. Simpson trial) or the even more recent arrival of the telephone in Tokelau. In the Caribbean, cable TV has replaced the cultural legacy of British colonialism with 'American monoculture' (Chapter 6), whilst the Internet has provided a forum for uniting Tongans worldwide, stretching and deepening social relations across space, creating a cyber-migrant network with the island at its hub (Chapter 12). A past history of migration and the establishment of diaspora communities hence places islands at key nodes in global society. Island populations become 'transnational socio-cultural systems' which themselves evolve along with the developing map of migration. Flows of people and information by telephone, post, Internet, video, visits and longer-term migrations move in many directions and create a spatially dispersed yet culturally cohesive community. Festivals, celebrations, family decisions and even day-to-day events are shared across this transnational space in a kind of 'in here/out there' multiple connectivity across increasingly permeable cultural and political boundaries (Held, 1995). Island communities are simultaneously being united in virtual space and becoming fragmented across the globe.

In the long boom of the 1960s and 1970s, tourism transformed many islands, as idyllic images were manipulated in the leisure era. Vast numbers of islands were increasingly perceived and conceptualized in this way alone. Simultaneously, the destruction of isolation, combined with greater longevity and the portability of improved pensions, brought a new romantic turn – a combination of return migration and retirement migration. Even seemingly unpropitious islands, such as isolated Wallis and Futuna, could be described as 'holiday islands and havens for the retired' (Roux, 1980, p. 174), whilst in the Mediterranean and around other developed countries, the new 'end game' transformed and revived island livelihoods, though not always without friction. Established residents often resented not only those who merely sojourned in the islands, but even those fellow-islanders who returned with new wealth and ways from distant metropolises.

In every era, and in different regions at different scales, there are diverse island images and realities; hence migration routes and destinies vary.

Whilst a definition of islands may be easy, as islands are simply places surrounded by water, their ends and limits are uncertain. As we have tried to show in this opening chapter, they are places of colonialism, neo-colonialism and tourism; they are places that control oceanic realms, reflected in the manner in which the Pacific has been conceptualized as 'a sea of islands' rather than 'islands in a far sea' (Hau'ofa, 1993); and they are places of emigration and immigration. In short, they have become places where 'local experience no longer coincides with the place in which it takes place' (Pred and Watts, 1992, p. 7). No longer little isolated fragments of geography, the once 'small worlds' of islands have become enmeshed in the opportunities and uncertainties of globalization. As islanders emigrate, return, circulate and commute, they lead 'global lives'. Mobility is both real and virtual, as diasporic islanders email their observations on island cultures and the migration experience. The identities of islands and islanders, and the meaning of migration, are constantly in flux.

Acknowledgements

Every effort has been made to obtain permission from copyright holders for reproducing song lyrics in this chapter.

Note

1. Alkire (1978, pp. 28–30) suggested a figure of 30–50 for coral atolls, and Williamson and Sabath (1984, p. 31) argued that the Marshallese abandoned permanent settlements on islands that would support less than 40–80 people. However, Aboriginals apparently abandoned Bass Strait islands (between Tasmania and the Australian mainland) that were too small to support populations of less than 300–450. Yet McArthur *et al.* (1976, p. 325) concluded that survival was possible for isolated communities of less than 30. The Scottish island of St Kilda was abandoned in 1930 when the population fell to 36.

References

Aldrich, R. (1993) *The Seduction of the Mediterranean*. London: Routledge.

Aldrich, R. (1995) 'From Francité to Créolité: French West Indian literature comes home', in R. King, J. Connell and P. White (eds), *Writing Across Worlds: Literature and Migration*. London: Routledge, pp. 101–24.

Aldrich, R. and Connell, J. (1998) *The Last Colonies*. Cambridge: Cambridge University Press.

Alexander, C. (1995) 'Rivers to cross: exile and transformation in the Caribbean migration novels of George Lamming', in R. King, J. Connell and P. White (eds), *Writing Across Worlds: Literature and Migration*. London: Routledge, pp. 57–69.

Alkire, W.H. (1978) *Coral Islanders*. Arlington Heights: AHM Press.

Ballantyne, R.M. (1857) *The Coral Island*. London: Blackie, 1903.

Beller, W.S. (ed.) (1987) *Proceedings of the Interoceanic Workshop on Sustainable Development and Environmental Management of Small Islands*. Washington, DC: US Department of State, Man and the Biosphere Program.

Bertram, I.G. and Watters, R.F. (1985) 'The MIRAB economy in South Pacific microstates', *Pacific Viewpoint*, 26(3), pp. 497–520.

Binder, P. (1977) *Treasure Islands: The Trials of the Ocean Islanders.* London: Blond & Briggs.

Boyle, P., Halfacree, K. and Robinson, V. (1998) *Exploring Contemporary Migration.* Harlow: Addison Wesley Longman.

Braudel, F. (1972) *The Mediterranean and the Mediterranean World at the Time of Philip II.* London: Methuen.

Brunhes, J. (1920) *Human Geography.* London: Harrap.

Carrier, J. (1984) *Education and Society in a Manus Village.* Port Moresby: University of Papua New Guinea, Education Research Unit Report 47.

Cassola, A. (1995) 'Migration in contemporary Maltese fiction', in R. King, J. Connell and P. White (eds), *Writing Across Worlds: Literature and Migration.* London: Routledge, pp. 172–9.

Chamberlain, M. (1995) 'Family narratives and migration dynamics: Barbadians to Britain', *Nieuwe West-Indische Gids*, 69(3–4), pp. 253–75.

Chapman, M. (1991) 'Pacific island movement and socioeconomic change: metaphors of misunderstanding', *Population and Development Review*, 17(2), pp. 263–92.

Cleland, J.G. and Singh, S. (1980) 'Islands and the demographic transition', *World Development*, 8(12), pp. 969–94.

Cohen, A.P. (1982) 'A sense of time, a sense of place: the meaning of close social association in Whalsay, Scotland', in A.P. Cohen (ed.), *Belonging: Identity and Social Organisation in British Rural Culture.* Cambridge: Cambridge University Press, pp. 21–49.

Condon, S. and Ogden, P. (1996) 'Questions of emigration, circulation and return: mobility between the French Caribbean and France', *International Journal of Population Geography*, 2(1), pp. 35–50.

Connell, J. (1988) *Sovereignty and Survival: Island Microstates in the Third World,* Research Monograph 3, Sydney: University of Sydney, Department of Geography.

Connell, J. (1990) 'Modernity and its discontents: migration and social change in the South Pacific', in J. Connell (ed.), *Migration and Development in the South Pacific,* Pacific Research Monograph 24. Canberra: ANU National Centre for Development Studies, pp. 1–28.

Connell, J. (1995) 'In Samoan worlds: culture, migration, identity and Albert Wendt', in R. King, J. Connell and P. White (eds), *Writing Across Worlds: Literature and Migration.* London: Routledge, pp. 263–79.

Connell, J. (1997) *Papua New Guinea: The Struggle for Development.* London: Routledge.

Connell, J. and Brown, R. (1995) 'Migration and remittances in the South Pacific: towards new perspectives', *Asian and Pacific Migration Journal*, 4(1), pp. 1–34.

Cruz, M., d'Ayala, P.G., Marcus, E., McElroy, J. and Rossi, O. (1987) 'The demographic dynamics of small islands', *Ekistics*, 54(323–4), pp. 110–15.

Cunningham, P. (1998) 'Blasket's case', *Independent on Sunday: Sunday Review*, 11 January, pp. 41–4.

Defoe, D. (1719) *Robinson Crusoe* (edited by A. Ross). Harmondsworth: Penguin, 1965.

Findlay, A.M. and Li, L.L.N. (1997) 'An auto-biographical approach to understanding migration: the case of Hong Kong emigrants', *Area*, 29(1), pp. 33–44.

Fuka, M.L.A. (1985) 'The Auckland Tongan Community and Overseas Remittances', unpublished MA thesis. Auckland: University of Auckland.

Galtung, J. (1971) *Members of Two Worlds: A Study of Development in Three Villages in Sicily*. Oslo: Universitetsforlaget.

Gmelch, G. (1987) 'Work, innovation and investment: the impact of return migrants in Barbados', *Human Organization*, 46(2), pp. 131–40.

Golding, W. (1954) *Lord of the Flies*. London: Faber & Faber.

Halfacree, K. and Boyle, P. (1993) 'The challenge facing migration research: the case for a biographical approach', *Progress in Human Geography*, 17(3), pp. 333–48.

Hau'ofa, E. (1993) 'Our sea of islands', in E. Waddell, V. Naidu and E. Hau'ofa (eds), *A New Oceania*. Suva: University of the South Pacific, pp. 2–16.

Held, D. (1995) *Democracy and the Global Order: From the Modern State to Cosmopolitan Governance*. Cambridge: Polity Press.

Hionidou, V. (1995) 'The demographic system of a Mediterranean island: Mykonos, Greece', *International Journal of Population Geography*, 1(2), pp. 125–46.

Hooper, A. and Huntsman, J. (1973) 'A demographic history of the Tokelau islands', *Journal of the Polynesian Society*, 82(4), pp. 366–411.

James, K.E. (1991) 'Migration and remittances: a Tongan village perspective', *Pacific Viewpoint*, 32(1), pp. 1–23.

Jones, H.R. (1973) 'Modern emigration from Malta', *Transactions of the Institute of British Geographers*, 60, pp. 101–20.

King, R. (1995) 'Migrations, globalization and place', in D. Massey and P. Jess (eds), *A Place in the World?* Oxford: Oxford University Press, pp. 5–44.

King, R. (1998) 'Islands and migration', in E. Biagini and B. Hoyle (eds), *Insularity and Development: International Perspectives on Islands, Planning and Core-Periphery Relationships*. London: Pinter, pp. 93–115.

King, R., Connell, J. and White, P., (eds) (1995) *Writing Across Worlds: Literature and Migration*. London: Routledge.

King, R., Iosifides, T. and Myrivili, L. (1998) 'A migrant's story: from Albania to Athens', *Journal of Ethnic and Migration Studies*, 24(1), pp. 159–75.

King, R. and Strachan, A.J. (1980) 'The effects of return migration on a Gozitan village', *Human Organization*, 39(2), pp. 175–9.

Lesourd, M. and Réaud-Thomas, G. (1987) 'Le fait créole dans la formation de l'identité nationale en République du Cap-Vert', in J.P. Doumenge (ed.), *Iles Tropicales: Insularité, Insularisme*. Talence: Centre de Recherches sur les Espaces Tropicaux, pp. 107–24.

Lowenthal, D. and Comitas, L. (1962) 'Emigration and depopulation: some neglected aspects of population geography', *Geographical Review*, 52(2), pp. 195–210.

Marcus, G.E. (1981) 'Power on the extreme periphery: the perspective of Tongan elites in the modern world system', *Pacific Viewpoint*, 22(1), pp. 48–64.

McArthur, N., Saunders, I.W. and Tweedie, R.L. (1976) 'Small population isolates: a micro-simulation study', *Journal of the Polynesian Society*, 85(3), pp. 307–26.

McElroy, J.E. and de Albuquerque, K. (1988) 'Migration transition in small northern and eastern Caribbean states', *International Migration Review*, 22(1), pp. 30–58.

McGrath, F. (1991) 'The economic, social and cultural impacts of return migration on Achill Island', in R. King (ed.), *Contemporary Irish Migration*. Dublin: Geographical Society of Ireland, Special Publication 6, pp. 55–69.

More, T. (1516) *Utopia* (trans. P.Turner). Harmondsworth: Penguin, 1965.

Moorhouse, G. (1997) *Sun Dancing: A Mediaeval Vision*. London: Weidenfeld & Nicolson.

Munro, D. (1990) 'Transnational corporations of kin and the MIRAB system', *Pacific Viewpoint*, 31(1), pp. 63–6.

Naidu, V. (1981) 'Fijian development and national unity: some thoughts', *Review*, 2, pp. 3–12.

Neale, T. (1966) *An Island to Oneself.* Auckland: Collins.

Nietschmann, B. (1979) 'Ecological change, inflation and migration in the far western Caribbean', *Geographical Review,* 69(1), pp. 1–24.

Nuttall, M. (1992) *Arctic Homeland: Kinship, Community and Development in Northwest Greenland.* London: Belhaven Press.

O'Sullivan, M. (1993) *Twenty Years a-Growing.* Oxford: Oxford University Press.

Potts, L. (1990) *The Global Labour Market: A History of Migration.* London: Zed Books.

Pred, A. and Watts, M. (1992) *Reworking Modernity: Capitalism and Symbolic Discontent.* New Jersey: Rutgers University Press.

Prentice, R. (1990) 'The Manxness of Man: renewed immigration to the Isle of Man and the nationalist response', *Scottish Geographical Magazine,* 106(2), pp. 75–88.

Richardson, B.C. (1986) *Panama Money in the Barbados 1900–20.* Knoxville: University of Tennessee Press.

Roux, J.-C. (1980) 'Migration and change in Wallisian society', in R.T. Shand (ed.), *The Island States of the Pacific and Indian Oceans.* Canberra: Australian National University, Centre for Development Studies Monograph 23, pp. 163–76.

Rubenstein, H. (1979) 'The return ideology in West Indian migration', *Papers in Anthropology,* 20(1), pp. 21–38.

Semple, E.C. (1911) *Influences of the Geographic Environment.* London: Constable.

Shankman, P. (1976) *Migration and Underdevelopment: The Case of Western Samoa.* Boulder, CO: Westview.

Stevenson, R.L. (1883) *Treasure Island.* Cleveland: World Publishing Company, 1946.

Thomas-Hope, E. (1985) 'Return migration and its implications for development', in R. Pastor (ed.), *Migration and Development in the Caribbean.* New York: Westview, pp. 157–77.

Underwood, R. (1985) 'Excursions into inauthenticity: the Chamorros of Guam', in M. Chapman (ed.), *Mobility and Identity in the Island Pacific (Pacific Viewpoint 26),* pp. 139–59.

Warnes, A.M. and Patterson, G. (1998) 'British retirees in Malta: components of the cross-national relationship', *International Journal of Population Geography,* 4(2), pp. 113–33.

Wessen, A., Hooper, A., Huntsman, J., Prior, I. and Salmond, C. (1992) *Migration and Health in a Small Society: The Case of Tokelau.* Oxford: Clarendon Press.

Williamson, I. and Sabath, M.D. (1984) 'Small population instability and island settlement patterns', *Human Ecology,* 12(1), pp. 21–34.

Wyss, J.D. (1812) *The Swiss Family Robinson* (translated by J. Lovell). London: Cassell, 1869.

2

From the Periphery of the Periphery: Historical, Cultural and Literary Perspectives on Emigration from the Minor Islands of Ireland

Stephen A. Royle

We're still here, still fighting, proud that because of the sweat and tears and heartache we expended, the annoyance we caused and the publicity we created and the fight we put up, no other island off the Irish coastline has been evacuated. (Ó Péicín, 1997, p. 137)

This sums up *Islanders*, the recent book about the experiences of Diarmuid Ó Péicín, a priest who served on Tory Island in the 1980s at a time when the official evacuation of Tory and some of the other islands off the coast of Ireland was viewed as being likely on economic grounds. He sought to change this view, fought to diversify and modernize his island's economy: the book's subtitle is 'The true story of one man's fight to save a way of life'. Since leaving Tory he has been active in establishing pressure groups which strive to resist the continuing economic imperatives for emigration from the small Irish islands and to promote a positive view of their significance to Ireland. This campaign has been at least partially successful because government policy is now to support the islanders in their homes, based on a new realization of the cultural significance and value of the islands and their communities (Government of Ireland, 1996).

In the past the small offshore islands of Ireland had rates of population loss remarkable even within an Irish context, for the island of Ireland itself was subject to a diaspora of its people of biblical proportions. This emigration dated from before the Great Famine of the 1840s, though it much increased at and after that period. In all 120 years of population decline saw Ireland's population fall from 8.2m in 1841 to 4.2m in 1961. At 5.1m in 1991 the combined population of the Republic of Ireland and Northern Ireland remained 37.6 per cent below the peak of 1841 (Table 2.1). Irish people spread throughout much of the world; Australia, Canada, New Zealand, South Africa and, especially, the United States of America together contain far more people of Irish descent than live today in Ireland. Perhaps 60 million people worldwide are of Irish descent. The scattering of its people makes Ireland a suitable subject for this book's 'global lives' theme, but Ireland, the world's twenty-first largest island, is not a 'small world', being of sufficient scale to have large cities. Instead, in this chapter, I concentrate on the Irish islands which are indeed small worlds and which have participated even more dramatically than Ireland as a whole in emigration.

This chapter is in three main parts. First, I set the scene by presenting some historical material on emigration from the Irish islands, including detailed population data. The second part sees a shift in focus to an exploration of the 'island experience' of migration through the lens of Irish island literature, notably the works of three authors from the now-abandoned Great Blasket Island. The third section surveys the islands in the modern era. Migration continues to be a relevant process, but official policies, tourism and other economic opportunities combine to make the demographic situation somewhat less bleak now than in the past, at least for many islands.

Historical Migration from the Irish Islands

In the past pressures on the economic or social systems of the small islands of Europe's Celtic fringe were normally met by out-migration. Sometimes an island proved unable to sustain a viable society at all and depopulation would result. One well-known example of this is St Kilda, the outermost of Scotland's Outer Hebrides, from where the population was removed in 1930 (Steel, 1975). Even where depopulation did not result, migration became a way of life. This can be demonstrated clearly from the Irish census statistics. Table 2.1 identifies an overall decline in Irish island residents of 74.6 per cent

Table 2.1   Population of Ireland and its islands, 1841 and 1991

	1841	*1991*	*% change*
Ireland	8,175,124	5,101,237	–37.6
Irish islands (211 inhabited islands in 1841, 66 in 1991)	38,138	c. 9,700	–74.6

Source: Censuses.

Note: The imprecision concerning the 1991 island population relates to the Northern Ireland Census which does not separately identify the population of the small islands in Strangford Lough.

between 1841 and 1991, compared to Ireland's overall decline of 37.6 per cent. A recent analysis of Irish island populations (Royle and Scott, 1996) postulated that poor accessibility helped explain the broad differences in the rates of decline between islands. There are some islands, such as Inishturk, where the population total has held up well in recent years despite isolation (Cross, 1996), but in no case has an island remained anywhere near its historic population total. Table 2.2 demonstrates this latter fact for all the islands of the Republic that were inhabited in 1971.

There is also much physical evidence of migration on the islands. All were more intensively farmed in the past than today and this can be seen in the form of abandoned fields and lazy beds (potato ridges) within fields now used only for grazing, if at all. Also obvious are the abandoned and ruined houses. Table 2.3 records the author's survey of several islands' housing stock in 1997 compared with historic data. There has been a decline in habitation in isolated islands such as Inishturk but also those with good access such as Inishnee, which has a bridge and is only a short distance from a town. The location of Inishturk, Inishnee and other islands mentioned in this chapter is given in Figure 2.1; for a more detailed map of all the minor Irish islands see Royle and Scott (1996, p. 112).

Sometimes small islands lose their people in dramatic fashion: the crisis caused by the volcano in Montserrat may yet necessitate further evacuation of that island, as Philpott discusses in Chapter 7. In the Irish situation, however, migration has usually been far from dramatic: a steady drip of individuals or families leaving for mundane reasons of economics. Small islands with their restricted opportunities offer in reality or perception a life less rich materially than alternatives on neighbouring or distant mainlands. Often the relative success of early migrants is communicated back, encouraging others to follow. A downward spiral of decline thus becomes established. Aalen and Brody in their study of Gola sum it up thus: islanders come to 'expect more than their small country communities can provide. As they leave, so such communities ... come to be able to provide less and less' (1969, p. 126).

Island economies were often stretched to the limit. This can be seen clearly in the economic history of the Aran Islands. Through the fortunate survival of 1821 census enumerators' books (a category of documents almost entirely lost in Ireland when the record office in Dublin was blown up in the Civil War in 1921), these islands' economies can be recreated in considerable detail (Royle, 1983). Here was a peasant society where the few resources available to the islanders were exploited to the utmost. Most worked in primary activities, creating a subsistence society that had little need for traders or service providers. Almost all household heads either worked their own landholdings as tenants (the islands were owned by an absentee landlord in Dublin) and/or were labourers. The land they farmed often had to be 'made'; soil was formed by spreading layers of sand and sea

Table 2.2 Population development of islands which were inhabited in 1971, Republic of Ireland, 1841–1991

Island	1841	1871	1901	1926	1951	1971	1979	1981	1986	1991
County Clare										
Canon	54	39	24	18	7	5	–			
Coney	145	90	35	41	23	6	2	2		
Green	9	5	3	7	5	1	1	1	–	
Inishloe	114	91	32	28	14	3	–			
Inishmacowney	142	95	44	41	22	18	–			
Inishmore (Deer)	133	71	45	44	24	15	3	3	2	1
Saints	–	18	14	6	1	1	–			
Scattery	65	140	99	100	56	2	–			
County Cork										
Ballycottin	–	11	3	3	3	3	7	5	3	4
Bere (Bear)	2,122	1,125	1,059	1,182	534	288	258	252	230	216
Clear (Cape Clear)	1,052	572	601	453	257	192	155	164	145	132
Dursey	48	197	205	162	96	38	25	19	20	20
Fastnet	–	–	3	3	3	3	3	3	–	
Garinish	13	–	5	87	3	2	2	2	2	1
Haulbowline	222	379	431	430	275	306	325	452	466	307
Inchydoney	235	79	n/a	n/a	n/a	59	114	89	106	109
Inishbeg	335	44	17	n/a	n/a	10	24	18	19	17
Hare (Inishdriscol)	358	279	317	216	139	79	54	35	19	22
Long	336	220	199	143	61	40	39	38	18	11
Ringarogy	597	243	166	134	97	75	70	69	75	71
Roancarrigmore	0	4	2	2	4	3	–			
Sherkin	1,131	452	350	248	146	82	82	70	87	93
Spike	202	1,310	617	442	299	185	185	130	60	113
The Bull	0	0	4	3	3	5	5	1	–	
Whiddy	729	n/a	259	215	104	111	57	54	41	34

Table 2.2 (cont'd)

Island	1841	1871	1901	1926	1951	1971	1979	1981	1986	1991
County Donegal										
Aranmore	1,431	1,174	1,308	1,390	1,249	773	825	803	735	596
Auginish	26	5	10	7	5	5	3	7	7	6
Cruit	258	273	245	237	151	69	84	83	89	77
Inishbofin	121	125	150	120	n/a	103	69	46	6	3
Inishfree Upper	137	181	190	153	89	27	7	11	17	19
Inishtrahull	54	53	65	39	4	3	3	3	–	
Island Roy	74	47	40	30	18	17	9	10	13	13
Owey	94	130	137	143	106	51	–			
Rathlin O'Birne	0	4	4	3	3	3	–			
Rutland	125	65	65	41	35	2				
Tory	399	343	335	250	257	273	213	208	136	119
County Dublin										
Lambay	89	88	28	45	21	24	8	12	10	8
County Galway										
Annaghvaun	125	166	109	69	76	73	75	79	90	103
Dinish	59	64	52	25	35	12	3	3	–	
Finish	66	134	100	81	49	10	5	5	–	
Furnace	155	154	87	77	85	68	66	65	64	56
Gorumna	1,910	1,417	1,540	1,451	1,440	1,108	1,122	1,120	1,080	1,082
Illaunamid	30	16	3	3	4	3	4	3	–	
Illaunamore	89	89	68	46	23	3	7	6	4	2
Inchaghaun	0	7	7	13	5	7	7	5	3	1
Inchamackinna	41	7	9	31	51	37	–			
Inishbarra	0	9	0	88	75	21	5	5	1	1
Inishbofin	121	125	150	540	291	236	203	195	177	181
Inisheer	456	495	480	409	388	313	257	239	255	270
Inisheltia	0	3	4	3	12	4	2	3	3	3

Table 2.2 (cont'd)

Island	1841	1871	1901	1926	1951	1971	1979	1981	1986	1991
County Galway (cont'd)										
Inisherk	49	36	28	31	29	13	7	7	–	
Inishlackan	126	130	117	54	27	2	2	2	–	
Inishmaan	473	433	420	380	361	319	237	238	236	216
Inishmore	2,592	2,110	1,768	1,363	1,016	864	883	891	848	836
Inishnee	455	409	271	174	145	79	83	86	74	47
Inishtravin	93	158	89	102	42	34	17	7	7	4
Inishturk	84	92	95	90	86	35	–	7	–	
Island Eddy	125	59	38	35	16	2	–			
Islands in North Sound	0	12	11	5	3	3	–			
Lettermore	844	718	801	721	843	625	597	582	542	508
Lettermullin	587	497	533	346	320	221	216	218	211	196
Mweenish	649	418	425	313	206	198	182	188	170	149
Omey	397	281	115	71	66	34	21	21	12	12
Roeillaun	17	38	28	13	7	1	–			
Rossroe	7	30	34	28	34	19	25	31	26	28
Rusheenacholla	53	45	32	13	18	9	6	6	7	4
Turbot	146	160	129	108	89	65	7	1	–	
County Kerry										
Begenish	84	22	44	26	14	2	2	2	2	5
Carrig	105	31	16	22	23	16	13	10	11	8
Dinish	11	7	9	4	5	6	5	6	6	5
Rossmore	0	101	79	42	32	11	9	9	–	
Samphire	0	0	9	2	2	7	–			
Skellig Rock, Great	0	0	6	3	3	3	3	3	–	
Tarbert	134	175	124	22	23	20	13	18	17	17
Tearaght	0	4	3	3	3	5	3	3	–	
Valencia	2,290	2,139	1,864	1,483	1,015	770	712	718	666	680

Table 2.2 (cont'd)

Island	1841	1871	1901	1926	1951	1971	1979	1981	1986	1991
County Mayo										
Achill	4,901	4,757	4,825	4,790	4,906	3,129	3,089	3,101	3,161	2,802
Annagh	0	15	11	10	6	5	–			
Black Rock	0	6	3	3	3	3	–			
Clare	1,615	494	490	378	278	168	132	127	140	137
Clynish	87	20	30	18	7	4	4	4	3	5
Dorinish More	40	14	8	0	0	1	–			
Eagle	15	17	3	3	3	3	3	3	–	
Illaunataggart	33	26	9	5	4	1	–			
Inishbiggle	67	154	135	162	123	112	97	89	69	51
Inishcottle	40	47	27	30	11	8	11	11	10	10
Inishgort	32	35	9	0	9	6	3	3	3	2
Inishlyre	113	73	47	21	14	18	12	10	7	6
Inishakillew	126	45	41	16	11	15	15	13	7	7
Inishturk	577	112	135	101	123	83	85	76	90	78
Knockycahillaun	41	23	21	14	8	4	–			
County Sligo										
Dernish	42	48	39	31	11	4	1	1	–	
Inishmulclohy (Coney)	124	93	64	47	25	5	7	7	6	5
County Wexford										
Tuskar Rock	4	6	3	6	3	3	3	3	3	5

Source: Censuses.

Table 2.3 Buildings and services, selected Irish islands, *circa* 1900 and 1997

| Island | Historic data | | 1997 data | | | | | |
| | Residential buildings | Services | Residential buildings | | | | | Services |
			Inhabited	Uninhabited	Ruined	Sheds	Being built	
Clare	104 (data 1915)	Church Police barracks Post office Schools (2)	59	5	25	48	3	Bed and Breakfast (4) Church Cycle hire (2) Community centre Hotel Lighthouse Post office Power station Public house Schools (2) Shops (3)
Dursey	69 (data 1898)	School	8 (3 holiday homes)	16	16	27	–	None
Inishnee	85 (data 1899)	School	31	1	21	21	–	None
Inishturk	54 (data 1915)	Church Post office School	19	6	9	30	2	Bed and Breakfast (2) Community centre Church Post office Power station School
Tory	87 (data 1909)	Church Lighthouse Post office School	60	16	8	64	11 (or restoration)	Bed and Breakfast Café Church Community centre Cycle hire Gallery Hostel Hotel Lighthouse Offices (2) Power station Shops (2)
Whiddy	68 (data 1896)	School	15	–	6	25	–	Museum Post office Public house Restaurant Shop

Sources: 1:10,560 Ordnance Survey Maps for the historic data; field survey for the 1997 data.

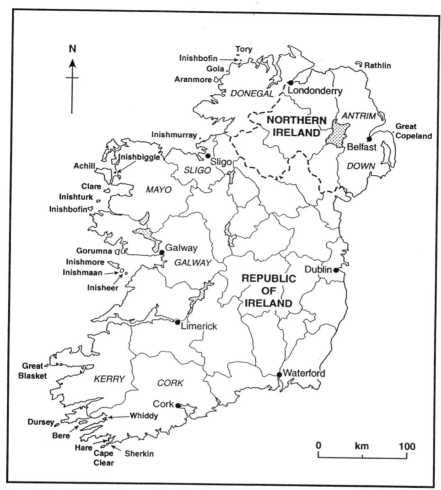

Figure 2.1 Irish islands mentioned in the text

weed onto the bare limestone surfaces of the islands, loose rocks being made into walls to protect the precious productive land (Conry, 1971). In addition, most men worked as fishermen or kelp-makers (kelp is seaweed which was gathered, dried and burnt to ash, from which iodine and other by-products were extracted). Kelp-making was an important feature of the Aran economy as it produced cash, needed to meet rental payments. Few people had only a single occupation – some full-time fishermen on Inishmore, labourers and a small number of widows who worked in textile or net production. Further activities not recorded as occupations but known to have existed included sealing, the taking of seabirds and illicit distilling. Wrack (material washed ashore) was avidly gathered – providing wood for building or fuel.

Otherwise on their treeless, thin-soiled islands which provided neither wood nor turf (the local expression for peat, the common fuel of rural Ireland), the islanders burnt *bualtrach* (cow dung), which would be dried on the walls. In short, the Aran Islands in 1821 displayed an economy which was stretched to its utmost with every feasible activity exploited. No wonder then that in 1822 when bad weather caused the potatoes to fail, there was a crisis on the islands serious enough to require outside help (Royle, 1984).

Such over-stressed economies were not capable of much expansion and were increasingly unable to satisfy people who during the nineteenth century became aware of the opportunities caused by industrialization in north-east Ireland and Great Britain as well as the real or mythical opportunities of faraway North America or elsewhere. Consequently, Aran and other islands became subject to considerable emigration, reflected in the statistics of Tables 2.1 and 2.2.

In the late nineteenth and early twentieth centuries, many islands were subject to outside help through the work of the Congested Districts Board. Published accounts are available of its activities on Great Blasket (Stagles and Stagles, 1980), the Aran Islands (Harvey, 1991) and Clare (Dwyer, 1963). Clare's farming landscape was remodelled from a fragmented pattern into 74 consolidated strip farms, with new houses, strung out along flatter coastal land, backed by a wall the length of the island to separate the farms from communal outfield pastures (Figure 2.2). Rental payments were replaced by purchase annuities to convert the farmers from tenants into potential owners. However, within a short time several new, expensive farmhouses were abandoned as islanders continued to emigrate – population fell 23 per cent during 1901–26 – the attractions of the outside world continuing to exceed whatever Clare could be reorganized to offer.

In tandem with the economic push factors were those related to insular society. One difficulty was in finding marriage partners. For example, on Great Copeland in 1911 when there were just three families, only two elderly couples were married; none of the fourteen people between eighteen and thirty was wed (Royle, 1994). Perhaps taboos on kinship marriage prevailed since all the islanders shared just two surnames. The young population drifted away to seek partners elsewhere leaving 'the older people to continue to wrest a living from the soil and to battle with the sea. Eventually this proved too much for their tired bodies and they were compelled to leave the islands' (Pollock, 1982, p. 62).

Sometimes the remaining islanders left as a group, as with Scotland's St Kilda. The depopulation of Inishmurray in 1948 and Great Blasket in 1953 are examples of the 'last boat out' situation, but usually there is a slow drift of people away until one winter there is nobody left, a process Aalen and Brody were observing on Gola prior to 1969. Even on Great Blasket one family did not take the 'last boat' and hung on until 1954. The historian of

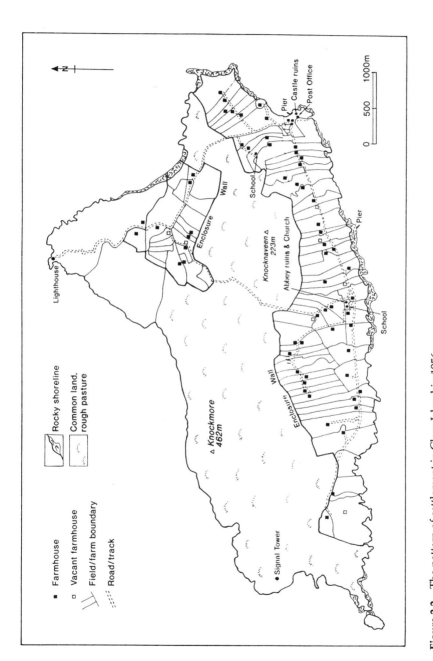

Figure 2.2 The pattern of settlement in Clare Island in 1956
Source: adapted from Dwyer, 1963; previously published in Royle, 1989.

Inishmurray, Patrick Heraughty, was one of those who were evacuated in 1948. He records that emigration had always been a way of life but had to some extent been balanced by return migration. Then 'in the 1920s and 1930s emigrants no longer returned. Life was better elsewhere' (1982, p. 71). By the 1940s the remaining islanders were petitioning the council to resettle them on the mainland. Eight houses were built on the coast opposite the island but in the event only six were needed as some petitioners had already migrated. There were just 46 people taken off Inishmurray on 12 November 1948, mostly very old or very young from an island which at its peak in 1880 supported 102 people.

On Achill, the largest Irish island, there was much seasonal migration, especially of 'tatie [potato] squads' to Britain, but migration soon became more permanent and the island's landscape shows clear evidence of this, especially in the deserted village of Slievemore (see McGrath, 1991, who also details more recent return migration). Slievemore's straggle of roofless, stone cottages is now a tourist attraction and the village has a preservation society. This is not the only example of cultural heritage emanating from migration, as will be demonstrated; although migration, of course, could also cause great damage, especially to language and other non-material cultural forms (Hindley, 1994).

'Twould be a bad place that wouldn't be better for you than this dreadful rock': Irish Island Culture and Emigration

Irish island culture was predicated both upon the 'islandness' of its origins and, especially within the outpouring of literature in the twentieth century, upon the tradition of out-migration. The separateness from mainland and mainstream society of small islands engenders a sense of their being special. The romance of this is exploited by the modern travel industry which markets, it seems, almost every island with an 'island of dreams' or 'paradise island' tag. Romance was hardly a factor on the Irish islands, but the idea of them being remarkable places inspired a number of islanders to write about their homelands. Some wrote consciously for posterity, aware that a way of life was slipping away. This feeling was reinforced by visitors who were not tourists in the modern sense but rather antiquarians conscious both of the quality of island culture and also of its fragility in the face of development and out-migration.

All three of the notable Great Blasket Island autobiographers I am going to focus on were inspired thus by outsiders (see Flower, 1944; Thompson, 1982). The three writers, who came from a population never more than about 160, were Tomás O'Crohan (1856–1937), author of *The Islandman (An tOileánach)*, his great-nephew Maurice O'Sullivan (1904–50), who wrote *Twenty Years a'Growing* and Peig Sayers (1873–1958) who married into the island in 1892 and published *Peig* in 1936. Sayers, a traditional storyteller, narrated her books to her son. This son, a schoolmate of Maurice O'Sullivan, also wrote about the island (O'Guiheen, 1982) as did Tomás O'Crohan's son (O'Crohan, S., 1992) and daughter-in-law (Ní Shúlleabháin, 1978). Both Tomás O'Crohan (1986) and Sayers (1978) also wrote other books about Great Blasket. There are other accounts of life on the Blaskets including that of the island nurse, Méiní Dunleavy (1876–1967), whose biography (Matson, 1996) is puffed on the cover as a 'worthy successor to *An tOileánach and Twenty Years a'Growing'*, though the biographer himself makes no such claim in his more modest foreword. This is the latest in what has been recognized as the 'Blasket library', a canon itself the subject of further books (Stagles and Stagles, 1980; MacConghail, 1987). Other islands have a less considerable output, but worthy of note is Conchúr Ó Síocháin's *The Man from Cape Clear*, when 'for the first time, a voice from another island spoke', to quote the foreword to the translation (1975, p. ix).[1]

Notable authors from outside the island realm who wrote on the islands include J.M. Synge, who visited Inishmaan regularly and wrote *The Aran Islands* (1907), and Deborah Tall, who acknowledged her debt to Synge in her not always flattering account of a period of residence on Inishbofin (County Galway), hardly disguised by its title, *The Island of the White Cow* (1986), that being a translation of Inishbofin. Novelists writing about the islands include Peadar O'Donnell (*Islanders*, 1927; *Proud Island*, 1975) and Liam O'Flaherty, whose powerful *Skerrett* (1932) is set in 'Nara', anagram for Aran, his home island of Inishmore.

The amount and quality of this literary output is remarkable testimony to the culture of the Irish islands. This is expressed also through music, language, crafts (including, but not entirely represented by, Aran sweaters) and painting, this last most notably with the Tory Primitive School, whose artists, inspired originally by English painter Derek Hill and islander James Dixon, produce scenes of their island home. The School has exhibited internationally and its work is important also to Tory's economy, some visitors coming specifically because of it (Scott and Royle, 1992; see also Fox, 1978).

There is not space here to do justice to all the books mentioned (see, instead, MacConghail, 1987 and Ó Conluain, 1983) but the three principal Blasket writers are considered. They demonstrate how even whilst celebrating an extant, culturally-rich and vibrant society, the authors were conscious of its transitory nature and their work is larded with references to emigration and to the passing of an already historic way of life.

The Islandman

Much of O'Crohan's book *The Islandman* details the way in which Blasket islanders earned their living: farming, fishing, trading with the mainland and with passing ships, gathering wrack, cutting peat. His own occupational pluralism – reminiscent of that revealed by the census for the Aran Islands generations before – extended to building his own house, the roofless ruins of which are extant, and to helping the Congested Districts Board build five houses on the island. *The Islandman* has bleak passages, particularly those on the death of most of his children through infection, cliff falls or drowning, but it details in a positive manner a vibrant, if always pressured, island community struggling to wrest a livelihood from an unfavourable ecumene. There is much in the book about the islanders' society and fellowship. O'Crohan strayed little distance from Great Blasket but though his life was 'narrow in its range ... within that range [it was] absolute and complete' (O'Crohan, 1978, p. viii). O'Crohan, helped in this belief by scholarly visitors to whom he taught Irish, thought his small world remarkable enough to set down before it changed completely: 'for the like of us will never be again' (p. 244). Much of this change was caused by emigration. Many of his family went to America leaving Tomás, the 'spoilt child' to 'keep the little house' going for his parents (p. 65). Later a brother returned, having been unsuccessful – 'anybody would have conjectured from his ways that it was in the woods he had spent his time in America' (p. 171) – but such occasional returns did not stem the tide of emigration. One man 'had a houseful of children then, [but] there are only three of them in my neighbourhood now, the rest of them are in America ... like so many others' (p. 188).

Peig and An Old Woman's Reflections

The second author by seniority, though not by publication date, was Peig Sayers. She was brought up on the mainland and had wanted to migrate to America, following her childhood friend, Cait-Jim. One of the most powerful scenes in *Peig* is an 'American wake', the party given to a departing migrant. The funereal element is a recognition that the person would probably never be seen again. However, Cait-Jim was unable to remit passage money to Sayers as promised, leaving her with 'two choices in the palm of my hand – to marry or go into service again' (1974, p. 151). She chose marriage, an arranged match with Blasket Island man, Peats Guiheen. Her first impression of Great Blasket acts as a counterpoise to what was to come, for the island seemed 'black with people' as she was welcomed to her new home (p. 152). Though comforted by the thought of her 'fine handsome man', she still hankered for America (p. 153):

> The blessing of God be with you, Cait-Jim, I said in my own mind, you were the lucky one. Whatever happens your feet will be planted on mainland clay. Not so with me! How lonely I am on this island in the ocean.

Her bounded horizons continued to fret her: 'I think this is a very confined place with the sea out there to terrorize me' (p. 159). In the last chapter, as a lonely widow living in one of the new cottages O'Crohan had built, she again laments her island life (p. 211):

The most of my life I've spent on this lonely rock in the middle of the great sea. There's a great deal of ... hardship in the life of a person who lives on an island like this that no one knows about except one who has lived here – going to bed at night with little food and rising again at the first chirp of the sparrow, then harrowing away at the world and maybe having no life worth talking about after doing our very best.

She certainly had a life worth talking about – literally, as her autobiography was dictated – but the necessity to 'harrow away' is ever present in all the Blasket biographies and this engendered the feelings about the inevitability of emigration. Migration is a constant theme running through *Peig*. In a sad chapter entitled 'Scattering and sorrow' she loses one son, Tomás, to a cliff fall and 'six months after this my son Pádraig hoisted his sail and went off to America' (p. 184), it having been remarked by Tomás on the morning of his death that 'isn't it time he [Pádraig] went' (p. 180). Another son, Muiris, was

deeply attached to his country and to his native language and he never had any desire to leave Ireland. But that's not the way events turned out for he, too, had to take the road like the others, his heart laden with sorrow. As soon as he had turned the last sod on his father's grave he made ready to go. The day he went will remain forever in my memory because beyond all I had endured, nothing ever dealt me as crushing a blow as that day's parting with Muiris ... [but] 'Twould be a bad place that wouldn't be better for you than this dreadful rock. (Sayers, 1974, pp. 185–6)

Muiris promised to return, but his mother recalled an emigration proverb: 'the city has a broad entrance but a narrow exit' (p. 186). Her other children also departed and Sayers was left with only her dead husband's blind brother, a 'feeble old man', for company. Thus when one son, not Muiris but Micheál, returns because of the 'hardship of the world', if 'he was sad, I was delighted' (p. 188).

This fine example of island literature is predicated upon migration – the American 'deaths' trouble her more, it seems, than the real thing – and upon her awareness of change: 'I suppose that never again will there be an old woman as Irish as me on this island' (p. 210). Further, in *An Old Woman's Reflections*, she concludes that 'the little house where I used to eat and drink, it's unlikely there'll be a trace of it there' (1978, p. 129). Actually her house remains and still has a roof, unlike those of the other two authors which are open to the skies, but her foreboding certainly came true regarding the people on her 'beautiful little place, sun of my life' (1978, p. 128), for she must have been one of those evacuated from the island, a few years before her death in 1958.

Twenty Years a-Growing

O'Sullivan's autobiography was by a young man. His grandfather told him that 'the life of man' has four parts: 'Twenty years a-growing, twenty years in blossom, twenty years a-stooping and twenty years declining' (1953, p. 86), and the book is about him a-growing. He left the island in early blossom, and was drowned in his mid-forties, not long into his a-stooping. Thus he does not end his book like O'Crohan, Sayers and, from another island, Ó Síocháin, detailing the years of decline. Instead his book is full of 'gaiety and magic' (Thompson, 1982, p. 44; see also Duffy, 1995, for another analysis of O'Sullivan's book in relation to migration). Like the others he tells of 'harrowing away', sometimes through the experiences of 'Daddo', his grandfather. He writes of hunting seals and birds and mentions having livestock, but his principal activity seems to have been fishing, an often precarious existence, for as a friend says (O'Sullivan, p. 206):

> 'The chief livelihood – that's the fishing – is gone underfoot, and when the fishing is gone underfoot, the Blasket is gone underfoot, for all the boys and girls who have any vigour in them will go over the sea'...
> 'And what will our parents do when they grow old?'
> 'It is my opinion that they will have to do without us.'

Chapter 18 is devoted to the American wake of his sister Maura, off to Springfield, Missouri, the destination of most emigrants from Great Blasket. 'God help the old people, there will be none to bury them', says their father, who remembered when there was 'no thought of America ... and they were fine, happy days' (p. 218). Maybe so, but they were days past, for Maura retorted, 'Don't you see everyone is going now and you will see me beyond like the rest of them' (p. 219). Thus 'a great change was coming on the island ... all the young people were departing across to America, five or six of them every year ... there was nobody left in the house but my grandfather, my father and myself' (p. 230). O'Sullivan did not join the rest of the young people for, although 'New Island [America] was before me with its fine streets and great high houses ... gold and silver out on the ditches and nothing to do but to gather it' (p. 230), instead he went to Dublin to join the civic guard (police). The book ends with his first visit back to Great Blasket after two years with the guards. His content at seeing his father and Daddo either side of the fireplace is contrasted to the changes wrought by emigration on his beloved island (p. 298):

> green grass was growing on the paths for lack of walking, five or six houses were shut up and the people gone out to the mainland; fields which had once had fine stone walls around them left to ruin; the big red patches on the sandhills made by the feet of the boys and girls dancing – there was not a trace of them no.

Today the footprints of the 'children of one mother' (p. 218), as O'Sullivan called the people of his island brought up so closely together, tread different paths and the last child born on the Blaskets is becoming elderly. There are paths on Great Blasket again, but they are trodden by the feet of strangers walking round the ruined village, stepping into the houses of O'Crohan,

Sayers and O'Sullivan which are marked out for them. Some write in a comments book; one with a Springfield address writes that she 'was glad to see my grandfather's home'.

The Minor Islands of Ireland in the Modern Era

Migration is still a factor in the experience of the Irish islands in the modern era (Table 2.2), but the situation is now somewhat easier compared to the nineteenth and early twentieth centuries. This improvement is in response to an alteration in a number of factors, as detailed below.

The attitude of islanders

Firstly there is the attitude of those who have stayed towards their island homes. Many still acknowledge their islands as special, as did the Blasket authors. However, whilst O'Crohan and Sayers could only lament the passing of their contemporaries and the era they represented and O'Sullivan became a migrant himself, modern islanders, through their efforts in economic and political spheres, can try pro-actively to ensure a viable future for themselves and their families on their islands. Outsiders may wonder why they bother, for life on a small west coast Irish island can never be made convenient or even particularly comfortable. However, there is a quality and a tranquillity to that inconvenient life that many find alluring. One resident of Dursey put it thus: 'we'll stick to the ship while there's rigging in the mast' (Durrell, 1996, p. 160).

The islanders help themselves through the work of their co-operatives. That on Cape Clear, for example, deals with the island's water and electricity supplies (there are three wind turbines), administers one of the two ferries, runs a pottery, restaurant and bar, campsite, fish farm, enterprise centre, community hall, shop, petrol filling station, machinery repair service, an Irish language college in the summer and is heavily involved in social activities (Cape Clear Community Co-operative, 1994). The Inishmaan co-operative is another that is particularly active, operating a textile factory that markets high-quality knitwear around the world (Scott and Royle, 1992).

In the political arena, the Irish islands have their own pressure group formed from islanders from different places, the Comhdháil Oileán na hEirann (the Council of the Islands of Ireland). This strives to see that official policies to support island life continue to be developed and implemented (Royle, 1986). At local government level, islanders are often active. For example, Mayo County Council has a council committee with a substantial budget to deal with island affairs. Each offshore island elects representatives to this committee and the islanders make cases for expenditure to ease their lot. Islanders' work on this committee at least partially explains the substantial improvements in recent years to the harbour on Inishturk, which enabled local fishermen to operate larger boats to the benefit of their industry and the island's population stability. County Cork also has a council committee for island affairs on which sit representatives from Bere, Cape Clear, Dursey, Hare, Sherkin and Whiddy.

Official support

The second development refers to outside support. The first part of this chapter noted outside aid to islanders dating back to the early 1820s and later support of the Congested Districts Board. More recently, however, such support has been made more secure and regularized by the two governments in Ireland. The authorities in the Republic of Ireland have come to recognize the intrinsic merits of island life, particularly with regard to their disproportionate contribution to Irish culture, including, in many cases, native use of the Irish language. There has been extra government aid available to sustain these bastions of traditional language use (Hindley, 1990), somewhat to the chagrin of English-speaking islands such as Clare and Inishbofin (County Galway).

Further, the European Union is aware of the special needs and qualities of its islands. Thus in the preface (p. iii) to its *Portrait of the Islands* (European Commission, 1994) the Commissioners state that:

> the true value of these islands ... lies in their terrain, which at times can be both severe and calming to locals and visitors alike, and the islanders, who are proud of their traditions and way of life whilst also able to offer a warm welcome to outsiders.

The EU has no specific programme to aid islands; instead support for these valued places comes through the operation of its normal funding programmes. All Ireland, north and south, is within the EU's highest priority bracket for aid and this is often made available for projects affecting islands. Thus, the Valoren programme made major financial provision to a wind-powered electricity generation scheme on Rathlin Island in the early 1990s (Royle *et al.*, 1994). In 1997 a new community centre on Clare was erected with the aid of the Irish exchequer, the local community and the Regional Development and Social Funds of the European Union. Sadly, further evidence of outside help was revealed regarding this community centre when a building worker who had fallen from its roof during the course of construction had to be airlifted to hospital by a rescue helicopter. All the major inhabited islands now have helicopter pads for such purposes.

Access to the islands by means other than helicopter is also often supported by government. Thus in Northern Ireland the mid-1990s finally saw the provision of a roll on-roll off car ferry to Rathlin, provided by the nationalized Scottish ferry company, Caledonian MacBrayne. In the Republic of

Ireland a ministerial committee was set up with the involvement of islanders represented in the Comhdháil Oiléan na hEirann to develop a policy for the offshore islands. It reported that its mission was:

> to support island communities in their social, economic and cultural development, to preserve and enhance their unique cultural and linguistic heritage, and enable the islanders to secure access to adequate levels of public services so as to facilitate full and active participation in the overall economic and social life of the nation. (Government of Ireland, 1996, foreword)

There is no threshold population for this support; the County Donegal Inishbofin with its 1991 census population of three is included. But to qualify for support, the islands must have a community that 'has been resident on the island for at least one generation', presumably to avoid the government having to support alternative lifestyle groups which have sometimes set up communities on abandoned islands. This new government initiative is in recognition that:

> The offshore islands of Ireland constitute a unique element in the fabric of Irish society. Island communities have a special inheritance, a way of life that is to be cherished and valued. In particular the Committee recognizes the important cultural and historic dimension that islands bring to the nation. (Government of Ireland, 1996, p. 12)

Some of the islands most noted for their rich culture are now empty of people, including Great Blasket (though echoes are kept alive in the newly erected Blasket Heritage Centre overlooking the island). At a meeting of the first all-Ireland island pressure group, Comhdháil na nOileán, the words 'Don't let what happened to the Blaskets happen to us' were heard (Royle, 1986). It is now government policy to put this plea into operation. One result should be a lessening in the rate of migration.

New economic opportunities

A third reason why life on Irish islands is rendered more feasible in the modern era relates to new economic ventures. These include productive enterprises such as fish farms, as on Clare and Bear; the opportunities provided to all islands and rural areas by the telecommunications revolution; small-scale industry; and, especially, tourism (this last will be considered below). Gorumna, Inishmore and Inishmaan are amongst islands which have small factories supported by the state agency Údaras na Gaeltachta, whose brief is to encourage economic development in the Gaeltacht, the areas of Ireland where Irish remains the first language. Some Irish-speaking islands such as Aranmore and Cape Clear also make money from running intensive courses in Irish, this being a compulsory school subject in Ireland.

Tourism and the island as museum

Tourism is well known to be a double-edged sword with regard to the contribution it can make to the economy, society and environment of small islands. It is certainly a useful way of producing money for resource-poor economies and it does have the 'benefit' of seasonality which enables an island to earn enough money in the summer to support its population through a more traditional and tranquil winter. However, there are problems associated with tourist pressures on an island's environment, for example upon water supplies, or upon its people, whose traditional society and culture may be transformed, even corrupted, by the outsiders. Thus, even on remote Tory Island (Figure 2.3), one of the last refuges of spoken Irish, it is now possible for visitors to make all their transactions in English. On the other hand, the island now has a series of service facilities such as a hotel, hostel, general shop, craft shop, gallery, club and café (Table 2.3). All these are island-owned and provide jobs for locals. Many of these services are patronized extensively by non-Irish-speaking tourists who thus provide the islanders with their business and employment opportunities. There are just a few sheep left in souvenir of the island's traditional economy, though fishing may revive once the new IR£6m (c. $9m) investment in the harbour is completed. Meanwhile the building of this is giving employment to several island men. In sum, despite its drawbacks, tourism, itself helped by the provision of a scheduled ferry service, has helped to revive Tory when a few years earlier there had been consideration given to the evacuation of the island (Ó Péicín, 1997; Scott and Royle, 1992). Thus, although Tory's population still fell 12.5 per cent from 1986 to 1991, it had fallen by 35 per cent in the previous five years (Table 2.2).

Only in certain special circumstances can tourism be fully contained. Thus Hare has a well-known restaurant, but, given its insular location, numbers can be entirely controlled by the owners' operation of a dedicated evening ferry for diners. There are more problems with the Aran Islands, especially Inishmore. Here consideration may have to be given to limiting the number of tourists, such are the pressures on the infrastructure, especially the water supply. Thousands of tourists arrive daily in summer, some by air but most by ferry to the quay at Kilronan, where they are greeted by a seafront chain of three bicycle hire shops, five craft shops, five bed and breakfast establishments, two restaurants, two bars, a guest house, a hostel, and the tourist information office. There are further tourist establishments in Kilronan and throughout the island. And at Inishmore's premier attraction, the mighty Dun Aengus prehistoric stone fort on its magnificent clifftop site, half the fort is now fenced off to keep tourists away; where they are allowed, a ranger is employed to keep the visitors from climbing on the fragile dry-stone walls.

Inishmore, once a traditional Irish island, has become a museum of itself.

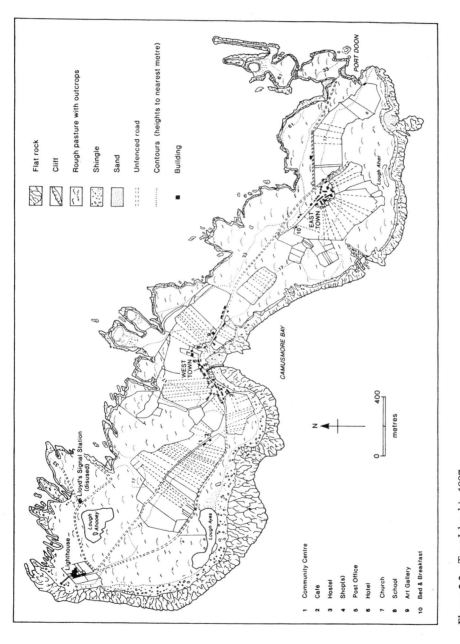

Figure 2.3 Tory Island in 1997

Source: adapted from a map in Scott and Royle, 1992.

This is nowhere more evident than in the Dún Árann heritage centre where the attractions include Dun Eochla (another stone fort), a lighthouse, a signal tower, a traditional thatched Aran cottage and, most poignantly of all, a patch of potatoes. The self-guided tour leaflet explains that 'potatoes and fish were the staple diet of the people for centuries'. Though Inishmore still supports some fishing boats, agriculture has been reduced to little more than a hobby and an exhibit. Inishmore might only be a museum of itself, but at least its population decline has been halted; as Table 2.2 shows, numbers have stabilized at around 850 since 1971.

Traditional islands

By contrast, there are other islands where things are less changed in social and economic terms. Bere has altered relatively little, still being dependent upon farming and fishing, though the fishing now includes fish farming. The island's economy continues to be supported by what could also be regarded here as a traditional use, as an army base. The British held Bere, the southern shore of the important naval anchorage of Berehaven, beyond Irish independence in 1921 and gave it up only in 1938. The Irish forces now use some of the British-built facilities as a training ground.

An even more traditional island is Inishturk to which, although there is now a ferry, access still retains an *ad hoc* flavour.[2] Inishturk interviewees expressed pride that their island has not sold out to tourism which they regarded as having blighted neighbouring Inishbofin, whilst the mere mention of Inishmore elicited gasps of horror, despite that island's relatively buoyant economy. Instead, Inishturk's now rising population, which reached 92 by 1997 from 78 in 1991, is supported in the traditional way by fishing and farming (Cross, 1996). There has been some outside help; a significant factor in the continuation of the fishing industry has been the ability to berth bigger boats in the newly extended harbour. By the harbour are sheds still used to house livestock before transportation to mainland markets, and there are still farms with fields of potatoes and oats. The only services are a post office, school, church and a community centre (Table 2.3). But even here life will change; the ferry service will become more reliable and two of the nineteen houses are already bed and breakfasts.

Islands in trouble

Despite all the available support, new opportunities and examples of islands which have maintained a traditional economy without excessive recent migration, there are some islands which have not been able to resist continued out-migration and population decline. One such is Inishbiggle, 1991 population 51. In 1997 there was only one family with children here and the whole family is moved by Mayo County Council in winter to adjacent Achill Island, to ensure that the children's education is not disrupted by their being cut off from the school there by storms. There are no services except a post office, and there is no ferry.[3] People on Inishbiggle want a cable car. Without some investment of this nature to improve access, there is no possibility of any tourism or other development in the economy. Complete depopulation must be a danger, for the island's population fell by 43 per cent during 1981–91.

Another island in difficulties is Whiddy. Unlike many islands Whiddy never had much focus to its settlement, no tiny urban place like West Town on Tory or Rerrin on Bere. Perhaps this is because Whiddy spent much of its history being used and abused by outsiders exploiting its strategic position at the head of Bantry Bay. Its landscape includes an early sixteenth-century castle, three nineteenth-century ramparts built by the British to threaten enemy (French) shipping in the bay and an American seaplane base from the First World War. Further, in 1969 fourteen large oil storage tanks were erected on its south eastern shore, a short-lived economy which probably brought little benefit to the island and which ceased after 1979 when the *Betelgeuse* exploded at the jetty with the loss of 50 lives. The tanks still blight the island, an unsightly reminder of the tragedy. Otherwise, Whiddy seems always to have been a collection of farms, most of which had their own jetty, there not even being the concentration of activity around a common pier as elsewhere. There have been some recent investments – a decent pier has been built; a licensed ferry plies there to a regular scheduled service from Bantry; passengers are attracted from the pier straight into a new bar, restaurant and shop with a children's playground attached. However, out-migration is still taking place, the population fell 37 per cent from 1981 to 1991 to 34 and it has fallen by a further 32 per cent to just 23 in 1997, with just six people under 50. The island has empty property; the school has been closed to be turned into a museum, one of its few services (Table 2.3). As a traditional, insular community, Whiddy's future seems in doubt.

Finally, let us consider the island of Dursey (Figure 2.4), which housed 358 people in 58 households in 1841 but in 1997 was down to six people in five households from what had been a stable population of 20 in 1991. Of its three villages, the most remote, Tilickafinna, is empty; the middle village, Kilmichael, had just one inhabited house from fourteen house sites visible in 1997 (there were 24 houses in 1841); whilst Ballynacallagh, closest to the cable car which is the island's lifeline, had four of its extant sixteen houses (28 in 1841) inhabited. Most Dursey islanders are elderly and it seems unlikely that when they die there will be people to occupy their houses.

In addition to the five houses inhabited by islanders, there are three occupied as holiday homes. However, seasonal habitation by outsiders cannot be regarded as maintaining an island's traditional community and Dursey is basically dying. This is despite the provision of the cable car in 1969 which removed the need to take a boat over the vicious waters of the Dursey

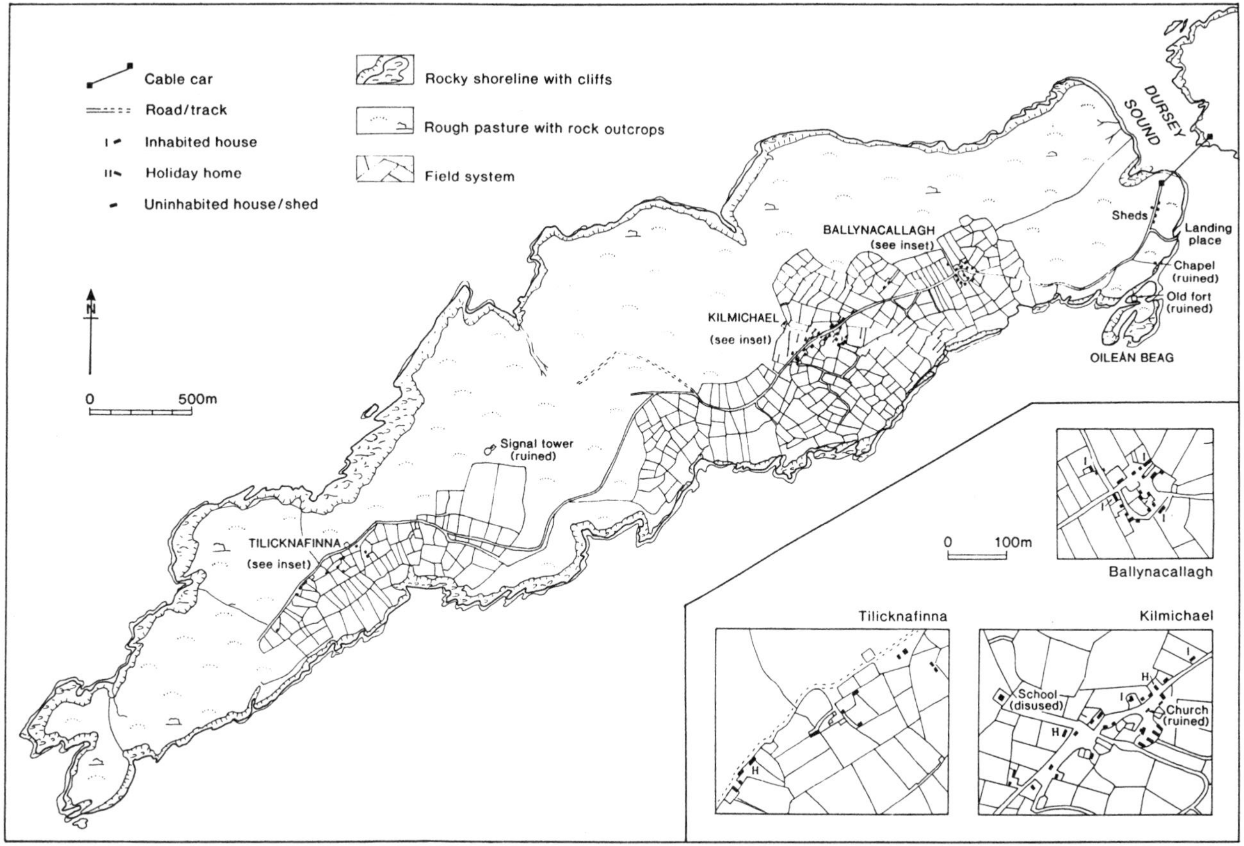

Figure 2.4 Dursey Island in 1997

Cable car

Road/track

Inhabited house

Holiday home

Uninhabited house/shed

Rocky shoreline with cliffs

Rough pasture with rock outcrops

Field system

N

0   500m

DURSEY SOUND

Sheds

Landing place

Chapel (ruined)

Old fort (ruined)

OILEAN BEAG

BALLYNACALLAGH (see inset)

KILMICHAEL (see inset)

Signal tower (ruined)

TILICKNAFINNA (see inset)

Ballynacallagh

Kilmichael

Church (ruined)

School (disused)

H

H

H

0   100m

Tilicknafinna

H

Sound, a short but often impossible journey. The cable car enables tourists to visit Dursey but they bring little economic benefit as the island is totally without commercial functions where money could be spent (Table 2.3). The cable car may even have accelerated Dursey's death throes for it gave guaranteed access to the island and thus enabled aspects of its economy to be run from the mainland. This is the case with a son of one of the island's last residents who lives away from the island and manages his land and stock from the mainland. Another son, a local fisherman, also operates from the mainland: had the cable car not been built, Dursey may have had the extensive pier improvements which have been provided on many offshore islands. This might have enabled this fisherman to have kept a modern boat on Dursey and stayed on the island. These two local out-migrants are part of a large family of nine children, all of whom have moved off the island to other parts of Ireland and to Britain. This family had provided some of the last pupils at the island school which closed in 1975, one year after the post office. Once an island has closed its primary school, families with young children seldom stay and the coming generation of islanders is lost. Depopulation becomes virtually inevitable. This was certainly the sequence on Inishbofin (County Donegal) where the school closed in 1981 when the population was 46; this island now usually has residents only in the summer.

Conclusion

In this chapter I have explored various dimensions of the economic, demographic and migration experiences of the minor islands of Ireland. We have seen that these islands represent a group of communities that have had to adapt their economies to the modern world, usually at a cost to their island traditions of economic subsistence and community cohesion. Where such adaptation has not taken place, unchecked out-migration threatens complete depopulation, as with Dursey.

My focus in the middle part of the chapter on island literature and the Great Blasket authors continues the lead given by Duffy (1995) in his more general survey of Irish emigrant writing and enables a more sensitive, personalized appreciation both of the intrinsic nature of small island societies off the west coast of Ireland, and of the human pain of emigration and the demographic erosion of these small but culturally rich communities. The irony of the Blasket stories is that their authors derive their material, and hence their fame, from their testimonies of a process of abandonment and disintegration.

The literary material of this chapter is sandwiched between an historical and a contemporary analysis which uses the more conventional investigative techniques of census compilation and field survey. Whilst the overall trend from the census data and from some of my fieldwork is one of progressive decline and landscape abandonment, the picture is not one of unmitigated gloom. Rates of population loss have stabilized in many islands and on a few there are even signs of small-scale population increase. The final section of the chapter details some of the efforts being made to reduce emigration, and the impact of new economic opportunities, notably tourism, has also been documented.

Population stabilization is not just about curtailing out-migration: return migration and immigration can be important. My 1997 field survey found clear evidence of return migration on Tory, Rathlin, Cape Clear and Clare. Elsewhere McGrath (1991) has described and mapped in detail the process and impact of return migration on the landscape, society and economy of Achill. Return is a hopeful sign for the future, particularly when it is associated not just with retirement but with an 'active return' and increased employment, for instance in tourism or fishing. Development of fish-farming on Clare was one factor which helped this island's population grow by 8 per cent between 1981 and 1991. On Inishmore there was confidence that the population was now stable, thanks to tourism. In other cases, there are the first signs that 'outsiders' are buying island property. Such 'blow-ins' do not usually add to the traditional island society – a helicopter parked on the lawn of a restored house on Sherkin in the summer of 1997 just emphasized the difference between that household and the farmers and fisherfolk of the island.

Finally, despite the various actions of the islanders themselves, and of policy initiatives at local community, national and European levels, it is clear that Irish islands are still subject to emigration pressures. Table 2.2 indicates that 27 islands (excluding eight lighthouse islands now unmanned) from the 1971 total of 94 had become uninhabited by 1991.

Acknowledgement

The author acknowledges, with thanks, a grant from Queen's University Belfast towards fieldwork expenses in July 1997.

Notes

1. It should be noted that in this review of the Great Blasket authors my references are to the English translations of books which were originally published in Irish some decades earlier. Full publication details of the English and Irish editions are given in the references at the end of the chapter.

2. In 1997 I was told I could get passage on a trawler from Roonagh Quay, County Mayo, then it was a ferry from Cleggan, County Galway. This was to sail at 16.30 but only arrived at 21.15 and did not depart until I had helped to unload its cargo of fish and then load up a return cargo of groceries. My return journey was not to Roonagh Quay as advised – itself an inconvenience, considering my car was in a different county – but was, thankfully, back to Cleggan, though neither in the boat nor at the time I had confirmed.

3. The advice potential visitors to Inishbiggle receive from the tourist office on Achill Island is to turn up at the quay around the time a boat normally comes over for the mail, then try to persuade the boatman to take you.

References

Aalen F.H.A. and Brody, H. (1969) *Gola: The Life and Last Days of an Island Community*. Cork: Mercier Press.

Cape Clear Community Co-operative (1994) *Short History and Review*. Cape Clear: Comharchumann Chléire Teo.

Conry, M.J. (1971) 'Irish plaggen soils: their distribution, origin and properties', *Journal of Soil Science*, 22(4), pp. 401–16.

Cross, M.D. (1996) 'Service availability and development among Ireland's island communities: the implications for population stability', Irish Geography, 29(1), pp. 13–26.

Duffy, P. (1995) 'Literary reflections on Irish migration in the nineteenth and twentieth centuries', in R. King, J. Connell and P. White (eds), *Writing Across Worlds: Literature and Migration*. London: Routledge, pp. 20–38.

Durell, P. (1996) *Discover Dursey*. Allihies: Ballinacarriga Books.

Dwyer, D. (1963) 'Farming an Atlantic outpost: Clare Island', *Geography*, 48(4), pp. 255–65.

European Commission (1994) *Portrait of the Islands*. Luxembourg: Office for Official Publications of the European Communities.

Flower, R. (1944) *The Western Ireland or the Great Blasket*. Oxford: Oxford University Press.

Fox, R. (1978) *The Tory Islanders: A People of the Celtic Fringe*. Cambridge: Cambridge University Press.

Government of Ireland (1996) *Report of the Interdepartmental Co-ordinating Committee on Island Development: A Strategic Framework for Developing the Offshore Islands of Ireland*. Dublin: Stationery Office.

Harvey, B. (1991) 'Changing fortunes in the Aran Islands in the 1890s', *Irish Historical Studies*, 27, pp. 237–49.

Heraughty, P. (1982) *Inishmurray: Ancient Monastic Island*. Dublin: O'Brien Press.

Hindley, R. (1994) 'Clear Island (Oileán Chléire) in 1958: a study in geolinguistic tradition', *Irish Geography*, 27(2), pp. 97–106.

Hindley, R. (1990) *The Death of the Irish Language: A Qualified Obituary*. London: Routledge.

MacConghail, M. (1987) *The Blaskets: People and Literature*. Dublin: Country House.

Matson, L. (1996) *Méiní, The Blasket Nurse*. Cork: Mercier Press.

McGrath, F. (1991) 'Emigration and landscape: the case of Achill Island', *Papers in Geography* 4. Dublin: Trinity College.

Ní Shúilleabháin, E. (1978) *Letters From The Great Blasket*. Cork: Mercier Press.

Ó Conluain, P. (ed.) (1983) *Islands and Authors: Pen Pictures of Life Past and Present on the Islands of Ireland*. Dublin: Radio Telefís Éireann and Mercier Press.

O'Crohan, S. (1992) *A Day in Our Life*. Oxford: Oxford University Press; first published as Ó Criomhthain, S. (1969) Lá Dár Saol. Baile Átha Cliath: Oifig an tSoláthair.

O'Donnell, P. (1927) *Islanders*. London: Jonathan Cape.

O'Donnell, P. (1975) *Proud Island*. Dublin: O'Brien Press.

O'Flaherty, L. (1932) *Skerrett*. London: Gollancz.

O'Guiheen, M (1982) *A Pity Youth Does Not Last*. Oxford: Oxford University Press; first published as O'Guiheen, M. (1953) *Is Trua Ná Fannan an Óige*. Baile Átha Cliath: Oifig an tSoláthair.

Ó Péicín, D. (1997) I*slanders: The True Story of One Man's Fight to Save a Way of Life*. London: Fount.

Ó Síocháin, C. (1975) *The Man from Cape Clear*. Cork: Mercier Press; first published as Ó Síocháin C. (1940) Seanchas Chléire. Baile Átha Cliath: Oifig an tSoláthair.

O'Sullivan, M. (1953) *Twenty Years a'Growing*. London: Oxford University Press; first published as Ó Súilleabháin, M. (1933) *Fiche Blian ag Fás*. Baile Átha Cliath: Clólucht an Talbóidigh.

Pollock, W.G. (1982) *Six Miles From Bangor: The Story of Donaghadee and the Copeland Islands*. Belfast: Appletree Press.

Royle, S.A. (1983) 'The economy and society of the Aran Islands in the early 19th century', *Irish Geography*, 16, pp. 36–54.

Royle, S.A. (1984) 'Irish famine relief in the pre-famine period: the Aran Islands in 1822', *Irish Economic and Social History*, 11, pp. 44–59.

Royle, S.A. (1986) 'A dispersed pressure group: *Comhdháil na nOileán*, the Federation of the Islands of Ireland', *Irish Geography*, 19, pp. 92–5.

Royle, S.A. (1989) 'Settlement, population and economy of the Mayo islands', *Cathair na Mart*, 9(1), pp. 120–34.

Royle, S.A. (1994) 'Island life off Co. Down: The Copeland Islands', *Ulster Journal of Archaeology*, 57, pp. 177–82.

Royle, S.A., Robinson, J. and McCrea, A. (1994) 'Renewable energy in Northern Ireland', *Geography*, 79(3), pp. 232–45.

Royle, S.A. and Scott, D. (1996) 'Accessibility and the Irish islands', *Geography*, 81(2), 111-19.

Sayers, P. (1974) *Peig: The Autobiography of Peig Sayers of the Great Blasket Island*. Dublin: Talbot Press; first published as Sayers, P. (1936) *Peig*. Baile Átha Cliath: Clólucht an Talbóidigh.

Sayers, P. (1978) *An Old Woman's Reflections*. Oxford: Oxford University Press; first published as Sayers, P. (1939) *Machtnamh seana-mhná*. Baile Átha Cliath: Oifig an tSoláthair.

Scott, D. and Royle, S.A. (1992) 'Population, society and economy on Tory Island, Co. Donegal', *Irish Geography*, 25(2), pp. 169–76.

Stagles, J. and Stagles, R. (1980) *The Blasket Islands: Next Parish America*. Dublin: O'Brien Press.

Steel, T. (1975) *The Life and Death of St Kilda*. Glasgow: Fontana.

Synge, J.M. (1907) *The Aran Islands*. Dublin: Maunsell.

Tall, D. (1986) *The Island of the White Cow: Memories of an Irish Island*. London: André Deutsch.

Thompson, G. (1982) *The Blasket That Was: The Story of a Deserted Village*. Maigh Nuad: An Sagart.

3

The Azores: Between Europe and North America

Allan M. Williams and Maria Lucinda Fonseca

Introduction: A Globalization Perspective

International migration from the Azores can be understood as part of the long process of globalization. The development of global interconnections has been integral to the expansion of the world economy and the rise of the modern state (Wallerstein, 1974). Over time, these interconnections have intensified and Held (1995), amongst others, contends that there has been a qualitative change in the later twentieth century related to technological, economic and political shifts. More importantly for this chapter, he also contends that globalization can be understood to denote 'the stretching and deepening of social relations and institutions across space and time such that, on the one hand, day-to-day activities are increasingly influenced by events happening on the other side of the globe and, on the other, the practices and decisions of local groups or communities can have significant global reverberations' (Held, 1995, p. 20). Therefore, while territorial boundaries demarcate the formal inclusion of individuals in decision-making affecting their lives, the implications of these decisions 'stretch' beyond national frontiers. Globalization is, thus, conceived of in this chapter as a process of linkage and interdependence between territories and of 'in here–out there' connectivity across increasingly permeable social and political boundaries. These interconnections have multiplied and become more complex over time; indeed, 'modern communications virtually annihilate distance and territorial boundaries as barriers to socio-economic activity' (Held, 1995, p. 20). While we are in broad agreement with these views, it is also a central contention of this chapter that the impacts of recent globalization are very much influenced by the longer-term evolution of social relationships across territorial boundaries.

The concept of globalization as a nexus of inter-relationships at different levels provides a useful framework for analysing international migration, as Kritz and Zlotnik (1992, p. 1) recognize:

> In the context of an increasingly inter-connected world, international population movements can naturally be seen as complements to other flows and exchanges taking place between countries. Indeed international migrations do not occur randomly but take place usually between countries that have close historical, cultural or economic ties.

This is particularly apposite for a study of the Azores. The external influences on emigration are not a set of abstract global processes; instead, they are shaped by the historically evolved links with particular places, especially established emigrant communities in specific regional contexts within North America. These links involve not only individual mobility, capital and the transfer of ideas, but also a particular ideology of emigration and (non-) return. In turn, social relationships within the Azores have shaped the imprint of the Azorean immigrant communities on the New World. This web of 'connectivity' is perhaps best understood in terms of the concept of migration networks (Gurak and Caces, 1992), set in the context of Held's notion of globalization. Above all, attention has to be paid to how the long evolution of Azorean emigration to North America created a set of social relationships which significantly mediated later globalization tendencies.

These general observations concerning international migration could be seen to apply to the migratory experiences of most island communities. They are, however, especially pertinent in the case of the Azores, given the scale of migration: for example, the number of emigrants between 1950 and 1975 was equivalent to about one half of the 1950 total population.

The particular character of Azorean emigration derives from two features. The first of these is the dispersed population distribution, with some islands having only a few thousand or even a few hundred inhabitants. This gives a particular intensity to kinship networks for, with the exception of some of the more populous islands, virtually all first-generation emigrants will be known personally by, and will know, those living in their home island.

Second and equally important is the islands' mid-Atlantic location (in physical, economic and cultural terms), so that they are 'neither of Europe nor of the Americas'. Not all commentators would concur with this view, and an alternative perspective, typified by Mota Amaral's statement (1981, p. 243), is that 'there exists a much more intimate link between the Azores and the Americas than with Europe'. The link is constituted not only by emigration but also by the considerable support from the Canadian and American governments at times of crisis, notably following natural disasters; this support has been fostered by both the interest-group activities of the Azorean émigré communities, and the strategic self-interests of the USA, which has important air bases on the island of Santa Maria. Much of the literature on Azorean emigration accepts, if only implicitly, this latter perspective, not least because most of the published research emanates from the Azorean émigré communities in North America (notably Alpalhão and Pereira da Rosa, 1980; Anderson and Higgs, 1976; Williams, 1982). This perspective is misleading, for the economic development and culture of the islands have been shaped by their evolution as constituent parts of the Portuguese state. These islands had never been settled before the arrival of Portuguese settlers in the fifteenth century. Thereafter they were governed and peopled from Portugal, and were key components in the emerging Portuguese colonial system. The islands did develop trade and migration links with North America from the eighteenth century which were largely autonomous of the Portuguese state, but state policies (especially the neglect of structural reforms) continued to influence these links. Moreover, there were also migration flows to the Portuguese mainland (mainly from the elites), even if at much lower levels than the nineteenth- and twentieth-century exodus of peasants and landless rural workers to North America (Chapin, 1992).

While the main emphasis in this chapter will be on regions and localities – whether the Azores, New England or Ottawa – the role of the state should not be underestimated. First, and as already mentioned, there is the role of the Portuguese state in the development, and the underdevelopment, of the Azores. Secondly, there is the mediating influence of North American immigration policies. For example, a long period of emigration to the USA was brought to a halt by restrictive legislation in the 1920s, only to be opened up again in the late 1950s following catastrophic volcanic eruptions on the island of Faial. Canada, long closed to Azorean emigration, also opened its frontiers after 1953, precipitating the rapid growth of new Azorean communities there. And thirdly, there was the impact on the islands of the 1974 military coup in (effectively, mainland) Portugal, and the subsequent transition to democracy.

Following these brief introductory remarks, the remainder of the chapter is divided into three main sections. Firstly, we provide a brief outline of the evolution of international migration from the Azores. Secondly, we consider the formation of the Azorean communities in North America, and thirdly we assess some of the impacts on the Azores of its long history of emigration.

The Azores: A Long-established 'Emigration Platform'

The Azores (Figure 3.1) are a group of nine islands, located about 1100 km from Portugal and 3200 km from New England. Strung out over 550 km, the islands are of various sizes and economic importance. In 1991 the least populated island was Corvo with just 393 inhabitants while São Miguel had 125,915 inhabitants, or 53 per cent of the total. These simple 'geographical' facts are central to an understanding of the evolution of emigration from the islands. Relative isolation from and weak social ties with Portugal reinforced the *relative* accessibility of the USA (at particular times), while the small populations of the islands intensified migrant networks based on friendship and kinship.

The history of these islands is in many ways a history of migration, for they had no human settlement before being discovered in 1427 during the period of the Portuguese *Descobrimentos*. Early settlement was difficult, and suffered a number of setbacks, but the population eventually entered a long period of growth, peaking at 263,305 in 1878 before declining. The population total revived after 1920 to reach a new peak of 327,400 in 1960, before entering another phase of decline (Silveira e Sousa, 1995, p. 33). These changes were largely brought about by emigration.

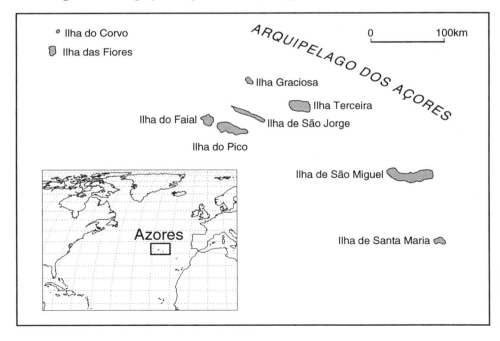

Figure 3.1   The Azores
Source: MPAT (1993).

The Azores had two functions in the early Portuguese empire. One was as a producer of foods and raw materials, and there were a number of well-documented cycles of production involving wheat, oranges and livestock (Pereira da Rosa, 1980). Migration from mainland Portugal was facilitated by the state to ensure that production levels were maintained. The Azores were also a source of recruits for the continuing project of colonization, particularly of Brazil. Ships en route to Brazil habitually called at the Azores, not only for supplies, but also for emigrants. While this 'emigration platform' function was to some extent managed by the state, it was also facilitated by recurring crises in the islands. All these factors contributed to what can be termed a 'culture of emigration', constructed around the expectation of emigration and the symbolic and actual importance of this as a channel for material and personal advancement.

Causes of emigration

Economic conditions have been the principal determinant of emigration from the Azores. The need for emigration as an economic 'escape-route' emerged relatively early, for by the sixteenth century the economic boom which had accompanied the Discoveries was drawing to a close (Pereira da Rosa, 1980, p. 64). Economic conditions subsequently continued to encourage, and often to compel, emigration. The root cause of the economic difficulties lay in the structure of agriculture, especially in the concentration of land ownership and power. In 1840, for example, 3 per cent of the population effectively controlled virtually all of the agricultural land in the islands (Williams, 1982, p. 64). Even as late as 1986, average farm size was still only 5 ha (MPAT, 1993, p. 118). The position was exacerbated by a system of perpetual leaseholds. Fixed rents were paid at the end of the year, irrespective of the size of the harvest, so that many farmers became locked into a cycle of indebtedness. Their children faced even greater difficulties, for farms were held on hereditary leases that could not be subdivided without the permission of the owners. Most of the children therefore faced the prospect of inheriting either nothing or a very small amount of capital, sufficient only to finance their emigration. Many islanders took the latter option in the face of rising land prices and a lack of farms to lease, given the intense pressures exerted by a growing population on a rigid and inefficient agricultural system. Not surprisingly, emigrants have tended to be drawn disproportionately from amongst the poorest members of what, until recently, was a largely agricultural society.

While emigration can be seen as only one component of household survival strategies, this was given an additional twist – especially in the case of temporary migrants – by the desire to accumulate sufficient wealth to enhance marriage prospects. While abroad 'they restricted their consumption ... in order to be able to send remittances and to accumulate economically with the aim of returning, marrying and reproducing the peasant farm' (Silveira e Sousa, 1995, p. 40). As in most peasant communities, land constituted the primary symbol, and source, of both wealth and status. In fact, most emigrants did not return and we consider this key point later in the chapter.

The almost constant pressures emanating from the mismatch between the agricultural system and strong population growth were exacerbated by recurrent but unpredictable crises. One of the major waves of emigration in the second half of the nineteenth century was partly triggered by what Williams (1982, p. 67) terms 'the arrival of two unwanted visitors' in the early 1850s: potato rot and odium (which devastated vine stocks). Earlier in the century, an orange tree disease had arrived by accident from the USA, and by the 1870s had spread to the main orange production area on São Miguel. More dramatic, and more localized, were volcanic eruptions and earthquakes, such as those in Faial in 1957 and São Jorge in 1964.

Finally, while economic conditions were the primary determinant of emigration, there were also social and cultural influences. Many emigrants were seeking to escape the isolation and restrictive social mores and customs of Azorean society. Another motivation was to evade military conscription to fight in distant places on behalf of a Portuguese state that had been of little material assistance or ideological significance to the peasant farmers of the islands. The Portuguese state's late concession of independence to its colonies in Africa (in 1975), after a bloody and ultimately futile colonial war, meant that the evasion of conscription remained a significant motivation until relatively recently.

While we have thus far written about emigration from the Azores in aggregate terms, there were also important inter-island variations. The overall level of emigration tended to be governed by rhythm of the economy, or immigration controls in North America (providing evidence of the pervading effects of globalization from an early date), but the experiences of individual islands were also shaped by localized economic failures or natural disasters. For example, the effects of odium were particularly severe in the 1850s in Graciosa, Terceira, Faial and Pico, while the spread of disease amongst the orange plantations brought about a crisis in São Miguel in the 1870s. There were also positive sources of variation, such as the links via whaling and, later, emigration between Faial and New Bedford (in Massachusetts). Some of these differences will be explored in the following section. However, inter-island differences should not be exaggerated, for the islands share a common and distinctive history of emigration that marks them out from mainland areas of mass out-migration, such as the Minho and Tras os Montes regions in northern Portugal. Not least, there is the sheer scale of emigration; for example, between 1866 and 1890 the Azores, with only 6 per cent of the country's population, provided 16 per cent of all Portuguese emigrants (Silveira e Sousa, 1995, p. 32).

The geography of Azorean emigration

The geography of Azorean emigration has been shaped by two features: economic opportunities, particularly the cyclical movements in the global economy which have seen changes in the hegemonic order; and the cumulative influences of established migration networks. Brazil was the first major destination of Azorean emigrants, and from the eighteenth century there was a steady flow to what was then the jewel in the Portuguese colonial empire. While independence disrupted these flows to some extent, Brazil remained the favoured destination of Azorean emigrants until well into the nineteenth century. For example, as late as 1870–4 some 46,000 Portuguese emigrated to Brazil and a large proportion were from the Azores. Even after the USA became the major focus of Azorean emigration, Brazil continued to have residual importance, especially for legal emigrants from São Miguel.

By the mid-nineteenth century the focus of emigration had shifted decisively to the booming economy of North America. The new geography of migration was highly specific, linking particular islands to the changing economic fortunes of particular American regions in particular time periods. As much of the emigration to North America was clandestine, it is difficult to construct an accurate statistical profile of these movements, but it is possible to outline the shifting geographical patterns. The early emigration was built on trade and fishing links with New England communities, such as New Bedford. Records show that, from the 1780s, American whaling boats were taking on Azoreans as crew and subsequently many settled for a period in the home ports of these boats. By the 1820s there was evidence of permanent settlement in North America. This pattern peaked in the later nineteenth century, and it was estimated that, in the 1880s, about one third of the crew of New Bedford whaling ships were Portuguese, drawn almost entirely from the Azores (Dias, 1982, p. 13). While fishing provided the initial economic link, later emigrants increasingly looked to the textile factories for work. Meanwhile, two new geographical foci of Azorean emigration emerged: California and Hawaii. These links too were initially forged by Azorean crew on American whaling ships, but in both cases agriculture quickly became the economic mainstay of the evolving communities. The Gold Rush also provided a brief if spectacular boost to Azorean settlement on the west coast of America.

Emigration to the USA, which had accounted for most of the international migration from the Azores from the late nineteenth century, was brought to a virtual halt by increasingly restrictive American immigration laws in the early twentieth century. The literacy requirement imposed on immigrants in 1917 was particularly severe on Azoreans who came from impoverished agricultural communities where little or no formal educational provision was available. These controls were further tightened by the National Origins legislation of the 1920s, while economic conditions in the 1920s and 1930s also served to deter immigrants and to foster return migration.

The North American emigration gates were only re-opened for the Azoreans in 1953, when Canada – faced with labour shortages – signed a bilateral migration agreement with the Portuguese state. There were 555 Portuguese immigrants in 1953, and annual flow numbers peaked at 16,333 in 1974. Soon after the Canadian initiative, a volcanic eruption on Faial in 1957 prompted the US government to ease immigration controls as part of an overall relief programme for the islanders. The subsequent Azorean Refugee Acts of 1958 and 1960 made more formal immigration exceptions for Azoreans suffering from the natural disaster on Faial, and these marked the start of a new phase of emigration to the USA. Generally less restrictive immigration legislation was enacted later in the 1960s. As a result, whereas only 10,752 Portuguese (mostly Azoreans) officially emigrated to the USA in the two decades before 1950, the number rose to 150,000 in the period 1950–76 (Williams, 1982, p. 96). This critical stage in the reinvigoration of emigration has, again, to be seen in terms of the existence of kinship networks in the context of globalization. The key factors were the labour requirements of the North American economy in its hegemonic phase, the pressures exerted on US policy makers by émigré interest groups, and what Gurak and Caces (1992, p. 166) term the 'momentum-maintaining role' of settled migrant communities.

Emigration from the Azores after 1960 is detailed in Figure 3.2. Legal emigration totals, which underestimate the true level, were running at about 3000 a year in the early 1960s but peaked at around 8000–12,000 per annum in the late 1960s, in response to the devastating seismic activity in 1964. Emigration then fell back gradually before levelling off at around 1000–3000 per annum in the 1980s. The contributions of the individual islands to the total varied, reflecting their particular economic circumstances and external ties, but the most populous island, São Miguel, consistently contributed the largest share, accounting for between 45 per cent and 80 per cent of the total. The higher rate of emigration from São Miguel derived from the particular economic, social and demographic conditions in the island which are different to those in the remainder of the archipelago. Property ownership is more concentrated in the hands of a small number of large proprietors and, consequently, there exists a great number of agricultural families who are dependent on a very precarious land holding system which does not provide the minimum levels of income commensurate with their needs or aspirations. For this reason, the rural population of São Miguel, peasants and tenant farmers alike, have long been obliged to look abroad in order to improve their conditions, whilst their children have been stimulated to seek non-agricultural professions either in the other islands or abroad (Serpa, 1978). The higher demographic density on São Miguel, associated with the existence of a younger age structure in comparison with the other islands,

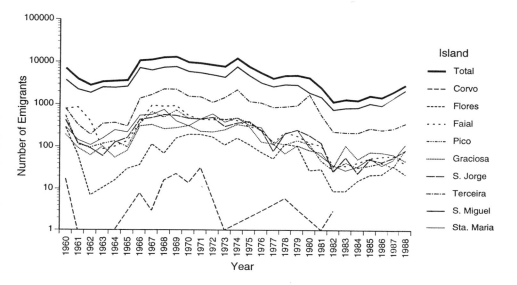

Figure 3.2   Emigration from the Azores, 1960–88
Source: Serviço Regional de Estatística dos Açores.

also serves to aggravate the extent of unemployment, which further increases the propensity to emigrate.

There are two remarkable features which emerge from this brief sketch of Azorean emigration. The first is the high and constant level of emigration over a long time period. The islands have accounted for a disproportionately large share of all Portuguese emigrants since the nineteenth century, and this has continued into the 1960s and 1970s. While emigration from continental Portugal fell from 538,000 in the 1960s to 292,000 in the 1970s, that from the Azores actually increased from 72,000 to 76,000. And while emigration declined throughout Portugal in the 1980s and 1990s as economic and political conditions changed, it remained relatively high in the Azores, even if the absolute level fell to 19,000 over the period 1980–88. To some extent, the persistence of emigration is due to the relatively strong employment growth in North America and the existence of well-organized receptor communities which eased the way economically and culturally for new arrivals. Not least there was a self-perpetuating economic system in which immigration was linked to social mobility. For example, in California's Central Valley, Azorean immigrants entered the labour market as hired hands, became tenant farmers and subsequently farm owners, at which point they hired relatives and friends from the Azores and started off a new cycle (Williams, 1982). Such a process supports the argument of Gurak and Caces (1992, p. 156) that migration networks influence decision-making by modifying the basic parameters of the migration system; their adaptive function reduces the 'costs' of migration, and so 'migration becomes an option for a broader segment of the origin society'.

The second feature of Azorean emigration is the extent to which it was geographically focused on North America. Unlike mainland Portugal, the Azoreans did not share in the major migrations to northern Europe in the 1960s and early 1970s (Lewis and Williams, 1986). The Azoreans also had little part in the major internal migrations (such as those from the Alentejo region) which fed the rapid metropolitan expansion of Lisbon after 1960. Instead, migration to the mainland was largely limited to the higher education of the middle and upper classes. As a corollary of these highly selective migratory flows, the Azoreans – despite coming from a region which now has only about 3 per cent of the population of Portugal – account for more than one half of all Portuguese immigrants in North America.

Azorean Communities in North America

Detailed consideration of the Azoreans in North America lies outside the scope of this chapter, but the individual communities have been well documented elsewhere (Alpalhão and Pereira da Rosa, 1980; Anderson and Higgs, 1975; Dias, 1982; Williams, 1982). However, a short account is necessary here, not only to understand the experiences of the Azoreans but also the impact of emigration (and return) on the islands. We begin with a brief review of the main communities, and then make some general observations.

In the USA there are three main foci of Azorean settlement – New England (especially Massachusetts, Rhode Island, Connecticut and, more recently, New Jersey), California and Hawaii; these six states account for more than 95 per cent of permanent resident Portuguese aliens.

- New England is the oldest established focus, and its origins lie in the links forged by American whaling ships picking up crew from the islands of Faial, Pico and São Jorge, when they called at the Azorean port of Horta. By the mid-nineteenth century, these initially port-based Azorean communities were already shifting to the expanding industrial towns of the New England region. The principal concentration, according to the 1870 census, was Massachusetts (which had 30 per cent of all Portuguese immigrants in the USA), especially Bristol County, and within this Fall River. By 1920 Portuguese-born residents alone (that is, excluding their descendants) constituted 19 per cent of the population of the latter city (Williams, 1982, p. 17). The textile industry was the economic mainstay of the Azoreans, and the post-war decline in the local industry led to the growth of a new community in New Jersey, based on employment in the garment industry and other semi-skilled and unskilled occupations. In this way, Essex County in New Jersey became the third largest county-level concentration of Azoreans on the east coast.

- Azorean settlement in California also developed initially in response to the spread of whaling, in this case to the Pacific. The Azoreans then moved into agriculture, where they were able to draw on their traditional island work experiences. The first focus was the production of vegetables around San Francisco and Oakland, but from about the 1880s there was a shift to dairy production in the Central Valley, particularly in San Joaquín. Tuna fishing, especially from San Diego, provided another focus for the Portuguese on the west coast. While most of the early settlers were from the more westerly islands, especially Faial, Pico and São Jorge, a significant proportion of later arrivals were from Terceira. The peak of California's relative importance was 1870 when it accounted for approximately 40 per cent of all the Portuguese in the USA. By 1890 this had slipped to 32 per cent, and it stabilized at this level for the next half century (Williams, 1982, p. 14). In more recent years, there has been greater economic diversification, with industries such as food processing, shipbuilding, and electronics being especially attractive to the Azoreans. There were an estimated 83,000 Azoreans in California in the 1970s, at which date 90 per cent of all the Portuguese in the state were from the Azores (Dias, 1982, p. 60).

- The third major focus was Hawaii. Whaling again provided the original link, for Hawaii was a base for reprovisioning the Pacific whaling fleets. Immigration increased markedly after an 1877 agreement whereby the Azoreans and Madeirans obtained free transport to Hawaii in return for working on the sugar plantations for a minimum of three years. At its peak, Hawaii accounted for 16 per cent of all the Portuguese – mostly Azoreans – living in the USA in the early twentieth century (Alpalhão and Pereira da Rosa, 1980, p. 53). The strategy of encouraging Portuguese emigration stemmed from the need to contract a low-cost labour supply for the sugar plantations. Chinese emigration, which was initiated in 1852, provided a temporary solution to these labour shortages. However, the indigenous population of Hawaii did not view favourably what was perceived as the excessive 'orientalization' of the islands, and the arrival of large numbers of men without families. For these two reasons, it was decided to promote the emigration of European families as a counterbalance to the growth of the Chinese population. Azorean (and Madeiran) families were the only major group who were inclined to accept the offer of work in the Hawaiian islands for ten dollars a month, plus their food, accommodation and medical assistance.

Other than these factors, the development of Portuguese emigration was also due to the favourable reputation of the Portuguese who had gone to the islands. A Commissioner of Immigration, William Hillebrand, wrote to the Minister of the Interior of the Kingdom of Hawaii on 6 June 1877 that

> In my opinion your islands could not possibly get a more desirable class of Immigrants than the population of Madeira and Azores Islands. Sober, honest, industrious and peaceable, they combine all the qualities of a good settler and with all this, they are inured to your climate. Their education and ideas of comfort and social requirements are just low enough to make them contented with the lot of an isolated settler and its attendant privations, while on the other hand their mental capabilities and habits of work will ensure them a much higher status in the next generation, as the means of improvement grow up around them. (Dias, 1983, p. 6)

Their role as contracted plantation labourers ascribed a very low status to the Azoreans and they had some of the poorest living and working conditions in the Hawaiian islands. This led, in contrast to the other Portuguese communities, to early attempts to shed their distinctive cultural identity. It also led to out-migration, mostly to California, in the early twentieth century, for example involving an estimated 2000 in the period 1911–14 (Williams, 1982, p. 56). Simultaneously, there were also internal migrations within the islands, with many of the plantation workers moving to Honolulu, following the parallel course of the earlier Chinese immigrants. In this city and, later, also in Hilo (the second city of the archipelago), the Portuguese tended to find work in the commercial and service sectors. Other members of the Portuguese community became independent agricultural workers, outside of the plantation system. As a result, 'The Punchbowl district of Honolulu became the residential core of the Portuguese islanders, who soon converted this arid rock into productive gardens' (Pap, 1981, p. 78).

- There were two main foci in Canada – Ontario and Quebec. Between 1960 and 1975, 51 per cent of all Portuguese immigrants to Canada were from the Azores, and 78 per cent of these were from São Miguel (Pereira da Rosa, 1980, p. 67). Recruited originally to work in rural areas, they quickly moved to urban centres where wages and working conditions were better. Taking emigrants and their descendants together, Anderson and Higgs (1976) estimated that there were 220,000 Portuguese in Canada in the early 1970s, while Teixeira (1995) estimates that this number had risen to 300,000–500,000 some twenty years later. Therefore, within a period of 40 years, Portuguese communities have evolved in places such as Toronto which are as visible, numerous and well-organized as those in California and New England.

These different communities have individual characteristics, reflecting the interaction of particular features of the locales of origin and destination as well as global changes, but they also share common features. Here we consider three of these: the economic success of the emigrants, their communal organization and the essentially permanent nature of emigration.

First, Azorean emigrants have enjoyed economic success. The relatively unskilled emigrant men mainly filled matching occupational niches in whaling, farming, factories and construction, while the women tended to work as domestic or office cleaners. Economic progress was probably greatest in California, where the emigrants' high propensity to save enabled them to become self-employed farmers. In the 1970s, an estimated 40 per cent of milk producers in California were of Azorean origin (Dias, 1982, p. 21). Some emigrants also achieved upward social mobility through investment in ethnic businesses, notably shops and travel agencies, serving their Azorean compatriots. The Azoreans, while generally achieving only modest social mobility, were able by dint of hard work and self-deprivation to accumulate substantial savings. The extent to which these benefited the islands is dependent on the temporary/permanent character of emigration.

The second feature of the Azoreans in North America is their retention of a strong cultural identity. This has been facilitated by the highly selective emigration flows from individual islands to particular areas of settlement. For example, not only are 87 per cent of all the Portuguese in Canada to be found in Ontario and Quebec, but within these provinces they are concentrated in small areas (for example, Kensington Market) within particular cities such as Toronto (Teixeira, 1995). There is also concentration in terms of islands of origin. For example, in San Diego most of the Portuguese are from Pico while in Arcata most are from Flores. Geographical concentration is facilitated not only by language barriers, lack of skills and of capital, but also by a desire to settle in close proximity to friends and relatives. The desire for cultural affinity is encapsulated in the popular saying that 'Azoreans always carry their islands on their backs'. According to Dias (1984, p. 112), 'the evocation of the past becomes an antidote to the strangeness of the new reality and acts as a shelter from the anguish of the first days of life in a new country.' The second and third generations have often moved out from the initial settlement areas but have not usually entirely desegregated, instead forming new suburban communities such as that at Mississauga in Toronto (Teixeira, 1995). The geographical dispersal of the Azoreans in the agricultural areas of California did not lead to any markedly greater acculturation, and indeed the relatively isolated nature of their economic activities reduced the need to learn English and to mix with other nationalities. Instead there has been a relatively high degree of institutionalization of national differences in all the major Azorean communities in North America, as indicated by the wide range of ethnic businesses, and social, cultural and religious associations (Dias, 1982). The main exception to this has been Hawaii, as noted earlier, although there is also evidence that the decline in immigration to California (perhaps due to reduced opportunities in agriculture) has weakened cultural revitalization in that state (Williams, 1982, p. 128). However, these informal and institutionalized networks have played a key role in mediating how emigration from the Azores has been shaped by the processes of globalization. As Kritz and Zlotnik (1992) have commented in a more general context, but with direct relevance to emigration from the islands, the functions of networks include linking communities at origin and destination, serving as channels for information, migrants and resources, and insulating migrants from the negative aspects of living in host societies, or aiding in their adaptation to them.

The third feature of the Azoreans in North America is the essentially permanent nature of their emigration. There have of course been returns and Williams (1982, pp. 79–80) writes that there were complex patterns of movement from and to the islands:

> It was common practice for young men who had immigrated to America to work until they could save enough money to return to the Azores and take a wife. Many returned immediately with their new brides; others remained in the islands, hoping to use their accumulated capital to become successful farmers or merchants. Some were successful; others watched their money dwindling away and reluctantly decided to return to the United States – only this time accompanied by a wife and children. Still others returned to marry a wife only to find that she refused to leave her family and were themselves forced to remain in the islands ... There were also those who returned to the Azores with the intention of staying and found that they could no longer adjust to the slow pace of village life and soon departed again.

The rates of return have fluctuated in accordance with economic and entry conditions in the USA and Canada, but have generally been rather low. The permanency of migration is reflected in emigration being essentially a family affair. Through chain migration, the Azoreans were able, remarkably quickly, to reconstitute their nuclear, and sometimes their extended, families in North America. The result, according to Alpalhão and Pereira da Rosa (1980, p. 30) in their study in Quebec, is that 'Some tens of thousands of Portuguese live in Quebec today. Their destiny is bound to two continents, two peoples and several governments'. This is a particularly strong illustration of the nature of globalization, and of 'the stretching and deepening of social relations and institutions across space and time' (Held, 1995, p. 20). What implications does this hold for the Azores? The most obvious is that the impact of the emigrant Azoreans is not primarily effected through remittances and investments in the islands. Instead, Azoreans' high propensity to save has been translated into very high property ownership rates in North America (Brettell and Pereira da Rosa, 1984, p. 91). This preoccupation with home ownership is embodied in the popular adage that *'quem casa quer casa'*, whoever marries wants a house.

Given the low levels of return migration, the main influence of these

North American emigrant communities has been ideological. The permanent emigrants have retained strong links with the islands through intense information and recruitment networks (Silveira e Sousa, 1995, p. 40), in which women have usually played a key role (Alpalhão and Pereira da Rosa, 1980, p. 137). These networks have sustained the ideology of emigration which, combined with the existence of strong facilitating communities, has provided a powerful stimulus for the continuation of emigration. As Silveira e Sousa (1995, p. 45) writes in respect of São Jorge island, the returnees from America played an essential part in linking the islanders to the outside world through 'the construction of the symbolic images and desired trajectories and life projects'. Whilst historically the networks and ideology mostly operated amongst the peasantry, there is evidence that from the 1970s they also involved the children of the elite (Chapin, 1992, p. 4). Above all, these images combined personal and family material advancement (evident in conspicuous housing consumption), with the fulfilment of family and marriage projects.

The Imprint of Emigration on the Azores

The persistent, large-scale and mainly permanent emigration from the Azores has left a strong imprint on the social and demographic structures of the islands. While these have been studied in less detail than in the areas of emigration in mainland Portugal, it is possible to identify significant impacts on population levels, ageing, investment/consumption and social structure.

The long history of emigration, and the persistently higher rates of out-migration compared to return, have had direct and, via their impact on the natural rate of increase, indirect effects on population levels in the Azores. From the late nineteenth century the population fell steadily, declining from 256,673 in 1900 to 231,453 in 1920. Thereafter, the reduced opportunities for emigration to North America stemmed population losses and the islands' population gradually increased to reach 327,400 in 1960. After this, the combination of population pressure in the islands and the renewal of emigration opportunities to North America saw the onset of a new migration wave. By 1991 the population had been reduced by almost one third to 237,795; the rate of decline was 12 per cent in the 1960s and a remarkable 16 per cent in the 1970s. No other Portuguese region experienced such a sustained loss of population in the post-war period, and indeed there are few parallels to be found anywhere in Europe in this period. By the 1980s, improved living conditions in the islands and tighter North American immigration controls had reduced net emigration, and the population decline was only 2.4 per cent in this decade. As ever, the experiences of the islands were varied (Table 3.1). Santa Maria had a population loss of 8.9 per cent during 1981–91 while, at the other extreme, Corvo and Terceira had population gains. Population

Table 3.1   The Azores: population change, 1981–91

Island	1981	1991	% change
Santa Maria	6,500	5,922	-8.9
São Miguel	131,908	125,915	-4.5
Terceira	53,570	55,706	+4.0
Graciosa	5,377	5,189	-3.5
São Jorge	10,361	10,219	-1.4
Pico	15,483	15,202	-1.8
Faial	15,489	14,920	-3.7
Flores	4,352	4,329	-0.5
Corvo	370	393	+6.2
Azores	243,410	237,795	-2.4

Source: MPAT (1993, p. 116).

changes are not related in any simple manner to the sizes of the islands, the economic opportunities they provided or the services which they have available, for the most populated island, São Miguel, had the second largest population loss in the 1980s. The key to these variations is the way in which the overall economic situation interacts with conditions and natural disasters in particular islands, and the individuality of their links with overseas communities on the North American mainland.

The effects of depopulation are to be seen in the gradual ageing of the Azorean population. Between 1981 and 1990 the proportion aged over 65 increased from 11.3 per cent to 12.5 per cent. While there is a trend towards an older population, the age structure in the Azores is little different from that of Portugal as a whole, and is considerably younger than that of the Alentejo and Central Portugal, which are also regions of out-migration (Figure 3.3). There are a number of reasons for this, including the relatively high birth-rates in the Azores, and the lack of migration of young people from the islands to work or be educated in the metropolitan centres of mainland Portugal. Historically, this lack of mobility towards the mainland has been due to the lack of opportunities in a relatively remote capital, but more recently there has been the pull of established ties in North America, and the establishment of a university in the islands. But the most important factor has been the nature of emigration from these islands. The permanent nature of emigration from the Azores has tended to involve the staged emigration of entire families, sometimes spanning three generations.[1] Therefore, the age-specific impact has been less pronounced than in some regions of mainland Portugal, where the out-movement of young people to foreign countries or to Portuguese urban regions has severely denuded the age–sex pyramids (Figure 3.3).

The permanent nature of emigration has also meant that transfers of

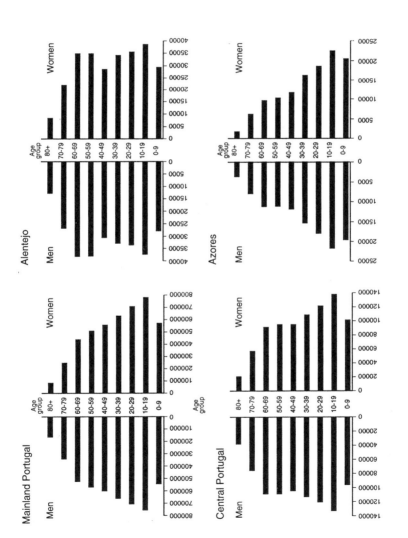

Figure 3.3 Age–sex pyramids in all of mainland Portugal, the Alentejo, Central Portugal and the Azores, 1991
Source: Serviço Regional de Estatística dos Açores.

accumulated wealth to the communities of origin were smaller than they would have been in the case of temporary emigration. Of course, there were emigrants who did remit all their savings to the islands and then returned, but they were outnumbered by the permanent emigrants. Although the latter also remitted some savings to support the families who remained behind, and there were examples of collective efforts among the North American emigrant communities to raise money for particular projects such as repairing a church or civic building, their first priority was to buy a home and, if possible, establish a business in their new communities. Therefore, there is less visible evidence in the Azores than there is in, say, rural northern Portugal of the return of the emigrants and of their conspicuous consumption in the form of imported architectural house styles and consumer durables (Lewis and Williams, 1986).

Despite the relatively low rates of return, the sheer scale of emigration meant that even the small proportions of returnees had an impact on the Azorean economy. Silveira e Sousa (1995, p. 43) comments that, on average, returnees had been abroad for about twelve years. On their return, their first priority was to invest in a large house and land that constitute the main positional goods in these rural communities. In this, there is little to differentiate their behaviour from that of returnees to mainland Portugal (Lewis and Williams, 1986). There is some evidence that they have effected agricultural innovations, and this reflects the strong complementarity between their working experiences abroad (in agriculture, especially in California and initially in Canada) and the economic environment they were returning to in the islands; this stands in marked contrast to the generally non-innovatory behaviour of returnees to mainland Portugal from northern Europe. A recent study of the island of São Jorge has shown that, in the last few years, there has been a gradual shift of investment from agriculture and livestock activities to small industries, commerce, personal savings, restaurants, cafés, bars and other similar establishments (Fonseca, 1999). It is probable that there have been similar trends in the other islands. However, in common with the mainland, the returnees were rarely able to accumulate sufficient funds, and to build up sufficient productive resources, to guarantee that the next generation would not also have to emigrate in due course (Silveira e Sousa, 1995, p. 20).

The strong Azorean presence in North America has also been an important factor in the development of tourism in the archipelago. The USA and Canada constitute two of the principal markets for tourists to the islands, representing in the course of the 1980s between 45 per cent and 70 per cent of all guests coming from abroad who are staying in tourist accommodation (Moniz, 1996, p. 57). The way in which this influences emigration has already been discussed, but it is important to stress the direct economic importance of such tourism flows.

Finally, emigration did have an important influence on the political and social structures of the Azores. Firstly, in common with many other regions of emigration, this was a process that served to defuse the social tensions inherent in underemployment, poverty and a profoundly unequal society. Secondly, the returnees, through the purchase of land and houses, were able to climb the social hierarchy and to challenge the power of the traditional elites in the islands, even if they were not able to dislodge them from the upper echelons of political power (Silveira e Sousa, 1995). Thirdly, the distinctive nature of Azorean emigration, and the symbiotic economic and cultural exchanges with the substantial émigré communities in North America, served to differentiate the islands socially from mainland Portugal. This contributed to the brief surge of separatist sentiment in the 1970s, when secession to the USA was mooted by some minority groups in the face of political instability and a leftwards shift in Portuguese politics following the 1974 military coup which overthrew the Salazar/Caetano regimes (Lewis and Williams, 1994). While this soon fizzled out, it did provide another reminder that the Azores do lie between Europe and North America in more than a narrowly geographical sense.

Conclusion

The latest major wave of emigration from the Azores, which began in the late 1950s, seems to have dissipated. In part this is due to changes in Portugal, related to both the institutionalization of democracy and economic development. Portugal has experienced a marked acceleration of economic growth from the mid-1980s, related in no small measure to its accession to the EU in 1986. Some of the resulting prosperity has diffused to the islands, notably in the form of state- and EU-funded infrastructural projects. This is symbolized by the fact that all the islands now have their own airfields, which has significantly improved accessibility to and within this geographically isolated and dispersed group of islands. In addition, the Azores have generated a modest growth in tourism, some of which is based on their ecological attractions, but also drawing on the vast potential for visits to friends and relatives from the large émigré communities in North America. The net effect of these various economic developments has been an increase in total employment in the islands from 77,320 to 89,200 jobs in the 1980s (MPAT, 1993, p. 117).

Although emigration has declined, it has not ceased, as is indicated by the 2.8 per cent decline in population in the 1980s. In 1987 – the last year for which reliable sub-national statistics are available – the gross emigration rate in the islands was still twice that recorded in *any* mainland district. Moreover, while the pace of economic development has accelerated, structural economic change has been limited. For example, male employment declined in the 1980s, despite the overall increase in jobs. The islands continue to be reliant on transfer payments from Lisbon and Brussels, on emigrant remittances, and on a potentially fickle external demand for tourism services and dairy products. A major reduction in any of these sources of income, or another natural disaster, could bring about a renewal of emigration. The distinctive Azorean émigré communities in North America would facilitate any such renewal of large-scale emigration.

As ever, then, the evolution of the Azores remains caught between North America and Europe, and it is emigration more than any other process that serves to cement relationships with the former. The islands' history also serves to illustrate the long-standing but changing nature of global–local relations, with emigration from any particular island being influenced as much, if not more, by events in distant émigré communities in North America as by policy changes emanating from Lisbon, or development in the main Azorean cities such as Ponta Delgada and Angra do Heroismo. It is equally true that the social and economic characteristics, and even the built landscapes, of particular sub-regions and city districts in North America have been shaped by their emigration links with the islands. Therefore, the Azores both influence, and are influenced by, events on both sides of the Atlantic.

Held's (1995) concept of globalization, with which we introduced this chapter, provides a useful perspective on emigration from the Azores. In this view, globalization constitutes a series of shifts which infiltrate the practices of everyday life and individual experiences of contemporary capitalism. As a result, globalization not only leads to a re-definition of relations between a global economy and national states but between a global economy and national and local civil societies. The interdependency between Azorean communities in particular islands and émigré communities in particular parts of North America provides strong evidence of such links between local civil societies, and we have also noted how global economic shifts and state interventions have mediated these. Held's 'stretching and deepening of social relations and institutions across space and time' are evident in a number of ways: income and property transfers, family division and re-unification, information channelling, and the creation and transfer of particular images and symbols. However, in the case of Azorean migration, many of these linkages had already achieved a high degree of intensity over a period of at least a century. They had created deep-rooted migrant networks which were modified by but not fundamentally transformed by the intensification of globalization in recent decades.

One of the main reasons for the persistence and intensity of such networks lies in the nature of islandness, and the way in which this relates to migration. Kritz and Zlotnik (1992, p. 6) emphasize that migrants are 'embedded in numerous formal and informal networks at origin and destination that affect migration outcomes'. These are particularly intense on islands where there will be relatively few competing networks. This is especially so in the Azores where many of the islands are isolated from each other, let alone from metropolitan Portugal, and where small populations can imply very high degrees of familiarity in what are island-centred migration chains. Politically a part of the Portuguese state, and territorially adjacent to it (at least in terms of intervening land masses), the particularities of the islands' economy, culture and location have meant that events in the Azores have – in an enduring manner – been significantly determined by decisions taken elsewhere (on the western side of the Atlantic) for a much longer time period than would seem to be implied by most recent theories of globalization.

Note

1. The permanent character and the family nature of Azorean emigration to the USA, since the nineteenth century, is due to – other than the economic factors already referred to – the high level of clandestine migration, especially of young adults seeking to evade military conscription.

References

Alpalhão, J.A. and Pereira da Rosa, V.M. (1980) *A Minority in a Changing Society: The Portuguese Communities of Quebec.* Ottawa: University of Ottawa Press.

Anderson, G.M. and Higgs, D. (1976) *A Future To Inherit: Portuguese Communities in Canada.* Toronto: McClelland & Stewart.

Brettell, C.B. and Pereira da Rosa, V.M. (1984) 'Immigration and the Portuguese family: a comparison between two receiving societies', in T.C. Bruneau, V.M. Pereira da Rosa and A. Macleod (eds), *Portugal in Development: Emigration, Industrialization, the European Community.* Toronto: University of Ottawa Press, pp. 83–110.

Chapin, F.W. (1992) 'Channels for change: emigrant tourists and the class structure of Azorean migration', *Human Organization,* 51(1), pp. 44–52.

Dias, E.M. (1982) *Açorianos na Califórnia.* Angora do Heroismo: Colecção Diaspora.

Dias, E.M. (1983) *A Presença Portuguesa no Hawai.* Lisbon: Ramos, Afonso & Moita.

Dias, E.M. (1984) 'Portuguese immigration to the East Coast of the United States and California: contrasting patterns', in T.C. Bruneau, V.M. Pereira da Rosa and A. Macleod (eds), *Portugal in Development: Emigration, Industrialization, the European Community.* Toronto: University of Ottawa Press, pp. 11–20.

Fonseca, M.L. (1999) 'A emigração na Ilha de S. Jorge', in M.V. Guerreiro *et al.* (eds), *A Ilha de S. Jorge: Uma Monografia Geográfico-Etnográfica.* Velas: Centro de Tradições Populares Portuguesas and Câmara Municipal de Velas, in press.

Gurak, D.T. and Caces, F. (1992) 'Migration networks and the shaping of migration systems', in M.M. Kritz, L.L. Lim and H. Zlotnik (eds), *International Migration Systems.* Oxford: Clarendon Press, pp. 150–76.

Held, D. (1995) *Democracy and the Global Order: From the Modern State to Cosmopolitan Governance.* Cambridge: Polity Press.

Kritz, M.M. and Zlotnik, H. (1992) 'Global interactions: migration systems, processes, and policies', in M.M. Kritz, L.L. Lim and H. Zlotnik (eds), *International Migration Systems.* Oxford: Clarendon Press, pp. 1–18.

Lewis, J.R. and Williams, A.M. (1986) 'The economic impact of return migration in Central Portugal', in R. King (ed.), *Return Migration and Regional Economic Problems.* London: Croom Helm, pp. 100–128.

Lewis, J.R. and Williams, A.M. (1994) 'Regional autonomy and the European Communities: the view from Portugal's Atlantic territories', *Regional Policy and Politics*, 4(1), pp. 67–85.

MPAT (Ministério do Planeamento e da Administração do Território) (1993) *Preparar Portugal Para o Século XXI – Análise Económico e Social*. Lisbon: MPAT/SEPDR.

Moniz, A.I. (1996) *O Turismo nos Açores: Estudo Sobre a Oferta de Alojamento Turístico*. Ponta Delgada: Jornal de Cultura.

Mota Amaral, J. (1981) 'As ilhas Portuguesas Atlânticas e a adesão de Portugal às Comunidades Europeias', in Intereuropa (ed.), *Portugal e o Alargamento das Comunidades Europeias*. Lisbon: Intereuropa, pp. 239–58.

Pap, L. (1981) *The Portuguese-Americans*. New York: Twayne Publishers.

Pereira da Rosa, V.M. (1980) 'Emigration and underdevelopment in a dependent society: the Azorean case', *Iberian Studies*, 19(2), pp. 62–8.

Serpa, C.V. (1978) *A Gente dos Açores: Identificação, Emigração e Religiosidade, Séculos XVI–XX*. Lisbon: Prelo Editora.

Silveira e Sousa, P.P. de (1995) 'Emigração e reprodução social no contexto açoriano: o case da ilha de S. Jorge na segunda metade do século XIX', *Islenha: Temas Culturais das Sociedades Insulares Atlânticas*, 17, pp. 31–49.

Teixeira, C. (1995) 'The Portuguese in Toronto: a community on the move', *Portuguese Studies Review*, 4(1), pp. 57–75.

Wallerstein, I. (1974) *The Modern World System*. New York: Academic Press.

Williams, J.R. (1982) *And Yet They Come: Portuguese Immigration from the Azores to the United States*. New York: Center for Migration Studies.

4

Emigration and Demography in Cape Verde: Escaping the Malthusian Trap

Annababette Wils

Introduction

Cape Verde, contrary to its name, is a dry country. During large parts of the year much of the land area is predominantly brown, set off against a blue sky and a blue sea. The country is an archipelago of nine inhabited islands located in the Atlantic Ocean 800 km west of Senegal (Figure 4.1). Altogether, about 400,000 people lived there in the middle of the 1990s. It was one of the world's poorest countries in the 1980s, but by 1994 it was classified towards the lower end of the United Nations' 'medium human development' band of countries, ranking 123rd (out of 175 countries) on the 'human development index' (an amalgam of per capita wealth, life expectancy and education), and 126th on the basis of real per capita GDP (UNDP, 1997).

The islands were colonized by the Portuguese from 1460 and the population evolved as a mixture of European and African descent. In early centuries, a buoyant colonial economy evolved which consisted mainly of slave trade. When this collapsed with the abolition of slavery, even the colonial elite on the islands degenerated into a state of chronic poverty (Carreira, 1982; Davidson, 1989). The majority of the population survived on subsistence agriculture as sharecroppers or renters, and some worked on plantations. The agricultural base crop of the islands was and is corn which, together with beans and fish, makes up the main elements of the diet. The climate prohibited then, and still hampers, the production of export crops

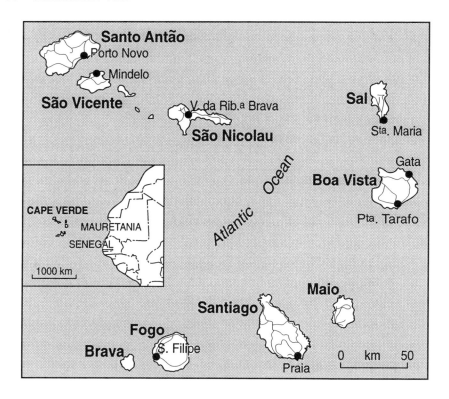

Figure 4.1   Map of Cape Verde

on a large scale. In the eighteenth and nineteenth centuries, Cape Verde acquired some foreign revenue by acting as a water and coal depot for long-range ocean ships. The effects of these poor trade relations were exacerbated by regular droughts and ensuing famines, which could wipe out up to 30 per cent of the population in a few years.

Throughout the present century, there has been remarkably little change in the domestic economy. Corn and beans for subsistence remain the staple crops, augmented by local fishery. Commercial fishing, in spite of foreign investments, has failed to grow. There has been the development of some domestic industry, such as beer, cigarettes, bread, furniture, and construction. Recently, tourism has mushroomed – achieving 17 per cent annual growth in the 1980s – and since 1990 the government policy has been to promote export processing industry.

Since independence in 1974, famines have been averted by a large influx of foreign exchange and food. About half the value of this flow has been in the form of international aid, while the other half has been remittances (not counting unofficial private transfers).

The dependence on remittances – and on the flow of migration which is the source of that money – has a long tradition on Cape Verde. An historical analysis of the population growth, economic situation, and emigration data leads to the hypothesis that emigration waves allowed the Cape Verde population to escape the Malthusian trap twice in the last century and a half. In these periods, emigration allowed the domestic population to increase because remittances gave the population the means to buy food from abroad in time of need. This hypothesis stems from a number of observations which will be described in more detail below.

My second hypothesis in this chapter is that, although the direct effect of migration appears to be the alleviation of economic distress and the promotion of natural population growth, the indirect effect of migration in recent decades has been to reduce population growth by lowering the fertility of Cape Verdean women. This hypothesis is also explored in more detail below. It requires some amount of statistical number juggling, involving estimates of migration flows and population trends, but I shall attempt to keep the statistical discussion to a minimum.

Two Cape Verdean Escapes from Malthus

Until recently, the majority of people on Cape Verde made their living from the sparse land. In good years this afforded a meagre existence, but in bad years meant death for large portions of the population. It is therefore not surprising that, from the beginning, Cape Verdeans have been leaving their country. In the sixteenth and seventeenth centuries, after the decline of the slave trade, most of the Europeans returned to Europe, but this was a very small group. Another small group of people went to the African coast where they earned a living as smugglers along the West African rivers, by-passing the Portuguese Crown's monopoly of trade there.

In this early period of Cape Verde's history, that is until the end of the eighteenth century, the population grew slowly, regularly set back by famines, but generally increasing and gradually spreading around to the nine presently inhabited islands. Figure 4.2, which charts the population size from 1582 to 1992, shows this slow increase. Then, from 1807 (an early census year) to 1867 there was a demographic stagnation. In 1807 the population was 58,000, and in 1867, just after a devastating famine, it was still only 67,000. It appears that during these six decades, Cape Verde was caught in a cyclical version of the Malthusian trap.

The Malthusian trap (Malthus, 1798) is a situation where the given land area is so crowded that only a meagre subsistence level of production is ensured. At this subsistence level, mortality rates are high, so high that they cancel out the tendency of high fertility to increase population size.

Population data and vital statistics for Cape Verde from the last century (Carreira, 1985) indicate that Cape Verde experienced an altered, cyclical

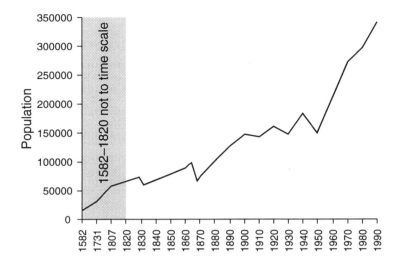

Figure 4.2 Population of Cape Verde, 1582–1990
Source: Wils (1996, p. 8).

version of this trap. The data show periods of high population growth (1.5–2.5 per cent annually) in the last and the present century, regularly interrupted by one or two years of severe attrition during a famine (loss of up to 30 per cent of the population). An interpretation of these numbers would show the following scenario. In good years, say at the beginning of the cycle, the land offers slightly more than subsistence to the population, which has a high fertility rate. Births exceed deaths and the population grows. The land becomes more crowded. Intermittently, a drought occurs, famine ensues and the population is reduced – to a level that can survive a drought. Then, the population is smaller than the hypothetical 'Malthusian crowded subsistence' size population, and the cycle starts again.

There are two main schools of thought on how to escape such a trap. One originates with Condorcet (1795) and suggests that by improving social conditions, particularly by educating women, fertility will decline and population growth will halt so that improvement of living conditions is possible. The second school is more recent (Boserup, 1968). It suggests that through high population density, a society is forced to introduce more productive technology thereby improving living conditions. There is little evidence that the poor or the elite of Cape Verde intervened in their Malthusian trap in either manner. This does not lead me to reject Condorcet's or Boserup's ideas as a whole; rather, it appears that there are situations, such as some periods in Cape Verde, where they do not apply. On Cape Verde, I believe, a third strategy was introduced to escape the Malthusian trap:

emigration. A detailed description of migration in the last century is found in Carreira (1982) and the following four paragraphs paraphrase his colourful account.

From the end of the eighteenth century, but increasingly in the nineteenth, North American sailors whaled in the profitable waters around the Cape Verde islands. The American whalers in the Cape Verdean waters would go ashore to cut, boil and treat the whales, and to obtain fresh supplies of food and water. There, they obviously came into contact with Cape Verdeans. It is uncertain when the first Cape Verdean men boarded the American ships to become sailors. Certainly, it became a large movement by the nineteenth century. A number of the Cape Verdeans disembarked from the ships upon arrival in Boston, New Bedford, and other towns in New England, and stayed in the United States. Some of the migrants remained in the USA, others returned with their fortunes. This is a story which has striking parallels with the early history of emigration from the Azores, as recounted by Williams and Fonseca in the previous chapter of this volume. During this period, there was also a revival of domestic activity as the British invested in the excellent natural harbour on one of the northern islands, São Vicente, as a coal depot for ships passing from Europe to the Americas.

Concurrent with the emigration and harbour activity, there was a sustained increase in the population, shown in Figure 4.2. The average rate of population growth from 1867 to 1898 was 2.4 per cent, as high as in some developing countries today. Compared with the earlier period of demographic stagnation, this growth indicates a significant improvement in living conditions. The activity on São Vicente in the harbour does not, however, explain the sustained population growth on other islands of Cape Verde. It seems likely that remittances from the USA played an important role in supplementing domestic subsistence activities. This is reflected in accounts from the period. For example the historian Barcelos writes about the years 1863-7 on the island of Brava:

> Many people left Brava for America. The crisis in Brava was more or less eased by remittances from their sons resident in the United States of America, crews of merchant and whaling ships, who managed by means of honest, steady work to procure regular sums of money to send to their families. (Quoted in Carreira, 1982, p. 49)

Around the turn of the century the harbour in São Vicente declined and the United States imposed strict immigration laws, so emigration was reduced. Thus, both activities which had carried the economy for some decades dried out simultaneously. From 1898 to 1947 the economic stagnation is reflected in a second period of demographic stagnation, this time at a higher plateau of about 140,000. It is not known why the islands supported a larger population in this recent Malthusian trap period than they had a century before. It is possible that there were better agricultural methods, although no convincing evidence of this can be found. We are left to conjecture that perhaps there was a constant flow of remittances from those migrants who had arrived in the United States earlier and remained there.

After 1948, which was the second year of a famine in which 17 per cent of the population died, Cape Verde emerged once again from its cyclical Malthusian trap. This time, it was because food aid came to the country, and also because emigration and remittances increased again (Carreira, 1982, 1985; Davidson, 1998; De Sousa Reis, 1989). The modern wave of migration has been predominantly to Europe, first to The Netherlands, then to Portugal, France, Luxembourg, Italy and Switzerland, as unskilled labour. In this period, women began to emigrate along with the men: whereas from 1900 to 1920 only 17 per cent of the emigrants had been women, during the period 1955–72 it was 38 per cent (Carreira, 1982, p. 83).

Figure 4.3 shows the rise in both emigration and immigration (i.e. mainly return migration) starting in 1930. From 1930 to 1946 there is virtually no external population movement. From the famine in 1946 migration starts, and increases continually until the latest data point in 1988. It is notable that the immigration curve follows the emigration curve very closely. Exceptions are brief periods during the 1946–7 drought, and the period 1970–4. This indicates that much of the migration from Cape Verde is temporary: people go abroad for a few years and return. The net migration is usually slightly negative, so there is a constant small demographic attrition from the country, and a growing stock of Cape Verdeans abroad.

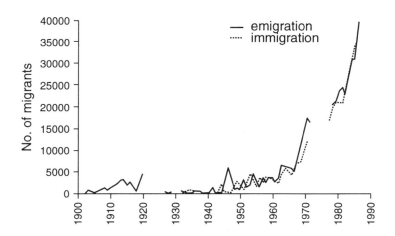

Figure 4.3 Emigration and immigration trends for Cape Verde, 1900–88
Source: Wils (1996, p. 21).

These migrants sent back substantial amounts of money – in some years remittances equated one quarter of GDP (World Bank, 1993). The upward trend of remittances closely follows the negative trend of net migration in the 1970s and 1980s. Elsewhere I have estimated how much each migrant sent back annually during this period. The range is from 67 to 245 US dollars per migrant in 1975 to a maximum of 379 to 1006 US dollars per migrant in 1980 (Wils, 1996). The range is due to the uncertainty of how large a pool of Cape Verdeans abroad is sending money – is it only those who emigrated a few years before, or does it include all Cape Verdeans who emigrated in this century and are still alive? The data show that the pattern of remittances is sensitive to hardship on the islands: during the crisis years shortly after independence in 1974, transfers per emigrant increased; and in the second half of the 1980s, when the economy stabilized and improved, they decreased (Figure 4.4).

Buoyed by remittances and foreign aid,[1] the population started to grow once more, as shown in Figure 4.2. The graph shows a small decrease in population growth from 1971 to 1974. This coincided with the Sahel drought, whose effects were also felt on Cape Verde. In these years, emigration was so high that it virtually cancelled out natural population growth. From 1974 the population rises. Overall, it is notable that a decline in population growth due to emigration is the exception; rather, the predominant effect of emigration is to allow the population to grow.

From all the above information, I conclude that there has been a correlation

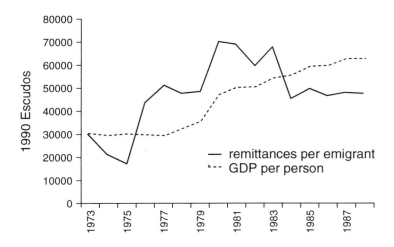

Figure 4.4   Remittances per migrant and GDP per capita, Cape Verde, 1973–88
Source: Wils (1996, p. 60); World Bank (1993).

between emigration and expanding population. This means that emigration is not mainly a mechanism to combat population pressure by reducing numbers of people. In the case of Cape Verde, it appears that emigration has been a strategy to reduce economic pressure because remittances improve the incomes of those who are left behind. With the better incomes, the average survival rates of the people on Cape Verde are improved. If this is the case, then the effect emigration has on lowering the population size is compensated by higher natural growth due to higher survival rates.

Some authors, e.g. Abernethy (1991), have used this reasoning, plus the argument that emigration reduces perceived limits of the natural and national system, to argue that emigration fosters population growth by keeping fertility high. In this view, emigration is not a sustainable strategy to solve problems of population growth and poverty, and what is more, ultimately exacerbates these problems. The logical conclusion of this line of argument is that emigration should be stopped, perhaps by limiting the immigration possibilities in the receiving countries. Such a conclusion would be callous for Cape Verde in the face of the obvious distress that the country would face if emigration were capped. Moreover, a closer look at the data show that emigration from Cape Verde may be a factor which reduces fertility and therefore, in the long run, slows natural population growth.

Migration as an Indirect Factor to Reduce Population Growth via Fertility Declines

There are three ways in which migration may reduce fertility. One is that partners are separated. There is an interesting time series data-set for Cape Verde which shows, in fact, lower births following years of high emigration during the period 1960–90. Besides such evidence of irregular dips in fertility, there may also be the longer-term effect of having a consistent, large portion of young women abroad temporarily who postpone their childbearing. Third, these women might bring back small-family-size values from their host country upon return. This section of the chapter shows the time series which correlates lower births to high emigration. It also presents the argument for the longer-term fertility-reducing effect. This argument is a complex one, not obvious from the data, and it requires a certain amount of mathematical patience from the reader. The idea that women abroad might postpone their childbearing stems from informal interviews I conducted on Cape Verde during April 1994, and is reinforced by passages from the literature, such as the following note about Cape Verde girls working temporarily as household helps in Italy:

> Note one special characteristic of this emigration – the girls go to earn money in order to marry boys of their own race. A few, insignificant in number, marry Italians. Generally, they mistrust Italian boys. They always go out together in a group to protect themselves ... (Quoted in Carreira, 1982, p. 85)

Figure 4.5 shows the annual number of births recorded by national data from 1960 to 1992. The number of births recorded by national data follows a rather irregular pattern. Although fluctuations in the number of births can be expected and are usual, more so in small populations like that of Cape Verde, the extent of the fluctuations is large. In 1975 for example, 10,196 births were registered, in 1976 there were 14 per cent more, and in the following year it was about 5 per cent less again.

Following an observation raised by Cape Verde's Secretaria de Estado da Cooperação e Planeamento (1984, p. 18–19), a correlation is made of births and emigration. The Secretaria noted that in the year following a period of large emigration, births fell, and after immigration, births increased. Figure 4.5 shows the annual number of births and the national figures on net migration, the former shifted forward by one year. The correlation of high emigration and low birth-rate is particularly apparent in the 1970s. In that decade, the years after high net emigration show low birth-rates. Migration is usually individual labour migration, so one can easily imagine the situation that one partner in a relationship emigrates to work and in the following period, until the migrant returns, or until a new relationship is started, there are fewer babies conceived.

Now we move on to the more complex argument that the overall level of fertility is reduced through migration. The strategy used for this argument is the following. First, an estimate of the average time abroad, and a detailed estimation of migration flows, is made for the years for which this

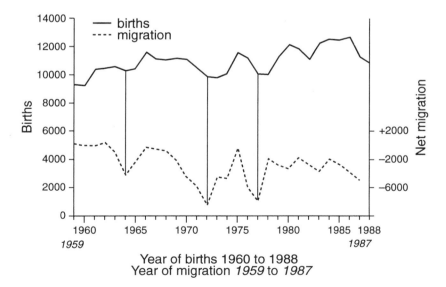

Figure 4.5 Births and net migration for Cape Verde, 1960–88
Source: Wils (1996, p. 27).

is possible, 1970–90. Second, this migration estimation is used to calculate what the fertility patterns of Cape Verdean women might have been during these decades when we include women temporarily abroad – a consideration which is omitted in the standard statistics which calculate fertility rates using only the births and women in the country.

It is difficult to estimate how long the average period abroad is and what percentage of the emigrants returns because these two factors – length of stay and percentage returning – confound each other in the estimate. However, the fact that the immigration curve in Figure 4.3 follows the emigration curve so closely shows that the stay abroad is short, and the percentage returning is high. A rearrangement of the data shows, for example, that if the average stay abroad is 10 years, then the percentage returning is 209, which is patently in contrast with the observed reality. More reasonable results come out if the average stay abroad is assumed to be a more realistic four years. Then the percentage returning is calculated as 88. Such a migration pattern could have a number of important implications. One would be that the ties to Cape Verde would remain strong, and therefore one could expect high remittances. Second, emigration would not be a way to reduce population size, rather it would be a way in which Cape Verde could increase its economic base. Third, the women who go abroad to work are more likely to postpone marriage and childbearing until their return to Cape Verde, thereby reducing their childbearing years and perhaps their total fertility – this is also mentioned in Carreira's account of female emigration in the 1960s (Carreira, 1982, pp. 84–5).

Estimation of the age and sex structure of migrants is not available directly from the published migration data. However, there are other population data which can be used to make an estimate of migration patterns in the past few decades. These other data are: age-specific population in census years (DGE, 1989 and 1993); a general knowledge of the age-structure of migrants; migration flows from Cape Verde by sex and four large age groups for the years 1986–9; and births and deaths (DGE, 1989). This information allows us to make an estimate of recent net migration flows from Cape Verde by age and sex in five-year blocks.

This method of migration estimation can be used in the situations where relatively reliable data exist on population age-structure (census years), age-specific mortality, and age-specific fertility or total births; but where data on migration are scant or unreliable. This situation is not rare. Migration is difficult to classify as it is not an unequivocal event such as a birth, death, marriage or divorce (see e.g. Courgeau, 1988; Wils, 1993). Even if precise data on migration exist – which commonly they do not – it is not certain when that movement should count as migration. After six months? After one year? After renting a house? Obtaining a job? For these and other reasons, migration registration is often of poorer quality than for births and deaths.

The data used for the migration estimation are: census counts in 1970,

1985 and 1990; emigration and immigration in total numbers; births and deaths by age; and migration by sex and by four large age groups for the years 1986–8. This information allows us to make an estimate of recent net migration flows from Cape Verde by age and sex in five-year blocks. To do this a method is used which is called the cohort component method of migration estimation (Wils, 1996, Chapter 3). Briefly, there are three steps. The first is to make a historical cohort component population projection from the beginning of the observation interval *0* to the end of the observation interval *t*, using existing data on the population age structure in time *0* and the age-specific death and birth rates or absolute births in the interval *0* to *t*. The historical population projection provides the population in time *t if there had been no migration*. The second step is to compare the projected population to the actual population in *t*. The difference between the projected and the actual population is assumed to be the result of migration movements in the interval *0* to *t*. The third step is to fit estimates of migration flows in the historical population projection from time *0*, in such a manner until the difference between the historical projection of the population and the actual population at time *t* is zero or close to zero. A variation on this method is to include absolute numbers of births throughout the historical observation period rather than age-specific birth rates. The former method provides the total effect of migration including the lost or gained births from migrants; the latter variation provides an estimate of the number of migrants without the birth effects.

The results for Cape Verde are shown in Table 4.1. Column one shows the five-year age groups and column two the age distribution of migration. Columns three to six show the estimated net migration for each five-year period. Column seven shows the estimated 1990 population using these migration estimates and the 1970–90 data for births and deaths. Column eight shows the actual population counted by the census in 1990 and the last column shows the ratio of the estimated to the counted 1990 population.

The accuracy of these estimations depends both on the quality of the data which are used, and on the quality of the age–structure estimates. The quality of the data can to some extent be checked by the consistency of different data sources with each other. The greater the across-the-board consistency, the greater our confidence that the data are correct. Overall, the consistency between data (census data, migration registration, birth and death registration) is within the 10 per cent range, representing a reasonably tight confirmation.

The age-structure estimates are fitted qualitatively, and there is a certain leeway possible. A number of different estimates were fitted by the author, with variations in the percentage of women migrating, the percentage of older migrants and the total emigrants. All estimates share two aspects which are, firstly, that the percentage of women among migrants is increasing and in 1985–9 was greater than 50 per cent; and secondly that migration

Table 4.1　Estimate of age-specific male and female migration, Cape Verde, 1970–90

Women	Distribution	Estimated migration				Estimated population	Census population	Population ratios
Age	1970–90	1970–4	1975–9	1980–4	1985–9	1990	1990	1990
0–4	0.00	0	0	0	0	27,138	29,934	0.907
5–9	0.00	0	0	0	0	25,670	25,448	1.009
10–14	0.00	0	0	0	0	22,263	21,250	1.048
15–19	0.27	3,015	2,518	1,739	3,800	17,259	16,971	1.017
20–24	0.30	3,350	2,798	1,933	3,722*	14,818	16,599	0.893
25–29	0.26	2,000*	2,425	1,675	3,659	13,944	13,833	1.008
30–34	0.03	335	280	193	422	11,344	10,763	1.054
35–39	0.03	335	280	193	422	8,285	7,871	1.053
40–44	0.03	335	280	193	422	4,288	4,143	1.035
45–49	0.02	223	187	129	281	4,726	4,453	1.061
50–54	0.02	223	149	129	281	6,140	6,361	0.965
55–59	0.02	223	187	129	281	5,946	5,675	1.048
60–64	0.02	223	187	129	281	4,576	4,846	0.944
65–69	0.00	0	0	0	0	2,927	2,981	0.982
70–74	0.00	0	0	0	0	2,416	2,623	0.921
75+	0.00	0	0	0	0	1,243	4,652	0.267

* indicates that the estimates deviate from the standard age distribution used elsewhere

Men	Distribution	Estimated migration				Estimated population	Census population	Population ratios
Age	1970–90	1970–4	1975–9	1980–4	1985–9	1990	1990	1990
0–4	0.00	0	0	0	0	27,661	30,032	0.921
5–9	0.00	0	0	0	0	26,190	25,652	1.021
10–14	0.00	0	0	0	0	22,747	21,107	1.078
15–19	0.25	4,886	2,798	1,933	4,478	17,079	16,587	1.030
20–24	0.30	5,863	3,358	2,319	4,374	13,933	15,077	0.924
25–29	0.20	3,909	2,239	2,010	4,583	12,457	12,318	1.011
30–34	0.03	586	336	232	537	7,923	8,178	0.969
35–39	0.03	586	336	232	537	5,234	5,076	1.031
40–44	0.03	586	336	232	537	2,656	2,660	0.998
45–49	0.02	391	224	155	358	3,502	2,634	1.329
50–54	0.02	391	224	155	358	4,056	3,935	1.031
55–59	0.02	391	224	155	358	3,831	4,003	0.957
60–64	0.02	391	224	155	358	3,388	3,937	0.861
65–69	0.00	0	0	0	0	2,079	2,401	0.866
70–74	0.00	0	0	0	0	1,461	2,063	0.708
75+	0.00	0	0	0	0	774	3,874	0.200

estimates are only slightly higher than the migration actually registered for 1970–84, and much larger for 1985–89.

The estimates show that the proportion of women partaking in the migration flows is growing. In the period 1970–74, which was the beginning of a large increase in the migration flows, women made up 40 per cent of the estimated migration flow. In 1985–89 women were estimated to have made up 55 per cent of the migration flow. This increase in female migrants is consistent with the rise observed by Carreira (1985) for the century as a whole (see above).

The estimate is compared to registered total migration, shown in Table 4.2. The estimates for 1970–84 are equal to the registered data; for 1985–9 the estimates are 60 per cent above the registered migration. Perhaps to some extent the higher numbers in the last estimation are due to incorrect choice on the part of the author, and perhaps to some extent these shifts are due to incorrect measures in the 1990 census. Also, there could be under-counting if the recent years include migrants who have not bothered to register.

With these migration estimates and caveats about their level of precision in mind, it is possible to move on to the fertility analysis. The hypothesis is that temporary migration of women since 1970 has reduced their fertility level and has thereby indirectly contributed to a slow-down of population growth. A number of assumptions are used. First, I assume that the four-year average stay and the 88 per cent return rate proposed above are accurate estimates of average male and female migration behaviour. For simplicity, let us use the rounded number 90 per cent, and call them temporary migrants. Second, if a temporary migrant woman has a baby while abroad, she brings the baby back to Cape Verde and leaves it in the care of a female relative or returns to Cape Verde herself. The offspring of temporary migrants should be counted as part of the long-term permanent population of Cape Verde. Third, practically no children under 15 migrate.

The migration estimate above provides an approximation of the number and the age structure of women who were abroad during 1970–90, excluding those who were abroad previous to 1970. However, there are no data on their fertility while abroad. Now, if these women behaved according to the

Table 4.2 Estimated compared to registered net migation, Cape Verde, 1970–89

	1970–4	1975–9	1980–4	1985–9
Registered	27,920	18,655	12,885	15,590
Estimated	27,920	18,655	12,885	25,590
Ratio	1.0	1.0	1.0	1.6

Source: DGE (1989) and author's calculations.

second assumption above, and brought their offspring back to Cape Verde, then, although these foreign births might not be recorded, the children would be counted in the censuses of 1980 and 1990. To obtain an estimate of the foreign births in period *t* is a simple multi-step process. One first takes the children of age *n* who were actually counted in period *t+n* and applies the age-specific mortality rates backwards to obtain the estimated births in period *t*. This is compared to the registered births in period *t*. The difference (if registered births are lower than estimated births) is attributable either to an under-registration of births, or to children born abroad who are brought to Cape Verde.

Table 4.3 shows the results of this estimation. The census data of 1980 are used to estimate births during 1970–4 and 1975–9, and the census data of 1990 are used for births in the 1980s. In 1970–4, there were 52,758 births registered on Cape Verde. However, the number of children aged 5–9 in 1980 would indicate that there had been close to 58,000 births in that period, a ratio of 1.10. The ratio is similar during the period 1985–9. In the interim, the ratio of estimated births to registered births is close to one. The closeness indicates that the data from the censuses and from the birth and death registration are broadly consistent with each other, which should give confidence in the accuracy of the data. For the estimation of fertility abroad, these figures indicate that if indeed assumption two above holds, then there were very few women who returned to Cape Verde with children who had been born abroad, in other words, that *the fertility of temporary migrant women abroad is close to zero*.

If this is true, then what was the fertility of Cape Verde women,

Table 4.3   Estimated births and registered births, and three estimates of total fertility rates of Cape Verdean women, 1970–90

Estimated total fertility of Cape Verde	*1970–4*	*1975–9*	*1980–4*	*1985–9*	*1990*
Estimated births	57,915	52,019	59,408	67,323	–
Registered births on Cape Verde	52,758	53,387	59,144	60,144	–
Ratio estimated: registered	1.098	0.974	1.004	1.119	–
TFR of women on Cape Verde and Cape Verde births only	6.15	5.95	5.40	4.45	4.30
TFR of women on Cape Verde plus Cape Verdean women abroad and estimated births, zero women abroad in 1970	6.00	4.65	4.20	3.95	3.50
As above, 9000 women abroad in 1970	5.60	4.00	3.80	3.70	3.30

Sources: DGE (1989) and author's calculations.

including those who were temporarily abroad? To estimate this, the historical population projection from above is repeated twice and compared. First a historical projection is made excluding the women who emigrated, and then, secondly, including the temporary migrant women who were abroad as if they were part of the resident population. The two historical projections give estimated populations from 1975, 1980, 1985, and 1990 (1970 is provided by the census of that year). Two alternative time series of total fertility are made. To estimate total fertility of a population, one takes the total number of births, distributes these by age of mother according to a fertility age distribution (age distribution from DGE, 1989), then obtains the age-specific birth rates with a division of the births by age of mother by the number of women in that age group, and adds the age-specific births rates to find the total fertility rate. In the two estimations, first the time series of women from the historical projection excluding the women who emigrated is used, and then the time series including women who emigrated.

The lower half of Table 4.3 shows three estimations for the period: 1) with women on Cape Verde only and births recorded on Cape Verde; 2) with women on Cape Verde plus the women abroad and births estimated and assuming no women abroad in 1970; 3) as 2) but with 9000 women abroad in 1970. The estimated fertility of Cape Verdean women in Cape Verde remained stable until 1980, and then began to drop quickly. By contrast, both estimates of the total fertility including the Cape Verdean women abroad show a declining fertility from 1970–4 onwards.

To the extent that the estimation procedures and assumptions are correct, the table shows two interesting results. First, the table indicates that the fertility of Cape Verdean women is considerably lower if we include the women who are temporarily abroad. Second, there could be another indirect fertility effect of migration, namely, that the low fertility of women who have been abroad diffuses, after a lag, to the fertility of women on Cape Verde. It may be that women who have been abroad bring small-family values with them from the countries they have visited. If both of these results are a true reflection of the fertility mechanisms in this island culture, then emigration, particularly temporary emigration, could be a significant factor inducing a fertility decline in the sending country.

Conclusion: Effects of Migration on the People of Cape Verde

The history of Cape Verde is intimately interwoven with the migration movements of its people, particularly in the past 150 years. Following a century of economic and demographic stagnation filled with droughts and famines, emigration in large numbers began around 1860. The migrants were men boarding American whaling ships and disembarking in New England. There, many of them stayed, but considerable numbers also returned to Cape Verde with their savings. The remittances and return funds appear to have been instrumental in fostering a period of lower mortality rates and higher population growth on the archipelago. Then, during the first five decades of this century, voluntary migration was low. The nadir of migration corresponded with zero average population growth. Migration and population growth did not pick up again until after 1946. From then onwards to today, the emigration and the immigration flows have increased consistently, reaching incredible highs of up to 100 per 1000 persons in the late 1980s. These extremely high values are of gross emigration and gross immigration per year and mean that, on average, one in ten Cape Verdeans either emigrated or immigrated (returned) each year in this decade. However, the net migration has been much lower as it appears that most Cape Verdean migrants return after a few years abroad.

This emigration has been economically vital. The migrants send remittances to Cape Verde and these have contributed up to one quarter of the country's GDP (Wils, 1996, p. 60). If emigration were stopped immediately, the economic consequences on Cape Verde would be grave. A part of these remittances have been consumed, but it appears from interviews I conducted in Cape Verde in April 1994 that a portion has been invested in housing and in private vehicles (used as trucks and buses on the islands). These vehicles, in turn, may be partly responsible for an increase in the production of fruits and vegetables for market, as farmers now have the means to transport their goods to urban areas (Wils, 1996). If migration were curbed, the economic shock might endanger the process of social development which is necessary for Cape Verde to have the chance to nurture its domestic economy to maturation.

The correlation between emigration and remittances on the one hand, with population growth on the other, leads to the hypothesis that migration has allowed Cape Verde to escape the Malthusian trap. The flip side is, of course, that the ensuing population growth makes the country heavily dependent on foreign income, unless there is very fast domestic economic development.

One could use this observation to say that emigration reduces perceived limits of the natural and national system, and to argue that emigration keeps population growth and fertility high. In this view, emigration is not a sustainable strategy to solve problems of population growth and poverty; and what is more, it ultimately exacerbates these problems. The conclusion is that emigration should be stopped by limiting the immigration possibilities in the receiving countries. In the case of Cape Verde, this reasoning would have been inhuman in the face of the real possibility of starvation if remittances were reduced. Furthermore, there are other arguments which actually support emigration from Cape Verde.

One could venture the hypothesis that emigration to low-fertility countries exposes the migrants to a culture of family limitation, which they might, particularly in the case of female migrants, take home with them as a new value. A second argument is that temporary labour emigration of women may directly lower total fertility. A careful analysis of the data from Cape Verde has shown that this might well be the case. The analysis shows that most of the women who go abroad to work are in their prime child-bearing ages, that they stay abroad temporarily, and that their fertility abroad is close to nil. If these findings are true, then the 'real' total fertility rate of Cape Verde is lower than is measured nationally, because the national measurements exclude the low-fertility period when female labour migrants are abroad.

The analysis shows that, if one includes only the women actually on Cape Verde, then the total fertility rates (average number of children per woman) have declined from around 6.0 in 1970 and 1980 to 4.3 in 1990. By contrast, if the women who are abroad temporarily, plus the offspring they bring back to Cape Verde, are included in the estimate, then the total fertility was below 6.0 in 1970, around 4.0 in 1980, and less than 3.5 in 1990. These numbers indicate that the real fertility of Cape Verdean women is lower than purely national data indicate. Also there may be a diffusion of low fertility values from those who have been abroad to those who have remained at home.

In summary, this chapter has demonstrated that the relationship between migration and population growth on Cape Verde has been an intimate one, and that this interaction has been quite complex when a full analysis is made of its component parts over the shorter and longer terms. The complex, staged nature of the linkages – namely that emigration may on the one hand increase population growth through the mechanism whereby remittances improve living standards and reduce mortality, but then also depress fertility through the absence of reproductively active females and a lower fertility behaviour imported on return – may well have relevance for the analysis of the demographic effects of migration in other islands and regions of the world affected by high levels of out-migration.

Note

1. There is one possible mechanism I should like to mention here; since it is very much an intuition, I relegate it to a footnote. The fact that Cape Verdeans have a relatively high level of education but no local university means that all students go abroad. The friends and contacts these students and graduates make when studying/working abroad in wealthy foreign countries may well have helped Cape Verde obtain the large amounts of foreign aid which have buoyed the economy as much as remittances since 1974.

References

Abernethy, V. (1991) *Population Politics: The Choices that Shape our Future.* New York: Plenum.

Boserup, E. (1968) *Population and Technological Change: A Study of Long-term Trends.* Chicago: University of Chicago Press.

Carreira, A. (1982) *The People of the Cape Verde Islands.* London: Hurst.

Carreira, A. (1985) *Demografia Caboverdeana: Subsidos Para o Seu Estudo 1807/1983.* Praia, Cape Verde: Instituto Caboverdeano do Livro.

Condorcet, J.-M. (1795, reprinted in 1976) *The Future Progress of the Human Mind.* New York: W.W. Norton.

Courgeau. D. (1988) *Methodes de mesure de la mobilité spatiale.* Paris: Institut National d'Etudes Démographiques.

Davidson, B. (1989) *The Fortunate Isles: A Study in African Transformation.* Trenton, New Jersey: Africa World Press.

De Sousa Reis, G. (1989) 'Die Politische Entwicklung auf den Kapverden von 1910 bis 1980', doctoral dissertation. Vienna: University of Vienna, Faculty of Philosophy.

DGE, Direccão Geral de Estatistica (1989) *Boletim Annual de Estatistica.* Praia, Cape Verde: Ministerio de Plano e da Cooperação.

DGE, Direccão Geral de Estatistica (1993) 2. *Recenseamento Geral da População e Habitação, 1990.* Praia, Cape Verde: Ministerio de Plano e da Cooperação.

Malthus, T.R. (1798, reprinted in 1976) *An Essay on the Principle of Population.* New York: W.W. Norton.

Secretaria de Estado da Cooperação e Planeamento (1984) *O Cresimento da População de Cabo Verde entre 1970 e 1980.* (Praia, Cape Verde: Secretaria de Estado da Cooperação e Planeamento.

UNDP (1997) *Human Development Report 1997.* New York: Oxford University Press.

Wils, A. (1993) *European Long-term Migration Data: Overview and Evaluation of Existing Data Collection,* Working Paper 93-28. Laxenburg, Austria: International Institute of Applied Systems Analysis.

Wils, A. (1996) *PDE–Cape Verde: A Systems Study of Population, Development, and Environment,* Working Paper 96-9. Laxenburg, Austria: International Institute of Applied Systems Analysis.

World Bank (1993) *World Tables 1993.* Washington, DC: World Bank.

5

Insiders and Outsiders: The Role of Insularity, Migration and Modernity on Grand Manan, New Brunswick

Joan Marshall

Introduction

Grand Manan, in the Bay of Fundy off Canada's east coast, is experiencing profound change in its economic structures, with important implications for social and community relations on the island. Insularity, by definition a characteristic of islands, describes the history of Grand Manan without addressing the complexity of the island's social, economic and political relationships. With the changes that have begun to dominate island structures in the past decade, the meaning of insularity for Grand Manan is changing and, indeed, is being challenged by the forces of globalization. Nonetheless, even as the media seems to accept the overwhelming ideological significance of state and corporate globalization, there is evidence that in unorganized but collective ways local communities are resisting. Globalization is not universally powerful, and responses to its encroachment have been mixed. In many island communities, resistance has occurred in relation to specific issues, or to particular initiatives that are seen to impact directly upon islanders. In this essay I describe two specific examples of issue-oriented resistance by the people of Grand Manan. I wish to suggest that community opposition to particular issues has meaning beyond the issues themselves, reflecting a collective will to protect an insular culture against external forces of change. While particular issues may focus community attention, resistance to perceived threats from exogenous forces may also be somewhat amorphous, and yet be equally effective in maintaining cultural norms and values that sustain island identity. Through the insider–outsider dynamic played out in the relationships between native islanders and residents 'from away', there is a more subtle, but nevertheless powerful type of resistance that will also be described. While the meanings of insularity for Grand Mananers may be transformed as the forces of global markets and new technologies become increasingly important, there are particular characteristics of Grand Manan that support resistance to outside forces of change. A profound sense of identity linked to historic and family lineage, underpinned by religious values, and combined with extraordinarily low rates of migration, is sustaining a strong sense of community, despite the challenges of incursions by the salmon aquaculture industry and, significantly reinforced by government initiatives, tourism.

Insularity, Migration and Modernity: Impacts on Local Culture and Community

The notion of insularity, while being partly defined by an island site, is not merely a geographical concept. It is a complex idea that incorporates distance, centre–periphery relationships, technology, political and economic decisions, external and internal information flows, physical characteristics of topography, soils and climate affecting the resource base, and social and cultural patterns. It is in fact a dynamic social construct that is not unproblematic. It may be contested. Insularity is as much a state of mind as it is an objective reality. While the concept usually carries negative connotations being used in reference to protectiveness of the status quo, defensiveness against change, and wariness in dealings with outsiders, it may also incorporate strong co-operative relations, survival skills, loyalty, and values of equity and justice.

The natural boundedness of islands, and the time–space distanciation that are implicated in this boundedness, are responsible for an insularity that both protects and defends, and that creates and defines distinctive local cultures. Shared understandings of individual experience evolve, rooted in taken-for-granted, habitual and repetitive activities by members of the community. Their daily, face-to-face relationships form layers of complex 'webs of significance' (Geertz, 1973) that persist over time, incorporating rituals, symbols and ceremonies that link people in common perceptions of their past. A sense of belonging is inextricably linked to a distinctiveness that can be described as 'local culture'. This connection between the formation of distinctive local cultures and insularity extends to and incorporates a separation between 'we' and 'they'. Community members define their localness with respect to family lineage and length of residence that excludes new arrivals. People 'from away' are defined as 'other', and excluded from decision-making opportunities and leadership positions.

While the notion of community is an 'untidy, confusing and difficult term' (Scherer, 1972, p. 1), its significance for people as a way to describe their sense of place and identity continues to be strong. Community describes daily relationships and existential life-worlds. In his exploration of the construction of 'self', philosopher Charles Taylor points out that 'the full definition of someone's identity thus usually involves not only his stand on moral and spiritual matters but also some reference to the defining community' (Taylor, 1991, p. 36). He goes on to suggest that our modern notion of the self is very much 'an historically local self-interpretation' (Taylor, 1991, p. 113). Under conditions of change, when life-histories are being challenged and social power eroded, the need for affirming identity will be more profound. The important linkages between history, memory and community have been recognized by a number of scholars (Lowenthal, 1985; Taylor, 1989, 1991; MacIntyre, 1981). The connections are crucial because of the challenges to identity and personal meaning systems implicit in challenges to community values that occur in the context of globalization and migration. In a similar vein, when the threat to personal and social identities becomes serious, it can often only be mitigated by a sense of community as a political enterprise. Communities themselves become the bases for political action. With similar concerns, Sagoff explored the ways in which environmental regulation and social policy are linked to community values, suggesting that 'people pursue these values not as individuals but as members of the group' (Sagoff, 1988, p. 100). It is shared experience that is at the basis of values and community solidarity.

The idea of community becomes concrete when there is a challenge to it. In his study of a blockade on the Shetland island community of Whalsay, Anthony Cohen showed that the ostensible purpose to protect the local environment may not entirely explain the protest. In fact it seems to have represented a confrontation with the outside world arising out of tensions created by the globalized elements of economic and political change (Cohen, 1987, p. 13). In another study, he points to the fact that 'Locality is anathema to the logic of the modern political economy' (Cohen, 1982, p. 7), and argues that this antipathy of locality and modernity is responsible for tension and conflict when forces of modernity are introduced. The result is that local communities respond to possible change by increased emphasis on the tightly structured intricacies of local social life, and an even greater wariness of outsiders.

Modernity is a complex concept that has been used to distinguish the present from the past, and is usually related to expectations and a world view of progress and development. Giddens points out that in the cause of modernity, social relations become disembedded from local contexts, and a modernist discourse of progress and economic development may be used to decontextualize local identity (Giddens, 1990). Three 'malaises' that accompany the challenge of modernity have been described by Taylor as loss of meaning, the eclipse of community responsibility, and a loss of freedom (Taylor, 1991). There is a tension inherent in the incursion of external forces of change that derives in part from perceived threat to community values that have been rooted in history, memory and shared understandings. Regardless of whether the threat is perceived to be from outside economic interests or from 'outsiders' who move into the community, there is a sense of invasion. Even as the prospects for new jobs may be welcomed, there is ambiguity in the response to those people who represent outside interests. The insider–outsider separation within the community becomes illuminated through different levels of resistance.

Grand Manan: The Setting

With a population of 2600 that has remained remarkably stable over the past 100 years, and very low rates of both in- and out-migration, the community of Grand Manan shows strong elements of insularity in its traditional values, resistance to change, and continuing strong influence of religion. While the distance from the mainland, one and a half hours by ferry, has certainly contributed to its sense of insularity, a more fundamental factor may be the low rates of migration clearly documented in the mobility statistics. As Table 5.1 shows, the percentage of 'movers' in the five-year period between 1986 and 1991 on Grand Manan (22 per cent), was half the national rate (44 per cent). In the one-year period (1990–91), it was less than one third of the national average. The data on Table 5.1 further show that amongst those who did move within the five-year period, only 300, or about a half, changed location from outside their census subdivision, i.e. moved over from the mainland. Of the people on Grand Manan in 1991, fewer than 12 per cent had moved there (or returned after an absence) since 1986. For a rural, resource-based economy it is notable that, while small, total population change from 1986 to 1991 was positive, increasing by 125 people, or almost 5 per cent. Taking natural increase into account indicates that net migration was approximately 90 people over the five-year period, or 18 people per year.

Migration involves the exchange of ideas and values; hence a low rate of migration might be an important factor contributing to a sense of insularity and community identity. Yet for Grand Mananers 'migration' has no real meaning. One cannot undo or leave behind a Grand Manan identity. One young woman said to me about a recent shopping trip on the mainland: 'I feel like I have Grand Manan written across my forehead!' She not only feels a strong sense of individual identity, she also has a profound sense that others perceive her as different.

Apart from the statistics, evidence of the low rates of migration and thus of stability of the family lineage within the community itself is quickly gleaned from the local phone book. Names leading back to the first Loyalist

Table 5.1 Grand Manan, New Brunswick province and Canada: mobility statistics according to the 1991 census

	Grand Manan	New Brunswick	Canada
Population > 1 year	2,610	702,440	26,430,895
One-year movers			
no.	130	87,350	4,322,225
%	5.0	12.4	16.4
Five-year movers			
no.	580	228,390	11,637,185
%	22.2	32.5	44.0
Five-year migrants			
no.	300	108,385	5,860,970
%	51.7	47.5	50.4

Note: 'Movers' are defined as all those over 1 year of age who changed place of residence over a one-year or five-year period. 'Migrants' are movers who, on census day, were residing in a different census subdivision five years earlier; for migrants the percentages are calculated on the basis of the total of five-year movers, not the total population.

settlers in the eighteenth century are repeated in columns: Green, Wilcox, Benson, Brown, Guptill, Ingalls and Ingersoll. Indeed, so common are certain names that a particular island strategy for naming wives has evolved. Whereas there might be several Joan Greens on the island, there was less likelihood that a husband's first name would be duplicated. Therefore it became conventional to call such a person 'Joan Ron', or 'Joan Adian', etc. For many families on Grand Manan this is still how the women are known. Moreover, it is not unusual for a Betty Brown to marry a Tim Brown, and after divorce to marry into a different Brown family and still retain her original birth name.

Even today low rates of out-migration are the norm. In the graduating high school classes, fewer than 50 per cent of the young people will leave. Of those who do, about half will return within two years. There is no question that a sensibility described as insularity is linked to the island identity defined in terms of family roots, historical continuity, and associated with low rates of migration.

In-migration has also been historically low, as noted earlier. Those people who are from away fall into three main groups: retired single women who see the island as a safe refuge for a simple, community-based lifestyle; retired couples; and spouses, mainly women, who marry onto the island. Conversations with these 'from aways', described by one lively ex-main-lander as 'choosies', all reflect a common thread of exclusion. They are acutely aware of an insider–outsider cleavage in Grand Manan society that defines social relations. One woman who moved to the island when she married an islander whom she had met on the mainland, now owns and operates a small business. She is convinced that there are some islanders who will not patronize her business because she is a non-native. She feels 'tolerated'. There is only one example of an entire family who moved to the island with young children, and they are mentioned by everyone as a unique addition. In the hopes of finding work, they had migrated from Newfoundland after the closure of the cod fishery. After three years, they are beginning to be accepted by many islanders, mainly because 'they are workers'.

The question arises as to the reasons for this remarkable degree of population stability. It is unlikely that mere distance from the mainland can explain the 200 years of continuous community existence and the stable population numbers over the past 100 years. The single most important reason must be associated with both the richness and the diversity of the resource base, which has allowed and encouraged flexibility of livelihoods as fish stocks or markets have waxed and waned. Insularity has incorporated a strong sense of community identity combined with a spirit of independence and flexibility. Today, however, both resource degradation and changing markets are threatening the foundation of diversity. The changing social and space relations on Grand Manan Island are occurring in the context of four related processes of change in the island economy: the resource depletion of the groundfish, herring and scallop stocks; the decreased market demand for smoked herring; the increasing importance of salmon aquaculture; and the growing importance of tourism. About 30 per cent of the labour force is engaged in the traditional fishery, incorporating lobster, scallops, herring, and groundfish; with clams, crab, sea urchins, quahogs (a type of clam), as recent additions. Dulse *(Rhodymenia palmata)* also has a small but important niche in this ocean-based economy.[1] The value of the traditional fishery is about 11 million Canadian dollars, with shellfish accounting for about $7 million of that and herring another $2.3 million (1994 figures). Salmon aquaculture was introduced to the island in 1978 by a local entrepreneur, and for many years the site at Dark Harbour was the only one on the island. In 1988 another two sites, owned by off-island interests, were established. By 1994 the value of the salmon industry was estimated to be $10.9 million, almost equal to the total value of the rest of the fishery; and by 1997, with 16 sites, the value of the salmon industry was estimated at $25 million. Within five years it had more than doubled in value, and is today worth more than double all of the traditional fishery combined. Described as 'farming' rather than 'fishing' (Boghen, 1995), aquaculture requires a very different set of skills and knowledge, with profound implications for changing social relations.

Evidence of the changing importance of the traditional fishery can be clearly seen in the cultural landscape, most dramatically evident in the demise of the smoke houses (only two operating in 1996 compared to over 200 thirty years ago) and in the increase in salmon sites. New systems of government-imposed quotas and licensing have severely restricted entry to the traditional fishery and are creating increased disparities between heavily capitalised fishers and those who can no longer afford to maintain multiple licenses and large boats (Davis, 1991; Finlayson, 1994; Sinclair, 1983, 1985). Salmon aquaculture, largely financed through mainland institutions and private interests, is an industry that offers jobs but little in the traditional relationships of family entrepreneurship and control and the learning of multiple skills. In the changes threatened by exogenous forces such as aquaculture technologies, outside capitalization, market demands for fresh salmon, and government–corporate consensus on the need for 'development', the insularity of Grand Manan is being challenged.

The growth in tourism has been another major factor changing the island landscape, especially since 1990 when a large new ferry was inaugurated, almost doubling the crossing capacity. During the short tourist season of June to September the population almost doubles. Services and facilities available to tourists include cottages, small inns, whale and sea-bird-watching excursions, and hiking trails. Evidence of non-congruence between provincial and local interests will be described below when I examine one of the ways in which islanders are resisting the outside forces of change.

In addition to the economic changes within the fishery and tourism, the island has been experiencing the institutional restructuring associated with municipal and school boundaries that has been common throughout North America in the drive for efficiency and increased cost-cutting. The redrawing of boundaries has removed school board representation on the island, with the result that islanders have less direct control over decisions in the educational system. Similarly, the amalgamation of villages into one municipality in January 1996 has institutionalized a new identity that ignores significant village-based identities across the island. All of these changes, economic and political, related to governments, globalization, new technologies, and resource degradation, have implications for the meaning of insularity. Social and community relations on Grand Manan are changing.

For, in the change from the traditional fishery to a high-technology enterprise, there is more involved than a switch from one source of income to another. Equally important are the social implications: who makes decisions and who controls the capital; the change in weekly and daily work-time patterns; the allocation and use of space on the sea and on the wharves; and changes in class, family and gender relations. All these will have a significant impact upon community relationships and individual and collective identities. The introduction of globalized fish farming and market forces is decreasing the extent of economic insularity in terms of flows of information, money, and material goods; but it is also affecting the psychic insularity of local culture. While the globalization of markets may create particular flows of money and goods at the macro level, in its local and regional impacts globalization is variable. It not only differentiates between regions in its impact, it also differentiates between those social strata and classes involved in the new relations and those which are not. Furthermore, the ways in which people are able to cope with change and with the globalization processes vary.

Just as the expression of globalization is not universal, neither are the responses to modernity. The changes being introduced to Grand Manan through the high-tech industrial structures of fish farming are creating a particular mix of modernism and tradition that are in tension and are reflected in the changing dynamics of social relationships. It has been suggested that one of the characteristics of postmodernity is the loss of a common historical past, and that globalization may be contributing to this loss through the greater rate of movement, 'interchange and clashing of different images' (Featherstone, 1993, p. 171). Nevertheless, what we see on Grand Manan is resistance and the selective absorption of elements of the modernist project.

Moral Community and the 'Rockweed Debate'

Community opposition to a state-approved rockweed harvest around the island of Grand Manan illuminates many of the issues involved in the modernist project, in particular as it involves a traditional, insular society confronted by external market forces. The case focuses on a still-unresolved local conflict, with the small fishing community resisting the plans of the provincial government and an out-of-province private company to harvest rockweed on the island. The state and industry seem to have joined their considerable forces to engage in a for-profit activity against the wishes of the local community. In the process, class cleavages within the community are illuminated and the ideological dominance of the market over social and environmental issues is made apparent.

In the decision-making about environmental choices, the privileging of economic values often obscures the intrinsically cultural nature of the process and the human interests that are at stake. And yet, for an island community that has historically been self-reliant, a threat to community control over resource assets can be the focus for community activism that forges solidarity. It is useful to consider for a moment the literature that links ideas of environmental sustainability, personal and collective identities, historical roots, and civic responsibility and activism as sources of community solidarity. In drawing upon these linked concepts I wish to underline the complexity inherent in the community protest. As Cohen argued in his case study of a blockade on Whalsay, 'collectivities become aware of themselves largely through their interaction with others' (Cohen, 1987, p. 12). Even in the context of a small island community there are diverse interests represented that need to be affirmed in terms of the collective sense of purpose. In the case of Whalsay, a blockade provided the means through which the community could acknowledge, not similarity of motivations, but rather a consensus regarding the superiority of local knowledge and the integrity of local interests (Cohen, 1987, p. 318).

The rockweed protest on Grand Manan can be explored with respect to several ideas that are highlighted in the literature, particularly the work of Joni Seager and Gerald Sider. The perception of the fishery resource as a community 'commons' has been an important source of collective identity and of cultural formation in fishing communities. As Seager (1995) has so clearly demonstrated, there is a relationship between economic objectives in the exploitation of environmental resources and the cultural milieux within which they are framed. The environment is fundamentally and crucially a construction of meanings, ideas and experiences directly associated with structures that are spatially and historically specific. These structures describe and define social and cultural patterns of meaning. Gerald Sider's work in Newfoundland showed that cultures that provide their own meanings and that delineate their own social relations may also be relatively helpless against the forces of domination (1986, p. 26). He showed that the domination by the merchant class at the point of exchange was not complete because it did not encompass production. The autonomy of the fishing village was rooted in the resource base, the sea being acknowledged as common property. This commons was 'part of the collective defense of the village against economic and political domination' (Sider, 1986, p. 28). On Grand Manan we see this tightly interwoven relationship between the environmental resource and cultural patterns that are reflected in daily activities that are the basis for collective identity.

While the fishery may be conceived in the abstract as a global resource, it is essentially locally based. 'Over the years, coastal communities have evolved a complex of perceptions regarding what is acceptable along *their* coastline' (Millar and Aiken, 1995, p. 621). A feeling that it is 'our water' is expressed through a complex system of community-based customs, social practices, and political responsibilities. Salmon farming precludes this collective defence because of its ties to leased sites, owned on the mainland, with the result that the potential for the market paradigm to overwhelm community identity is real. Implicated in this challenge is the separation of social relations from economic institutions, and the inevitability of new social relations that ultimately redefine the meaning of 'community'.

After several years of negotiations, the New Brunswick government, based on only one information meeting on the island in 1993, licensed an out-of-province private corporation, Acadian Seaplants, to harvest rockweed around the island as a three-year pilot harvest. The news reached islanders just as the harvest began in June 1996. A community protest was organized within hours, attended by about 100 people. Despite this, and an intensive letter-writing campaign, the company continued to harvest. The company hired several young local men for the harvest, paying them a minimum hourly wage. In all cases they were men who had been marginalized because of government quotas and restricted licensing in the traditional fishery, and who saw this as an opportunity for work. Opposing the harvest was a broad spectrum of inshore fishers who relied upon a diverse, seasonally varied, fishery resource, as well as highly capitalized fishers who were worried about the potential loss of spawning and incubation beds for many fish species. By late July the company began to move the rockweed off the island, originally intending to load trucks at the wharfside. A blockade, using their own trucks and cars, by local fishers, representatives of the Fisherman's Association, and the Municipal Council, failed to stop the shipments, however. The company merely changed its plans, instead loading the harvest directly into large boats for the transfer to the mainland. The harvest continued through 1996 and 1997.

Issues of resource sustainability, local control and jobs were all discussed by the community in the debate over the rockweed harvest. But even those involved in the protests acknowledged that the problem was more complex than the specific issue of local control and an out-of-province company. The protest reflected concerns about cultural values and a way of life; questions about threats to the dulse and periwinkle niche fisheries; about who would have accessibility to the commons, who would be allowed to benefit, and, ultimately, who would control access. Accessibility is a matter of both control and the nature of the extraction process itself. Very little capital investment is required for harvesting dulse, periwinkles or clams. If these industries were to be affected, even in the short term, then not only 'jobs', but also a lifeworld of social and cultural meaning would be threatened. The particular fishery sectors that would be most directly affected (dulse, periwinkles, clams, sea urchins) are those in which the least powerful members of Grand Manan's society are participants. The diversity of fishery activity on the island, and the parallel diversity of capitalization, are reflected in class disparities that have historically been accommodated within existing social norms. The introduction of an outside, wage-based producer changes the internal dynamics. As Davis showed in his study of Digby Neck, the potential for 'gear conflict' – between different levels of capitalization that correspond to equity and class differences – is always in precarious balance. It 'concerns the very basis of solidarity within the occupational and social communities' (Davis, 1991, p. 93). While social norms have reflected a collective identity and sense of solidarity over generations within the traditional fishery, the incursions of outside interests can jeopardize the equilibrium of social relations. Individual identities can be brought into tension in relation to the community. The few individuals who 'signed-on' to become harvesters represent the most marginalized members of the community who have sought the 'security' of an hourly wage as against the production-based sustenance of the traditional niches. The conflict reflects a view of equity and community support that addresses the precariousness of the under-capitalized social groups. The niches provided by these smaller fisheries are no less important to the island community than the more lucrative, capital-intensive sectors. They are embedded in the complex interrelations and daily activities of the wider community.

In the case of Grand Manan, Acadian Seaplants is not only an out-of-province, and 'from away' company. It also represents a take-over of the commons and a threat to the diverse niches, daily activities, and social relations that contribute to the culture of Grand Manan. More than a loss of local control, the proposed rockweed harvest represents profound change in how people are able to sustain themselves economically, and in the playing-out of social relations. For some islanders the harvest of rockweed represented a direct threat to niche fisheries in which minimal capitalization guaranteed the possibility of broadly-based access. The threat to sustainable livelihoods in the context of a multi-faceted resource and, ultimately, to a source of individual and collective identities, seems to have been crucial in this conflict. The rockweed harvest was not being perceived merely in terms of a physical resource nor of an economic proposal. Resistance to it was profoundly social and cultural. As Cohen showed for the people of Whalsay, for Grand Mananers too, the intrusion of external forces was 'felt to be an impugnment of the integrity of local expertise', breaching their 'self-confident insularity' (Cohen, 1982, p. 305).

The idea of a 'moral community', in which co-responsibility and mutually acknowledged obligations guide social relations, has always been an attractive ideal. While its reality will always be contested from within because of an inevitable heterogeneity of class, gender and other interests, Harvey has suggested that it is a concept that may have relevance for forging and sustaining community solidarity in the face of environmental conflicts (Harvey, 1997, p. 179). As long as Grand Manan islanders were able to remain relatively aloof from mainland institutional structures, they did not need to articulate community values, nor even to debate their defining objectives. Such values and objectives were taken for granted. But the many challenges to island insularity in tourism, aquaculture and the rockweed harvest are all contributing to a growing sense of external threat, and it may be that local participatory democracy will be increasingly crucial to the islanders' survival as a distinctive culture.

The Ferry and Tourism: Ambiguous Relationships and Resistance

The inauguration of a new and larger ferry in 1990 was a significant factor in changing perceptions of insularity, particularly in relation to the possibility for growth in tourism. However, the link between tourism and the ferry is less important today, in the late 1990s, than that between the ferry and the new aquaculture industry, local perceptions notwithstanding. For several years prior to 1990, islanders had complained that tourists were usurping their rightful spaces on the former 32-car ferry. Without a system of reservations, there were no alternatives to waiting for the next boat, four hours later, and resentment against tourists was a common reaction. An inventive system of renting car wrecks to 'save' a place in line, bringing some revenue into the pockets of a few entrepreneurial spirits, is still described with glee by islanders. The new ferry killed that small business, doubled the crossing capacity in terms of numbers of cars, but more importantly allowed access to the island for large 18-wheeler trucks and equipment such as paving trucks and cement mixers. At first the new ferry facilitated the movement of people to and from the island and was unanimously seen as beneficial. In particular, the emerging aquaculture industry could not have grown and thrived without the new ferry. Frequent deliveries of feedstock, shipments of tonnes of smolts onto the island in October, and in the spring the daily truckloads of fresh killed salmon that leave the island for distant markets, are all dependent on the ferry.

Despite the unanimous praise for the improved service after 1990, the relationship of the islanders to the ferry is somewhat ambiguous. Because of the ever-expanding demands of the aquaculture industry for transport capability, there is again in 1997 a concerted effort by the Transport Commission of Grand Manan to increase the crossing capacity. The Commission, chaired by the mayor, is an *ad hoc* group of the Municipal Council, representing commercial interests that include the Chamber of Commerce and Tourist Association, the trucking companies, two fish plants and the aquaculture industry. What is significant here is that while it is tourism interests that are perceived to be most affected by the new ferry service, representation on the Commission is dominated by fishery interests. Furthermore, the companies who are lobbying for increased capacity have co-opted the tourist representatives to lobby on behalf of the cause, primarily because tourism has a higher profile and community 'understanding' than the business interests associated with trucking and aquaculture. The discourse during a meeting held in July 1997 was almost entirely around the issue of marketing fresh fish and encouraging growth of the salmon industry (owned off-island), at the same time as it relied upon arguments for a growing tourist sector. The trucking representative pointed out that 'right after the new ferry in 1990, the number of cottages tripled, but now tourism growth is stagnant', implying that tourism would benefit by a new ferry.[2] Referring to the island companies that run whale-watching tours, he argued: 'Dana Russell guarantees whale sightings, and he needs the flexibility of a late ferry.' But it was the mayor, Joey Green, who fully acknowledged the non-tourist-related needs: 'Tourism runs for a maximum of four months; it is really only a fill-in.' Then he went on to argue that the government's plans were for a commercial fishery dominated by a few large companies, and that Grand Manan had to be ready for large-scale operations, only possible with an improved and expanded ferry service. Whereas in 1990 it was tourists who might be blamed for filling the ferry, today it is large transport trucks with priority because of their perishable loads that may become the '*bête noire*'.

Nevertheless, in the community perception it is tourism that is growing and that a new ferry would support. But the irony is that tourism is not unambiguously welcomed. Indeed, there are many who would like to curtail its growth because of the alleged increased traffic, 'strangers walking across private fields', and because tourists are not seen to contribute to broadly based economic benefits. One fisherman complained to me about the new kayaking group, owned by someone 'from away', which relies upon a public ocean access platform, paid for by the fishermen. As he pointed out, 'I can't get at my boat when they're here; and they don't pay anything for the upkeep.' So truckers use the name of tourism in their lobby efforts for the ferry because they see it as a benign and accepted activity; at the same time it is the demands of the salmon industry, not tourism, that have substantially increased traffic on the roads all year long. As incursions of high-tech, capital-intensive operations percolate through Grand Manan society, relations of power and decision-making are shifting. As the flows of capital and goods intensify, protective barriers of insularity will inevitably weaken. Many Grand Mananers who are not involved in tourism, trucking or as owners of the fish plants (and most are not) are not enthusiastic about the prospects for an increased ferry service.

One source of externally generated change has been the provincial government with its plans for a more 'upscale' tourism. There is a beautiful Provincial Park on the island, that encompasses a sand beach stretching several kilometres along the ocean front, and a variety of forest, bog and dune habitats and spectacular rocky cliffs. For Grand Mananers, the park belongs to the community, being used for daily walks, berry picking, family outings, and church picnics. Very much part of the daily activities of island life, the Anchorage Park and Campground is seen as intrinsically part of their common heritage. In November 1996 the government announced plans for more intensive tourist development that would eventually replace the tent sites with cabins and a conference centre. There was also reference to the possibility of instigating daily usage fees. Like its earlier marketing campaign that promotes the 'Day Adventure', the government's objectives were clearly contrary to island sensibilities. Traditionally Grand Manan has attracted visitors who have stayed for several days or weeks, engaging in bird-watching, hiking, and beach-combing activities.[3] The campground is one of the most popular facilities for long-term visits, especially for young families. Almost a third of 1997 visitors stayed at the campground, making it the single most popular type of accommodation. Since 1994 the provincial government has indicated interest in a more aggressive type of tourism, with higher commercial value and a different type of tourist. In this case, the island-born Park Manager, only a few years from retirement and in the position for over twenty years, was summarily dismissed and told he would be relocated on the mainland. The government deemed it necessary to have a more marketing-oriented, tourism-trained, mainlander as Park Manager. Furthermore, while the precise details of the plans were never released, it was announced that the 'upgrading' and 'modernizing' would bring in higher revenues. Community resistance was strong and swift, causing the government to reconsider its strategy, at least in relation to the park. Letters and phone calls to the government department, the local member of the Legislative Assembly, and the media, were finally effective in forestalling the plans. Resistance to the modernist project of the government created a solidarity around an issue that was perceived to affect every islander.

Island Identity: Insiders and Outsiders

The two examples described above, both involving collective resistance to projects that threaten to change a way of life, suggest community solidarity around specific issues that threaten a shared understanding of island identity. There is a sense of 'we' defending our community against 'them'. Both episodes, it might be argued, reflect an insularity that is merely resistance to change. On the other hand, they can be seen through the prism of sustaining a culture that values environmental sustainability and livelihood flexibility, family continuity and mutual responsibility. Nevertheless, the apparent unity expressed in the events of resistance belies the complex internal 'webs' of relationships that are in constant tension and occasional conflict.

None is more pronounced nor more significant than the cleavage between native islanders and those 'from away'. As Cohen has pointed out, 'Belonging is the almost inexpressibly complex experience of culture' (1982, p. 16). One problem for people who migrate to the island is that they have not lived the experiences that create the 'nebulous threads' of island culture. They cannot grasp the 'subterranean level of meaning' that allows them to truly belong (Cohen, 1982, p. 11). Earlier I described three main groups of 'from aways': single women, retired couples, and spouses who marry onto the island. There is an interesting difference between the lived experience of the first two groups and that of the third. For the older people who choose to move to Grand Manan, there is an uneasy alliance with native islanders, but it is an alliance nevertheless. This older group tends to stay together and not to mix with 'true' islanders. One of them said to me: 'Somehow we just gravitate to each other.' They share understandings across a broad spectrum of social and cultural activities, such as book clubs, bird-watching, organic gardening and music, that are not part of the island way of life. At summer concerts in the local church hall, organized by a young spouse from away, the audience is invariably only tourists and residents from away. Not only do their interests differentiate them. For many of these 'new' residents there is also a feeling of not being welcomed when they volunteer for various committees – on the Museum Board, Chamber of Commerce or Rotary Club, for example. Indeed, it has been suggested that the separation of archival duties from those of the Museum Curator is a direct reflection of island protectiveness. As a person from away, the Museum Curator is not permitted to have responsibility for the archives that have been mandated to a native islander. Nevertheless, for the most part, the 'newer' residents seem to have accepted that there are certain positions that will never be ceded to a non-islander.

For those who have married onto the island, experience, social relations and perceptions are quite different. In the vast majority of cases it is the men who have returned with off-island spouses. Women who go away to university tend to marry away and not return to the island except for family reunions and special celebrations. While the reason for the gender split is undoubtedly related to the lack of jobs on the island for women, it is also related to the tendency for women to follow the jobs of their husbands, rather than the reverse. In a survey of island households defining those 'from away' and those perceived to be native, one 65-year-old islander clearly separated the two categories. While pointing to 31 households of 'mixed' marriage, she classified 947 as islanders (83 per cent), and 193 (17 per cent) as 'from away'. Significantly, she included as islanders several couples which included women from away, but who had a birth connection through grandparents to the island. Lineage and family history are crucial factors for integration into the community even after decades of residence. One woman who married onto the island almost 50 years ago described her sense of still being treated as an outsider. She has been asked if she will 'go home' if her husband dies. No conversation about someone on the island begins without an introductory outline of family lineage. For those from away the introduction is brief: 'She's from away.'

Two families illustrate the strong gender bias in migration associated with marriage. In one case there are six children, three boys and three girls, aged 30 to 45. All three girls live on the mainland and all three boys have married away and brought their wives back to the island. Somewhat exceptional is that all six have married spouses from away, as did their father. In the second family are three boys (now aged 40–50) all of whom 'were sent away to university so they could meet other girls', according to one island commentator. Their wives, all from small rural communities on the mainland, still feel excluded from the inner meanings of Grand Manan society, even after more than twenty years living on the island. One said that when she first arrived 'it was a dream come true', that she 'loved the isolation'. But she went on to acknowledge that since then she has come to be more aware of being separated from much of the life on the island. Another of these women went further: 'I took to it like a duck to water', she said of her first year, 'but then there was a complete turnaround. ... I sensed a betrayal by the people we were socializing with, they were drawing conclusions, thinking they knew my business'. These young women did not arrive to comfortable homes. They encountered many of the same struggles that have defined the lives of native island women. Living in a tent trailer without electricity or running water for six months, her university-educated husband working mending twine for the weirs, one of these women earned money gathering dulse on the low tides for several years. Talking about the culture shock of their adjustment, one of the women referred to the custom of naming wives with the husband's first name. 'When I realized how they were naming me, my hackles really went up!' The interviews with these women led to subsequent discussions with their husbands, who were surprised by some of the feelings and perceptions that their wives described. The husbands, as islanders who are totally comfortable with the meanings, experiences and values of the culture, do not fully comprehend the extent to which their wives will always be outsiders. The insularity of this island culture, naturally protected by time and space, is also actively defended by the social norms of Grand Manan. The question that the mothers from away are now asking themselves is, 'What about our children?' One woman was adamant that her son would be sent away after high school. But, by all accounts, most continue to return if they have grown up on the island. Said one mother: 'I feel we've brought up our children differently (from native-islanders), but I wonder! They grew up here, and the messages at home and in the community are different. It's a major concern.'

Conclusion

As the globalized economic forces of change impact upon Grand Manan, bringing new types of jobs in the aquaculture industry, increasing the links to mainland capital and investments, intensifying and changing the nature of tourism, there will inevitably be effects on the island's society. To date, islanders have shown an unorganized but effective ability to resist both the active intrusion of government and private enterprise and the more ambiguous changes suggested by the arrival of new residents. The degree of psychic insularity in times of mass communication and access to the Internet has been remarkable, but especially so when seen in the context of constantly flexible livelihood strategies that have repeatedly adapted to changes in resource and market realities. Grand Manan is an island rich in tradition and community stories. Whether or not the people will be willing to share these traditions with others and allow themselves to be moulded by new circumstances and new arrivals is an open question. The tensions and occasional conflicts that are experienced by the insider–outsider dynamic will be heightened by the impending changes, and individual and collective identities will be transformed.

As Cohen (1982) noted, modernity introduces tensions and potential conflicts into indigenous cultures. Resistance is a response that may be effective in the short term in relation to specific issues such as the Anchorage Campground; or, in a more covert way, in the segregation of insiders and outsiders. Low rates of migration have protected islanders from threats to their values and mores. However, globalization brings with it demands for a level of technology, communications, investment and efficiency that may not be addressed within the context of the traditional society. The knowledge and astuteness required to become part of a global network challenges this island. While values of equity and co-operation are acknowledged and affirmed, class divisions are growing with the impact of government regulations and the incursions of outside capital. A few islanders are able to increase their investments in equipment and larger boats, and to increase their stake in the traditional fishery. For salmon aquaculture, however, where the outlays and risks are much higher, the major financial returns will accrue to off-island owners and the few islanders who can afford expensive leases for sites. Aquaculture is attracting new people to the island, who have special skills such as welding (for the manufacture of the cages), or capital (for the net company). It also offers the prospects of low-paying but steady jobs, such as in the feedplant. The barriers that have existed in the past to employment in the traditional fishery where family-learned skills and boats were crucial, do not apply to the business of high-tech fish farming. On the other hand, the types of jobs that are available do not offer opportunities for advancement and the accumulation of individual stakes in the economy. While corporate and government planners may encourage the migration of people and dollars to the island, the history and culture of Grand Manan suggest a level of resistance that may continue to ensure its insularity.

It may be that for Grand Manan this period is not one of 'transition' from one state to another, but rather a reconfiguring of relationships within island society itself. Supporting Cohen's thesis of the incompatibility of locality and a modern political economy, the experience on Grand Manan suggests that the community may continue to maintain its distinctiveness even in the context of the homogenization of globalized markets. Insularity for Grand Manan will continue to mean a strong collective identity and community values associated with strength and flexibility, as well as determination to define their own future.

Notes

1. Dulse was discovered and promoted as a food source in the nineteenth century by James MacDonald at Dark Harbour on the west side of the island, still the 'world capital' of the dulse industry. This edible seaweed is picked at the lowest tides and dried on rocks for later sale either as a snack food or ground up for health food tablets and soup additives.

2. The tripling of tourist facilities was, in fact, largely due to a government pro- gramme through the Atlantic Canada Opportunities Agency that provided interest-free loans that would be written off after five years of continuous busi- ness.

3. In a tourist exit survey that I conducted in the summer of 1997 (sample of 275 returns, representing approximately 780 people), there was an overwhelming consensus that the island should 'stay natural', 'not change', remain 'non-com- mercial'. To the question about what they had enjoyed the most, over 40 per cent of visitors said it was the 'unique atmosphere', the 'peace'.

References

Boghen, A.D. (ed.) (1995) *Cold-water Aquaculture in Atlantic Canada*. Moncton, NB: Canadian Institute for Research on Regional Development.

Cohen, A. (ed.) (1982) *Belonging: Identity and Social Organization in British Rural Cultures*. St John's, NF: Memorial University, Institute of Social and Economic Research.

Cohen, A. (1987) *Whalsay: Symbol, Segment and Boundary in a Shetland Island Community*. Manchester: Manchester University Press.

Davis, A. (1991) *Dire Straits. The Dilemmas of a Fishery: The Case of Digby Neck and the Islands*. Cambridge: Cambridge University Press.

Featherstone, M. (1993) 'Global and local cultures', in J. Bird, B. Curtis, T. Putnam, G. Robertson and L. Tickner (eds), *Mapping the Future: Local Cultures, Global Change*. London: Routledge, pp. 169–87.

Finlayson, A.C. (1994) *Fishing for the Truth: A Sociological Analysis of Northern Cod Stock, Assessments from 1977–1990*. St John's, NF: Memorial University, Institute of Social and Economic Research.

Geertz, C. (1973) *The Interpretation of Cultures*. New York: Basic Books.

Giddens, A. (1990) *The Consequences of Modernity*. Stanford: Stanford University Press.

Harvey, D. (1997) *Justice, Nature and the Geography of Difference*. Oxford: Blackwell.

Lowenthal, D. (1985) *The Past is a Foreign Country*. Cambridge: Cambridge University Press.

MacIntyre, A. (1981) *After Virtue*. Notre Dame: Notre Dame University Press.

Millar, C. and Aiken, D.E. (1995) 'Conflict resolution in aquaculture: a matter of trust', in A.D. Boghen (ed.), *Cold-water Aquaculture in Atlantic Canada*. Moncton, NB: Canadian Institute for Research on Regional Development, pp. 617–45.

Sagoff, M. (1988) *The Economy of the Earth: Philosophy, Law and the Environment*. Cambridge: Cambridge University Press.

Scherer, J. (1972) *Contemporary Community: Sociological Illusion and Reality*. London: Tavistock Press.

Seager, J. (1995) 'Feminism and the environment: What's the problem here?', in L. Quesnel (ed.), *Social Sciences and the Environment*. Ottawa: Ottawa University Press, pp. 55–65.

Sider, G.M. (1986) *Culture and Class in Anthropology and History: A Newfoundland Illustration*. Cambridge: Cambridge University Press.

Sinclair, P. (1983) 'Fishermen divided: the impact of limited entry licensing in NW Newfoundland', *Human Organization*, 42(4), pp. 307–14.

Sinclair, P. (1985) *From Traps to Daggers*. St John's, NF: Memorial University, Institute of Social and Economic Research.

Taylor, C. (1989) *Sources of the Self*. Cambridge, MA: Harvard University Press.

Taylor, C. (1991) *The Malaise of Modernity*. Toronto: Anansi.

6

Migration as a Way of Life: Nevis and the Post-war Labour Movement to Britain

Margaret Byron

Introduction

Migration is an integral element of the history, culture and socio-economic life of Caribbean people. The island of Nevis is no exception to this (Byron, 1994; Olwig, 1995; Richardson, 1983). Whereas the focus is on Nevis in this chapter, many of the points can be applied to the Commonwealth Caribbean in general. Nevis is a small, almost circular island with an area of 57.6 sq km. It is situated in the Leeward Islands of the Caribbean archipelago and lies directly south-east of St Kitts (Figure 6.1). A 3.2 km-wide sea channel separates the two islands. The total population in the island at the 1991 Census was 9130 (St Kitts and Nevis Population Census Office, 1992). The island of Nevis is part of the independent state of St Kitts (St Christopher) and Nevis.[1] The state became politically independent from Britain in 1983. In 1991 a total population of 41,826 was recorded for the islands, a decline from 44,404 in 1980 (St Kitts and Nevis Population Census Office, 1992). Migration has created a marked impact on the population of these two islands. They are distinguished within the Commonwealth Caribbean for the highest rate of net emigration as a percentage of natural growth rate and the most rapidly shrinking population (Mills, 1988). This is despite a high birth-rate and, since 1950, a low mortality rate.

In 1988 McElroy and de Albuquerque applied the concept of 'migration transition' to a selection of small island states in the Caribbean. St Kitts–Nevis were placed in the 'pre-migration transition' group; in other

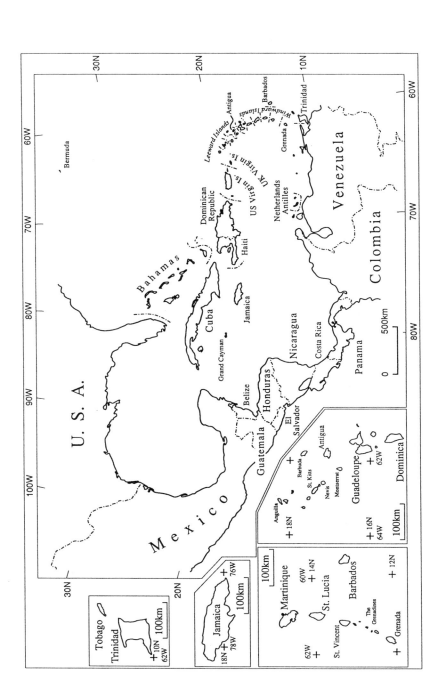

Figure 6.1 The Caribbean region including migrant destinations prior to World War II

words, they were still net exporters of labour. A decade later, St Kitts and Nevis are perhaps closer to the 'migration transition' stage. Tourism development and the growth of an international offshore financial sector have proceeded apace and, for St Kitts at least, the levels of out-migration are slowing. Annual tourist stayover arrivals for the state peaked in 1994 at nearly 100,000 while cruise-ship and yacht arrivals exceeded 110,000.[2] Since then arrivals have dropped slightly, perhaps due to the impact of hurricanes on the islands' tourism industry. The industry and related construction sector employ a significant proportion of the population. Since 1993 at least a quarter of total employment in Nevis has been in tourism and the sector has contributed up to 18 per cent of the island's revenue. Although the population decline which commenced in the 1960s was still evident in the 1991 census, the rate of decline has dropped. Clearly, the expansion of tourism has increased employment opportunities in Nevis, reducing out-migration and attracting potential returnees.

The relatively tiny size and population of the island of Nevis means that the migration process has affected the entire island community. In nearby Montserrat, very similar in area and population to Nevis, one-third of the population left the island for Britain in the post-war period, as Stuart Philpott shows in the next chapter. In such cases, island societies become saturated with migration information. The process permeates everywhere and the impact is more profound and lasting than it may be in a larger, more economically diversified society. Every Nevisian has been, or is, either a migrant or a close relative of a migrant. As children mature they become aware of the higher standard of living experienced on the island as a result of migrant remittances and the increasing presence of modern homes built by potential or actual return migrants. Few do not include migration in their range of possible survival or life improvement strategies and non-migrants muse on what might have been had they, too, 'gone away'.

Since the end of slavery, Caribbean populations, via migration, have become increasingly incorporated within the widening international division of labour. This position has become internalized by Caribbean societies and migration has become an accepted household income strategy for the majority of people. By the end of World War II labour migration flows from the region were well established. For the British colonies, the colonial power became the main migrant destination after 1948, when the British Nationality Act conferred citizenship and the right of abode in Britain on the people of the colonies.

The historical context to this post-war migration to Britain is an important aid to interpreting the contemporary Caribbean diaspora and its impact on the islands of the region, as we shall see in more detail in the next section of the chapter. The migration culture, heavily reinforced and expanded by the size and density of the flow of migrants to Britain, had been established in the previous century, starting with the emancipation of slaves in 1834.

Britain became the 'next available destination' and Caribbean migrants became part of the 15-million-strong migrant labour force serving the needs of post-war European capital, in this case as a direct result of colonial links with Britain (Cohen, 1987).

A feature of the British Caribbean colonial societies was the extreme insularity which resulted from existing as separate colonies supplying British markets and governed by Britain. Trade and cultural exchange were between the colonial power and the individual territories rather than within the region. Competition rather than co-operation prevailed. Migration turned out to be a force against this insularity. As early as the immediate post-emancipation years, Nevisians moved to Trinidad and Guiana to work. Some settled there while others returned and introduced these locations to the 'life-worlds' of those Nevisians who remained at home. This was followed by the Dominican Republic, Bermuda and the Netherlands Antilles. Knowledge that groups of Nevisian migrants and their descendants remained in these countries linked the destinations into the social field of the islanders. In addition to a view of the actual destinations, Nevisians brought back knowledge of, and a growing affinity with, their fellow Caribbean islanders. This was the beginning of a wider 'West Indian-Caribbean consciousness' which migration, within and outside the region, was to nurture and perpetuate (Sutton and Makiesky-Barrow, 1987).

The Migration History of St Kitts–Nevis from Emancipation to the Second World War

In the British colonies of the Caribbean, migration of freed labour began in 1834, the year in which the Act of Emancipation came into effect, freeing slaves in the British West Indies. This movement had been pre-empted by the redistribution of slaves in the region according to labour requirements in the various colonies. British planters in the newly-acquired colonies of Guiana and Trinidad in the southern Caribbean had their labour supplies cut when the slave trade was abolished by Britain in 1807 (Eltis, 1972; Higman, 1984). The subsequent shortages were partly alleviated by the illicit purchase of slaves from the labour-rich colonies of the northern and eastern Caribbean. This movement of labour continued after emancipation with negotiations then taking place between freed slaves and recruitment agents from the southern colonies.

Ex-slaves from St Kitts and Nevis were among the first workers to be recruited for plantation labour in Trinidad and Guiana (Richardson, 1980, 1983). Both islands contributed hundreds of labour migrants annually between 1839 and 1845 (Figure 6.2). The conditions of labour varied in the two islands: St Kitts being ideally suited in soil and climate to sugar production, while Nevis had a lower annual rainfall and stony soil. Due to their

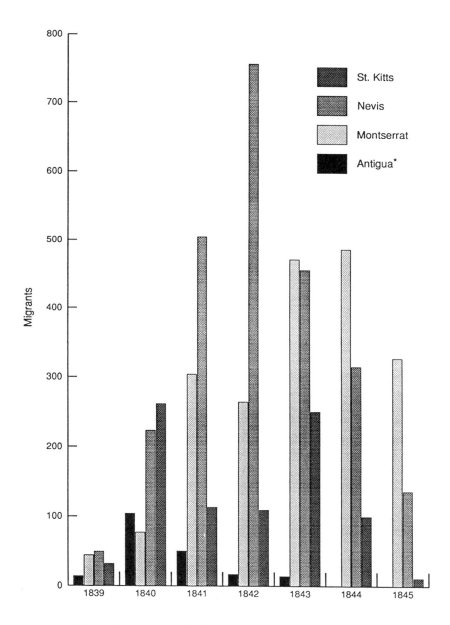

* No Antigua data available for 1844 and 1845

Figure 6.2 Migration from the Leeward Islands, 1839–45
Source: Hall (1971, p. 41).

relatively small areas (St Kitts 100 sq km, Nevis 60 sq km), neither island had much cultivable non-estate land so ex-slaves were obliged to live and work for their former owners on their estates. Planters in St Kitts strongly resisted the ex-slaves' attempts to purchase land,[3] fuelling a resentment which rendered them more amenable to the offers of the Trinidadian and Guianese recruitment agents. The following summary of the position of former slaves in a neighbouring island is equally applicable to St Kitts and Nevis:

> The human degradation of slave status had been removed, he was no longer driven to work, if wronged he could seek out the magistrate, his women and children were safe under one roof: but the roof was not his, nor the land, nor the freedom to avoid the daily routine of estate employment. There were two possible means of employment: he could leave the estates if he could find some other place to live, or he could leave Antigua. (Hall, 1971, p. 40)

In Nevis, migration offered the only alternative to continuing to work for the planters who had enslaved them. As the island suffered from stony soils and severe erosion, planters there were relatively poor and wages were lower than those in neighbouring territories (Hall, 1971; Richardson, 1983). Consequently, despite having a much smaller population than St Kitts, Nevis lost more than twice as many migrants over the six-year period shown in Figure 6.2. This movement to Trinidad and Guiana was seldom permanent. Richardson (1983, p. 90) observed that migrants 'usually returned home, once more braving the hazards of the inter-island passage'. Thus, virtually from the start of the post-emancipation period, 'spending some time out' was established as one way in which the working classes of this territory sought to improve their life chances.

The attraction of the sugar plantations in the southern Caribbean colonies of Trinidad and Guyana diminished with the introduction of the 1846 Act of the British Parliament which halved the duties on the importation of sugar from sources outside the Caribbean colonies. From this point the British West Indies lost the struggle for monopoly of the British sugar market (Parry and Sherlock, 1956; Williams, 1970). Consequent reduction of wages in Trinidad and British Guiana resulted in migrant labourers gradually trickling back to St Kitts–Nevis where they joined their fellow islanders in suffering the economic depression that pervaded the Caribbean at the time. For a few decades the population movement out of the islands ebbed and growing resentment at the local economic despair culminated in riots in the colony in 1896 (Richardson, 1983).

The out-migration that did occur during this period was in a north-western direction from the Leewards (Figure 6.1), and on a smaller scale to that following emancipation. It formed the precursor of the large-scale exodus to the Dominican Republic that was to feature over the ensuing three decades. Relative proximity and transport routes meant that labour from St Kitts and

Nevis was much more likely to serve the needs of American and European investments in the Dominican Republic and Cuba than those of the Canal Zone and banana plantations of Panama and Costa Rica. The British colonies of Jamaica, Barbados and St Vincent and the French colonies of Guadeloupe and Martinique were the main labour sources for developments in Central America (Koch, 1977; Newton, 1984; Thomas-Hope, 1993).

Other destinations for hundreds of Kittitian (people from St Kitts) and Nevisian labour migrants at the start of the twentieth century were Bermuda, where a floating dry dock was being constructed (Richardson, 1983), and cities of the eastern seaboard of the United States.[4] On the basis of information that the vast majority of the money remitted from the labourers on the docks went to St Kitts–Nevis, Richardson (1983, p. 119) concluded that the migration from the Caribbean to Bermuda during this period was 'essentially a migration of Kittitians and Nevisians'.

Between the last quarter of the nineteenth century and the economic depression of the 1930s, thousands of mainly male workers left St Kitts and Nevis for the Dominican Republic (Richardson, 1983). They worked as seasonal labourers in the sugar industry, mainly cutting sugar cane. Although the migration was organized and intended to be a seasonal labour movement, differences between the emigration and immigration figures indicate that some migrants did not return, as there was a net loss of several hundred in most years. It was clear that, for many migrants, the seasonal migration was not a success. Some remained in the Dominican Republic after the six month 'crop' period unable or unwilling to fund their return trip, which was not usually paid for by their employers. Others managed to get home but arrived with little more than the clothes they wore. The following case study, based on the oral history of an elderly Nevisian, illustrates some of the conditions experienced by the Dominican Republic's migrant labourers.

Mrs Henry was 83 years old at the time of this interview with the author in Nevis in 1990. She went to the Dominican Republic as a child with her older sister who had become her guardian upon the death of their mother. They found accommodation and work on the Consuelo estate which consisted of a sugar mill and a village of mainly labourers' barracks and 'sugar cane fields going back as far as you could see in the countryside'. They lived in the village of barrack-type accommodation for the plantation and factory workers, doing the laundry and cooking for male workers. Although under twelve years old, Mrs Henry helped her sister with her work from the time of their arrival in the village. Migrants who performed services such as these for the estate employees were the most vulnerable. They were entirely dependent on the incomes of their patrons who were themselves merely seasonal workers.

We didn't save anything, we just made out. For when crop done, for six months we doesn't make two dimes. The people we used to wash for, as they were not working they washed their own clothes. So we got no money. I was so glad when crop start and we get some work.

Despite wishing to escape the difficult conditions, the poverty which faced many migrants prevented a return to their homeland. Later, Mrs Henry and her husband had to devise a scheme for their return which would involve free passages. Migrants on the Consuelo estate were not assisted with passages home after the crop was harvested and many remained there throughout the off-season because they couldn't afford the return passage. The Henry family moved to La Romana estate at the beginning of the next season, as they had learnt that the owners were repatriating their workers at the 'crop over' stage. At the end of the harvest, Mr Henry approached the estate owners and expressed his desire to return home and was rewarded with passages to St Kitts for himself and his family. They travelled to St Kitts–Nevis in 1936 with no savings and for some time were dependent on their relatives in both islands. Mrs Henry left her first child with her childless older sister in Consuelo. Neither of them ever returned to Nevis. Her two Dominican-born offspring who accompanied their parents to Nevis subsequently re-migrated: one to Britain and one to the US Virgin Islands.

This account illustrates the misery which many labour migrants experienced and were virtually powerless to alleviate. Many migrants never returned, either through shame at not achieving the success expected of a migrant or simply due to their inability to afford the passage home.[5] For Nevisian migrants, sojourns abroad such as this one were undertaken primarily to alleviate their economic conditions. This objective was not necessarily achieved. Further evidence of the continued presence of a St Kitts–Nevis enclave in the Dominican Republic is found in the 1991 Commonwealth Caribbean Population and Housing Census. At the time of the census, 100 people who were born in the Dominican Republic were living in St Kitts–Nevis. More than 250 applications for citizenship were received between 1988 and 1991 by immigration authorities of St Kitts–Nevis from Dominican-born descendants of migrants.[6] In a literal reversal of fortunes, today the economic opportunities and living standards of the majority in St Kitts–Nevis are better than those in the Dominican Republic. Young Dominicans whose parents or grandparents were born in St Kitts–Nevis use this relationship to claim citizenship there.

The last major destination to attract workers from St Kitts and Nevis prior to the Second World War was the Netherlands Antilles. Here, United States investment in the form of oil refineries brought labour from the eastern Caribbean region during the period between the 1920s and the 1950s. Again, male migrants were in the majority, but female domestic workers were also sought to serve in the homes of the senior staff of the refineries. Frucht (1968) makes the point that most of the people who migrated to this destination were skilled artisans from the upper section of the working classes in Nevis. Their impact on the Nevisian landscape upon return was significant as they built relatively large houses and frequently opened village shops. Skilled though these workers were, they too were susceptible to repatriation in times of recession or when, due to illness, they were unable to work at their full capacity.

Demographically, the history of pre-war migration from Nevis is evident in the marked imbalances in the population structure. Figure 6.3 reveals a depleted young male population in Nevis over the age of 14 in censuses taken between 1911 and 1946. This is most noticeable in the 1911 and 1921 data when migration to the Dominican Republic was at its peak. By 1946, restrictions on movement to this destination and to the Netherlands Antilles during the economic depression of the 1930s had reduced out-migration of male migrants. The gender structure of subsequent labour migrant streams has reflected the demands of the international economies involved. The migration to Britain which is discussed in the ensuing pages was almost gender-balanced as there were many employment opportunities for women (Byron, 1998).

By the 1930s, when global economic depression closed the doors of the destination countries, there had already been a century of labour migration from Nevis. The various destinations were mapped in the islanders' perceptions of the world as successful migration ventures or as poor or disastrous experiences. Subsequent migrations were to be evaluated against this historical record. Social unrest was an evident result of the depression throughout the British colonies of the Caribbean (Hart, 1988). It was felt that a major contributor to this unrest was the presence of the young, active and politicized section of the population, many of whom would normally have been working abroad. This demonstrates how tightly and dependently the Caribbean territories were integrated into the global economy via the migration process. Labour had become a major export commodity and the lack of self-sufficiency in the island colonies meant that the global depression hit them particularly hard.

After the Second World War, two destinations emerged as important for the people of Nevis: the United States and Britain. The seeds of the movement to the United States had been set at the turn of the century as migrants moved to the cities of the Eastern Seaboard. However, most of this post-war migration was recruited as labour for agriculture in the southern states of the USA and in the US Virgin Islands (Byron, 1994). The agricultural workers' contract system was tightly monitored and migrants seldom managed to remain in the US beyond the length of their contracts. Non-recruited labour migration to the US was severely restricted from 1952 by the introduction of the Walter–McCarren Act. So, despite the proximity of the US and the links established by previous migrants, it was rapidly superseded by Britain as the major migrant destination at that time.

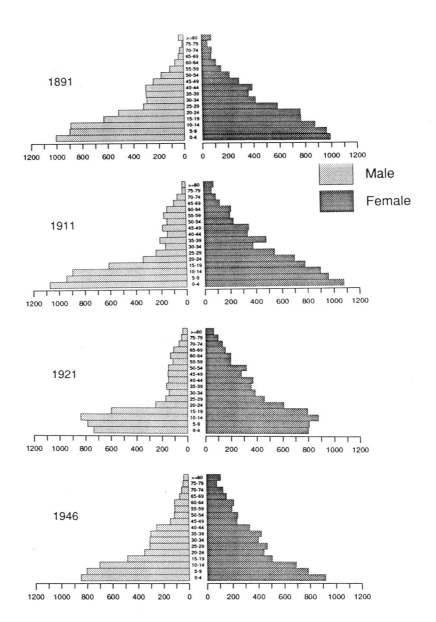

Figure 6.3 Age and sex structure of the Nevisian population at the 1891, 1911, 1921 and 1946 censuses

Post-war Caribbean Movement to Britain

The labour migration from the Caribbean to Britain was triggered by the movement of several thousand volunteers from the British colonies in the Caribbean to serve in the armed forces or work in munitions factories in Britain during the Second World War (Patterson, 1963). Upon their return to the Caribbean after the war many were disillusioned with the stagnant or declining economies of the colonies. This, coupled with the evidence of labour shortages in the manufacturing and service sectors they had witnessed in Britain, led many of this group of returnees to make a second voyage to Britain: this time in search of civilian employment (Davison, 1962). These pioneer migrants were followed by relatives and friends in a classic case of chain migration (Byron, 1994). Their status as citizens of the UK and Colonies, following the 1948 British Nationality Act, meant that they had rights of entry and abode in Britain.

Formal recruitment of labour in the Caribbean by British concerns occurred on a relatively small scale and focused on one territory in the region. State-controlled transport and health services in Britain and hotels and restaurants in the private sector co-operated with the government of Barbados in the recruitment of workers from that island (Davison, 1962). Like the post-war return to Britain of many Caribbean volunteers, this recruitment proved a popular example for potential, non-recruited migrants. It thus contributed more to the post-war migration than is indicated by the actual number of recruits (Byron, 1994).

The Caribbean-born population in Britain grew from 17,218 in 1951 to peak at 304,070 in 1971, according to the British censuses. In the two decades since then, this population has decreased in size to 295,179 in 1981 and to 264,591 in 1991 as a result of a growing out-migration from Britain and a greatly reduced in-migration from the Caribbean. The dwindling in-migration from the Caribbean was the result of the imposition of immigration restrictions and the withdrawal of citizenship rights from potential migrants in the Caribbean territories commencing with the 1962 Commonwealth Immigrants Act (Peach, 1968, 1982).

Britain as Major Destination for Nevisians

Between 1954 and 1961, 810 labour migrants were recorded leaving the colony of St Kitts–Nevis–Anguilla for the United States and the US Virgin Islands compared with 8297 leaving for Britain.[7] While the post-emancipation history of the Anglophone Caribbean shows migration developing into a major strategy of survival for low-income households, it is not always clear how particular destinations became salient for migrants at specific times. Undoubtedly, the international political economy determined where Caribbean labour migrated to. Meanwhile, however, the migrants themselves rationalized the emergence of particular destinations and expressed preferences.

The proximity and familiarity of the USA undoubtedly made it the first choice for potential migrants. Several migrants in Britain whom I interviewed during the late 1980s emphasized their preference for the USA after working there on short labour contracts but noted the difficulties of extending contracts or re-applying successfully once back in Nevis (Byron, 1994). The immigration restrictions which accompanied the 1952 Walter–McCarren Act and the rigid quota system of the US Department of Agriculture minimized access to this destination. This left Britain as the only freely accessible destination. In fact it was the first time that migrants were citizens of both their country of origin and the potential destination. When a sample of Nevisians living in the British city of Leicester was asked why Britain was the choice of destination, the response of the majority was: 'England was the only place "open" at the time'. For this society, the certainty was that a section of the population would migrate. 'Where to?' was the variable (Byron, 1994).

The migration from Nevis to Britain presents a good illustration of the operation of a labour migration that was unrecruited. In the absence of the infrastructure of a recruitment system, this migration was organized around the social networks of the migrants. These structures determined where migrants settled in Britain, provided initial accommodation and assisted in the search for employment. Early arrivals often subsidized the subsequent migration of relatives. In the city of Leicester in the British Midlands, the migration of all 113 Nevisians located in the city in 1988–9 was linked to three Nevisians who arrived in Leicester in 1954 (Byron, 1994). While most later migrants did not associate their presence in the city with these individuals it was possible to locate all migrants within the chain migration which commenced in 1954 with three vital links. Similar examples can be found across Britain from St Albans to Leeds. The mechanism of chain migration meant that elements of populations in specific villages in Nevis re-presented that locality in the new context of urban Britain.

The Impact of Migration on Nevisian Society and Economy

Frucht (1967) noted that it was only from the 1950s that migration became the dominant source of a cash income for the working classes of the population in Nevis. Prior to this the majority of the population depended on small-scale farming for subsistence, supplying domestic markets and producing the cash crops of cotton and sugar cane for export. This farming occurred either on a share-cropping basis with the land owned by a large estate owner to whom one third of all crops harvested was delivered, or on land leased or rented from the state by peasant farmers as part of land settlement schemes (Momsen, 1986). Lifelong objectives of most Nevisians were the purchase of land and home improvement. Within the constraints of the Nevisian economy, saving towards this was a very slow process. Migration proved one way of accelerating land acquisition. Returnees from earlier migrations to the United States and the Netherlands Antilles bought land and built impressive properties (Frucht, 1967, 1968; Richardson, 1983). Money sent home by migrants abroad was invested similarly. This set a precedent for subsequent migrations.

Gender and migration to Britain

After the War, potential migrants saw the movement to Britain as offering the latest opportunity to improve their living conditions by migrating abroad. For people in the Commonwealth Caribbean there were no restrictions on who could enter Britain at that time. The migration to Britain demonstrated that women were as available as men for labour migration. For the first time a major migration of labour from the Caribbean was almost gender-balanced. In earlier movements, within and outside the region, women migrated mainly as auxiliary workers to perform household tasks for managers and supervisors or to provide various services ranging from laundry to prostitution for the male migrants themselves (Richardson, 1983). The move to Britain therefore offered women the opportunity to appropriate what had been a male-dominated activity and to participate in wage generation in Britain *independently* of the male migrants (Byron, 1998). For this migration, in addition to and sometimes instead of men, women mobilized to form the migrant labour force as illustrated in the following quotation from a Nevisian woman interviewed in Leicester:

> One day my mother said that she was going to send me to England to help the family, I am the eldest of her children. She had a brother in Leicester so I came here to him. He helped my mother to get me here, it was his way of helping out his sister after he came to England. I then got a job and helped bring Mother and the younger kids over.

In the 1991 census of Great Britain, women made up 53 per cent of the Caribbean-born population. Given the central role played by females in maintaining the Caribbean household, it is not surprising that as increasing numbers of women left the islands, remittances became a major element in the income of many households (Momsen, 1986; Philpott, 1973). Perhaps the most lasting impact of the movement to Britain was the mobilization of the female labour force as a migrant group. While economic activity rates of women have traditionally been high in the Caribbean (Senior, 1991), recruited migrants were predominantly men. Since the migration to Britain, migrations have continued to include a high proportion of women and from the late 1970s women have often exceeded men in the migratory flows

(Henry, 1987; Mills, 1988). While this undoubtedly reflects changes in the industrial structure of the destination countries, societal attitudes in the Caribbean to female labour migration were certainly influenced by the movement to Britain.

Remittances and dependency

For the majority of labour migrants from Nevis this century, trips abroad were undertaken to accumulate enough money to invest in a higher standard of living at home. Migrants rarely expected their sojourns in North America or Britain to exceed five years. In reality this has not been the outcome for many. Between thirty and forty years on, the vast majority of the post-war migrants to Britain from Nevis and the other Caribbean islands are still there. Over this period millions of pounds in remittances have reached the Caribbean islands. For the smaller islands such as Nevis, with a population of less than 12,000, no family has remained untouched by the money and parcels of food and clothing which flowed in from abroad.

Accurate data on remittances are scarce due to the incoming foreign currency being exchanged at a range of banks and post offices. Richardson (1983) gave figures ranging from EC$ 0.75 million to 1.5 million from the UK alone annually exchanged at the St Kitts–Nevis post offices in most years between 1962 and 1975. By the 1970s as much again would have been exchanged at banks. The government decision to impose a tax on foreign currency transactions in the 1970s reflects the potential revenue from remittances for the state. Today, foreign currency receipts are rising significantly due to the payment of pensions from the UK to retired returned migrants.

Remittances have had two major effects on the society in Nevis. First, during the post-war period, many non-migrants came to rely on this source of income as an alternative to farming as their major source of income (Momsen, 1986; Richardson, 1983). The production of two cash crops on the island declined rapidly: sugar cane ceased to be produced by the 1970s and sea island cotton became a marginal crop. Ironically the income from these crops had contributed to financing the migration to Britain of many Nevisians (Byron, 1994). Secondly, the continuous arrival of money and other gifts created an image of prosperity abroad which exceeded what people had experienced in Nevis. The younger generations thus directed their goals overseas (Olwig, 1987).

The decline of the small farming sector in Nevis can partly be linked to the growing importance of migration in household income provision. The migration to Britain removed much of the young population of the island, the modal age group of migrants being 19–24 years (Byron, 1994). This had both immediate and longer-term impacts on productivity. While the seasonal labour force was depleted from the start of the migration, in the longer term many farmers had few or no children remaining in Nevis to assist and, eventually, to succeed them on the family farm. A survey early in the 1980s revealed that more than 65 per cent of farmers in the island were over 55 years old (Lowery and Lauckner, 1984). Meanwhile, the flow of remittances into Nevis reduced household and family members' dependence on farming. Momsen (1986) notes a mother who, having remained on the family farm in Nevis to raise her family, 'spends her old age living off the remittances of her children ... watching the weeds grow in her old vegetable patch by her newly painted house'. Gender is a significant determinant of remittance receipts. The high incidence of female-headed households in the Caribbean, the dominance of men in the migrant stream until the 1970s and the tendency for female migrants to leave their children with their mothers, sisters or other female relatives, meant that women were more likely to receive remittances than male non-migrants.

Regular remittance flows undermined the small farming economy of the island and, it is argued, thus made it more dependent on the global economy (Momsen, 1986; Richardson, 1983). Non-migrants received a cash income from abroad and often spent it on imported food. Remittances simultaneously presented and perpetuated an insidious image of a far more prosperous 'overseas' compared to the island environment. This was particularly attractive to the younger generations in the island society who internalised migration as the means to development and progress. Whereas the post-emancipation movement of Nevisians to Trinidad essentially commenced as 'assertions of freedom' (Richardson, 1983), the labour migrations since then, particularly those of the post-war period, have been explicitly livelihood strategies. The incomes earned abroad were always significantly above what the migrant could earn in Nevis. This was relayed to non-migrants via remittances, the prosperous appearance of visiting migrants and the higher standard of living maintained by returnees from abroad. Between 75 and 96 per cent of school children interviewed in Nevis in the early 1980s had received remittances from relatives abroad during the year of the interview. Not surprisingly, 64 per cent expressed a desire to emigrate from the island (Olwig, 1987). Remittances simultaneously subsidize the Nevisian household economy and intensify a culture of migration.

Return migration

Return has been a feature of Nevisian labour migration throughout the post-emancipation period and contributes significantly to the 'migration ideology' which prevails in the island. Today evidence of 'successful migrations' is present in large concrete houses, small businesses and regular inflows of foreign currency in the form of pension payments from North America and Britain to retired returnees (Byron and Condon, 1996). As mentioned earlier, images of successful and unsuccessful migrations, based on the apparent wealth of returnees, influence the goals of potential migrants.

Table 6.1   Main migration destinations of returnees present in Nevis, 1991

Country	no.	Country	no.
Britain	182	Netherlands Antilles	32
USA	108	Dominican Republic	11
British Virgin Islands	83	Other Leeward Islands	74
US Virgin Islands	286	Windward Islands	13
St Maarten	158	Jamaica	17
Canada	32	Barbados	16
		Trinidad	18

Source: Caribbean Community Regional Census Office (1994).

The higher numbers of returnees in Table 6.1 represent the most recent labour migrations which have been dominated by movements to Britain, the United States, US Virgin Islands, British Virgin Islands and St Martin. In the 1960s, coinciding with the immigration restrictions imposed by Britain, the United States and Canada eased their immigration laws by removing the often racially discriminatory quota systems by country of origin (North and Whitehead, 1991). Consequently, migration to these destinations from the Caribbean increased sharply. The US Virgin Islands, which have well-developed tourism industries, are popular destinations as are the nearby British Virgin Islands and St Maarten which have similar touristic economies (McElroy and de Albuquerque, 1988).

Amongst the ageing population of Nevis there are a few remaining representatives of the migrants to the Dominican Republic and the Netherlands Antilles. Some of these migrants, particularly those to the Netherlands Antilles, later migrated to Britain. The main goal of labour migrants from Nevis has been the improvement of their living standards in the home country. In the case of Nevis this has been realized by building new homes or improving existing housing and, in the case of pre-retirement returnees, starting a business. In 1991, 45 per cent of returnees from Britain to St Kitts–Nevis were of pre-retirement age. Self-employment is seen as the ideal situation by most returnees (Byron, 1994). Their experiences in Britain as employees, often in relatively low-skilled jobs, have fuelled the desire for the economic independence of self-employment in the Caribbean.

The economic impact of pre-retirement returnees has generally been to expand the service sector. In Nevis, most employment opportunities for returnees have arisen in the tourism industry or in linked service sector industries. Although some returnees have sought jobs as employees within hotels, many have entered self-employment within the service sector providing accommodation, transportation, boutiques and bars. Many of these cater specifically for the tourism industry. Several returnees shipped minibuses and vans to the Caribbean to use as taxis (Byron, 1994).

Recent return from Britain has made relatively little impact on the agricultural sector. Migrants who returned within 10 to 15 years, by around 1975, often continued in their pre-migration roles as small farmers. Their migration savings were used for improvements to their homes and for the purchase of land which they had previously leased from large land owners (Byron, 1994). However, in the 1990s, small farming is mainly the part-time occupation of retired returnees. Nonetheless, two egg farms on the island are owned and run by returnees from Britain. Their main markets are the hotels on the island.

Despite the heavy presence of migrants in the manufacturing sector in Britain, return migration has made a negligible contribution to this element of the Nevisian economy. Such a situation results from a combination of the low-skilled and specific positions held in Britain and the lack of a comparable manufacturing sector in Nevis. Investment in such a sector would exceed the capital assets of most returnees from Britain.

Remittances and the eventual return of labour migrants have undoubtedly raised the living standards of Nevisian society. The currently expanding tourism sector keeps employment levels high. Meanwhile farming, always a necessity as opposed to a chosen occupation, is more marginalized than ever. Through remittances and later return, migration provided both a source of externally produced food and hardware and the cash to purchase imports of these goods. This further tied Nevis into a dependency on the global economy and reduced any semblance of self-sufficiency in food that existed after the Second World War.

In the immediate post-war era, Britain was perceived as the 'Mother country'; it was the colonial power, and economic and cultural influences were still strong. When migration to Britain was curtailed by restrictive legislation in the early 1960s, potential migrants from the Commonwealth Caribbean looked again to the USA and its Caribbean territories as migrant destinations. From 1965 entry to the USA became easier due to changes in immigration legislation and Nevisians joined in this new movement. Many made the much shorter journey to the US Virgin Islands where there were labour shortages in the expanding tourism and construction sectors.

Migration, cable television and the all-encompassing effect of American culture

In Table 6.1 the large component of returnees from the USA and USVI is evident. Recent migration to these places has exceeded the migration to Britain and, although fluctuating in size over time, is still an active flow. The proximity of the USA and USVI has meant that communication has been more frequent. There have been two-way flows of people and goods, and people have often spent shorter time periods than were common in the migration to Britain; hence the younger average age of returnees from the USA recorded in the 1991 Census (Caribbean Community Regional Census Office, 1994).

Today the cultural environment in the Caribbean is heavily influenced by the American way of life. The considerable role played by migration in shaping cultural trends in Nevis is reinforced by cable television which brings 24-hour US television into 95 per cent of Nevisian homes. Olwig (1995) calls for an appreciation of the cultural complexity which characterized the post-emancipation Caribbean, arguing that such complexity is more commonly associated with 'the modern world of transnational relations'. Olwig's discussion of the complex and often contradictory patterns which have emerged in the post-emancipation Caribbean is very valid and builds on the points made by Frucht (1967) on Nevisian society in the 1960s. Today, however, the processes of migration and penetration of Caribbean island society by the US media have undermined diversity within and between Caribbean societies. Contrary to the image of complexity, the modern Caribbean increasingly seems to be overtaken by an 'American monoculture'. Everyone consumes American television, music, food and manufactured products. Even the local currency (EC dollar) is tied to the US dollar at the fixed rate of EC$2.67 to US$1.00. Indeed, the USA seems to have become the standard by which everything is measured.

Not surprisingly, returnees from Britain are compared to the more popular and all-pervasive image of a 'returnee from America'. In an interesting irony, the former emphasize their contribution to the creation of a home society which offers at best a limited welcome to returning migrants. As one Nevisian returnee told me in 1996, 'holidays were OK but once you come back for good they do not want you. We were only of use to them while we could send back money and parcels'. While individual perceptions vary, there is a general opinion among returnees of a pampered society for which they are partly responsible.

A Transnational Society?

While return migration is a major contributor to the migration culture of Nevis, it is important to consider the role of the non-returnees and non-migrants in the overall migration process. A useful model for inclusion of these elements is the concept of the 'transnational socio-cultural system' applied by Sutton (1987) to interpret the dynamism of Caribbean societies in New York. Sutton writes of the reconstitution of Caribbean life in the New York context 'by means of the mutual interaction of happenings in New York and the Caribbean' (1987, p. 20). 'Happenings' would range from the personal and family level such as funerals and weddings to the community level such as carnivals and national-day celebrations in the Caribbean and

New York. The constant bi-directional flow of people and information has resulted in this transnational system which re-forms and further 'creolizes' Caribbean culture in the migrant destinations and in the Caribbean.

The study of the impact of the migration process on Nevis is facilitated by extending the geographical and cultural space or 'social field' (Manners, 1965) to include all migrant destinations. We might suggest that a typical Nevisian non-migrant had seven siblings: two in Britain, one in Canada, two in the USA and two in the USVI. The same person also had an aunt in Bermuda and cousins in the Dominican Republic, descendants of the migrants at the turn of the century. Modern forms of communication have resulted in a social system which spans all of these destinations. Communication between people within the Nevisian social field occurs via a variety of media: telephone, postal services, the Internet, visits and more long-term migrations. It is multidirectional and impacts are experienced in all locations. Migration, arguably more than any other process, has firmly integrated Nevisian society into the global economic and cultural system while reinforcing its dependence on the metropolitan powers.

Conclusion

Caribbean societies as they exist today are largely descended from populations formed by the forced migration of slaves and, later, the importation of indentured labour. From a position of labour shortage when sugar production was at its height, the region has been transformed into a labour reserve which has served the needs of first a regional and later the global economy. Destinations have varied over time as labour needs changed and the streams of islanders have ebbed or flowed depending on the performance of the international economy.

For Nevisian people migration evolved as the most common improvement strategy for the individual or the household. Due to the small size of the island, all households were aware of migration as an option and the working classes became increasingly dependent on this strategy for their survival. At least one fifth of the entire population of the island left in the post-war migration to Britain. Yet the migration of labour to Britain in the 1950s and 1960s was simply the latest element in a trend of out-migration which was over a century old. It was unique in that, for over a decade, migrants were allowed unrestricted movement to the destination and it consequently had profound impacts on the demographic structure of the islands involved.

There is no doubt of the enormous changes wrought by migration on the culture and economies of the sender islands. In Nevis, the different receptions encountered by returnees from Britain and the USA illustrate the ways in which the migration process has contributed to local values and preferences. The current dominance of the USA as destination and trend-setter has been heavily supported by the introduction of cable television in the 1980s. 'Media are technologies which alter the cultural field they enter: this is the manner in which they mediate' (Rowe and Schelling, 1991, p. 106).

Destinations, while determined by the international political economy, are assessed by potential migrants in terms of the success of returnees and familiarity with returnees' culture. In Nevis today, the majority's perception of success is closely linked to the dominant ideology of the time. Migration has proven to be a critical agent in the construction of local ideologies, particularly in the context of small Caribbean islands.

Finally, the concept of a transnational system is potentially useful yet contradictory in the study of migration and Caribbean island states. It permits a more dynamic and inclusive approach to the analysis. In particular, the contribution Caribbean migrants may make to cultures of the destination countries is acknowledged, and changes which occur in all aspects of the social fields of migrants become potential foci for study. Transnationality is not, however, equitable in outcome and the dominance of the USA in the current migration era illustrates the need to include global power relationships within the analytical framework.

Notes

1. Due to political and economic ties between Nevis and St Kitts, records for the two islands are often aggregated and presented for the union. Consequently, some of the migration data presented in subsequent sections are for St Kitts and Nevis.

2. Source: St Kitts and Nevis Department of Tourism, March 1998.

3. British Parliamentary Papers 1842/XIII, 'Report from the Select Committee on West Indian Colonies', testimony of George Estridge, p. 232.

4. Interviews with return migrants in Nevis, 1988–90 and 1996.

5. A Colonial Office report for the years 1943 and 1944 commented as follows:

   Owing to the nationalist policies and restrictive legislation against aliens, as well as the discriminatory taxation which has been imposed on them, the conditions under which West Indians are living in both Cuba and the Dominican Republic are much less favourable at the present time than those which prevail in the colonies from which emigration took place, and steps have had to be taken by means of grants from the Imperial Exchequer to alleviate hardship. A considerable number of West Indians in Cuba and the Dominican Republic, especially the older and sick people, would welcome repatriation, but any large-scale repatriation is impracticable in view of the existing population problems in the West Indies (Stockdale, 1945, p. 2).

6. Interview with officials at the Ministry of Home Affairs, St Kitts, April 1991.

7. No data on migration to Britain was available in the Colonial reports for 1957 and 1958. It is likely that more than 1000 persons left in these two years, thus bringing the total for the period to over 9300.

References

Byron, M. (1994) *Post-war Caribbean Migration to Britain: The Unfinished Cycle*. Aldershot: Avebury.

Byron, M. (1998) 'Migration, work and gender', in M. Chamberlain (ed.), *Caribbean Migration: Globalized Identities*. London: Routledge, pp. 217–31.

Byron, M. and Condon, S. (1996) 'A comparative study of Caribbean return migration: towards a context-dependent explanation', *Transactions of the Institute of British Geographers*, 21(1), pp. 91–104.

Caribbean Community Regional Census Office (1994) *Commonwealth Caribbean Population and Housing Census 1991*. Trinidad: Caribbean Community Regional Census Office.

Cohen, R. (1987) *The New Helots: Migrants in the International Division of Labour*. Aldershot: Gower.

Davison, R.B. (1962) *West Indian Migrants*. London: Oxford University Press for the Institute of Race Relations.

Eltis, D. (1972) 'The traffic in slaves between the British West Indian colonies, 1807–1833', *Economic History Review*, 25(1), pp. 55–64.

Frucht, R. (1967) 'A Caribbean social type: neither peasant nor proletarian', *Social and Economic Studies*, 13(3), pp. 295–300.

Frucht, R. (1968) 'Community and context in a colonial society: social and economic change in Nevis, British West Indies', *Anthropologica*, 10(2), pp. 193–208.

Hall, D. (1971) *Five of the Leewards, 1834–1870*. Aylesbury: Caribbean Universities Press/Ginn and Co.

Hart, R. (1988) 'Origin and development of the working class in the English-speaking Caribbean area: 1897–1937', in M. Cross and G. Heuman (eds), *Labour in the Caribbean*. London: Macmillan, pp. 43–79.

Henry, F. (1987) 'Caribbean migration to Canada: prejudice and opportunity', in B. Levine (ed.), *The Caribbean Exodus*. London: Praeger, pp. 214–22.

Higman, B.W. (1984) *Slave Populations of the British Caribbean, 1807–1834*. Baltimore: Johns Hopkins University Press.

Koch, C.W. (1977) 'Jamaican blacks and their descendants in Costa Rica', *Social and Economic Studies*, 26(3), pp. 339–61.

Lowery, J. and Lauckner, F.B. (1984) *A Profile of Small Farming in Nevis*. Trinidad: Caribbean Agricultural Research and Development Institute.

Manners, R. (1965) 'Remittances and the unit of analysis in anthropological research', *Southwestern Journal of Anthropology*, 21(2), pp. 179–95.

McElroy, J.L. and de Albuquerque, K. (1988) 'Migration transition in small Northern and Eastern Caribbean states', *International Migration Review*, 22(3), pp. 30–58.

Mills, F.L. (1988) 'Determinants and consequences of the migration culture of St Kitts–Nevis', in P. Pessar (ed.), *When Borders Don't Divide: Labour Migration and Refugee Movements in the Americas*. New York: Center for Migration Studies, pp. 42–72.

Momsen, J.D. (1986) 'Migration and rural development in the Caribbean', *Tijdschrift voor Economische en Sociale Geografie*, 77(1), pp. 50–8.

Newton, V. (1984) *The Silver Men: West Indian Labour Migration to Panama, 1850–1914*. Jamaica: Institute of Social and Economic Research, University of the West Indies.

North, D.S. and Whitehead, J.A. (1991) 'Policy recommendations for improving the utilization of emigrant resources in Eastern Caribbean nations', in A.P. Maingot (ed.), *Small Country Development and International Labor Flows: Experiences in the Caribbean*. Boulder, CO: Westview Press, pp. 15–52.

Olwig, K.F. (1987) 'Children's attitudes to the island community: the aftermath of out-migration on Nevis', in J. Besson and J. Momsen (eds), *Land and Development in the Caribbean*. London: Macmillan, pp. 153–70.

Olwig, K.F. (1995) 'Cultural complexity after freedom: Nevis and beyond', in K.F. Olwig (ed.), *Small Islands, Large Questions: Society, Culture and Resistance in the Post-Emancipation Caribbean*. London: Frank Cass, pp. 100–20.

Parry, J.H. and Sherlock, P.M. (1956) *A Short History of the West Indies*. London: Macmillan.

Patterson, S. (1963) *Dark Strangers: A Study of West Indians in London*. London: Tavistock.

Peach, C. (1968) *West Indian Migration to Britain: A Social Geography*. London: Oxford University Press.

Peach, C. (1982) 'The growth and distribution of the black population in Britain 1945–1980', in D. Coleman (ed.), *Demography of Immigrants and Minority Groups in the United Kingdom*. London: Academic Press, pp. 23–42.

Philpott, S.B. (1973) *West Indian Migration: The Montserrat Case*. London: Athlone.

Richardson, B.C. (1980) 'Freedom and migration in the Leeward Caribbean, 1838-48', *Journal of Historical Geography*, 6(4), pp. 391–408.

Richardson, B.C. (1983) *Caribbean Migrants: Migration and Human Survival in St Kitts–Nevis*. Knoxville: University of Tennessee Press.

Richmond, A.H. (1954) *Colour Prejudice in Britain*. London: Routledge & Kegan Paul.

Rowe, W. and Schelling, V. (1991) *Memory and Modernity: Popular Culture in Latin America*. London: Verso.

Senior, O. (1991) *Working Miracles: Women's Lives in the English-Speaking Caribbean*. Barbados: Institute of Social and Economic Research, University of the West Indies.

St Kitts and Nevis Population Census Office (1992) *Preliminary Results of the 1991 Population and Housing Census*. St Kitts: St Kitts and Nevis Population Census Office.

Stockdale, F. (1945) *Development and Welfare in the West Indies, 1943–44*. London: HMSO.

Sutton, C.R. (1987) 'The Caribbeanization of New York City and the emergence of a transnational socio-cultural system', in C.R Sutton and E.M. Chaney (eds), *Caribbean Life in New York City: Socio-Cultural Dimensions*. New York: Center for Migration Studies, pp. 15–30.

Sutton C.R. and Makiesky-Barrow, S.R. (1987) 'Migration and West Indian racial and ethnic consciousness', in C.R Sutton and E.M. Chaney (eds), *Caribbean Life in New York City: Socio-Cultural Dimensions*. New York: Center for Migration Studies, pp. 92–130.

Thomas-Hope, E. (1993) *Explanations in Caribbean Migration*. London: Macmillan.

Williams, E. (1970) *From Columbus to Castro: The History of the Caribbean 1492–1969*. London: André Deutsch.

7

The Breath of 'The Beast': Migration, Volcanic Disaster, Place and Identity in Montserrat

Stuart B. Philpott

On 18 July 1995, Montserrat's Soufrière Hills volcano re-awoke after at least 400 years of dormancy, sending a cloud of superheated steam and ash 20,000 feet into the air and triggering mudflows and rockslides. Continuing volcanic activity soon buried the southern half of the island, including the island's capital, Plymouth, in ash and some villages in mud. The consequent evacuation of the area forced the inhabitants into shelters in schools and churches in the northern part of the island, into living in cars or with relatives, or into leaving the island altogether. The volcano, which had no name in village parlance (though it was the 'Soufrière' in official and tourist literature), became widely known as 'The Beast'.

After two years of frequent eruptions, ash clouds, and spreading volcanic debris, the 'safe zone' in the north had shrunk to less than one-third of the island's land mass. Then, on 25 June 1997, a major 'pyroclastic flow' roared eastward down the mountainside and in minutes totally destroyed several villages, burned many houses on an expatriate housing estate, and continued on to the now-defunct airport. This avalanche of searing gas and rock also snuffed out the lives of 19 islanders who had entered the restricted zone to tend animals or obtain belongings from their evacuated homes. Since the onset of the volcanic eruptions, the population of the island has steadily dropped from 11,000 to about 3200.

The tragic loss of life, social dislocation, deterritorialization of much of the island's resident population and the threatened evacuation of the rest, touched off an ongoing controversy with Britain over relocation and redevelopment efforts and heightened the political consciousness and sense of island identity among Montserratians elsewhere. Behind this lurked major questions as to the viability of Montserrat's continued existence as a society and whether the dramatic disruption and depopulation is only the latest – or the last – chapter in the history of Montserrat's long and intimate involvement with migration.

This chapter examines the manner in which migration has become incorporated into the island's social fabric, the increasingly 'transnational' character of its society and, particularly in the light of recent events, the changing construction of Montserratian identity. Although some of these issues are given a rather sharp focus by the recent sequence of volcanic events, the account builds on much earlier work on the impact of mass emigration on the island which I carried out in the 1960s and 1970s (see Philpott, 1968, 1973, 1977).

Migration History

Early migration

The positioning of Montserrat in the 'global economy' is nothing new. From the outset, Montserrat's social and economic history has been a continuing response to changing overseas demand for a series of tropical agricultural commodities and, ultimately, for labour itself. A British colony was first established on the island, apparently uninhabited at the time, in 1632. Except for two brief periods of French occupation, the island has remained a British possession (more recently, a 'dependent territory') up to the present day (Fergus, 1994). The earliest settlers were mainly Irish, either dissidents from other colonies or political exiles. However, despite some contemporary tourist and media depiction of Montserrat as 'A Touch of Old Ireland in the Caribbean' and some recent historical work on the Irish period (Akenson, 1997), the island quickly became a society based on European masters and African slaves. Like most other West Indian colonies Montserrat evolved a creolized culture and language.

The island's first crops were provisions and such export commodities as indigo, tobacco, cotton and ginger which could be economically produced by white smallholders and indentured servants. As North American tobacco became cheaper and more desirable on the British market, however, agricultural production throughout the Caribbean shifted to sugar. Increased capital outlay for grinding and boiling mills and other equipment required by sugar production concomitantly led to the concentration of land in larger holdings and an increased demand for labour. After various unsuccessful attempts to mobilize an adequate supply of European labour, planters throughout the island purchased African slaves. The first slave arrived in 1664 and by 1678 the number had increased to 992. At this same time, the white population reached a peak of 2682 from which it progressively declined until the mid-1960s.

When the island's sugar production reached its highest point in 1735, Montserrat was 'esteemed a possession of great importance to England' (Gipson, 1960, p. 225). The slave population of the island was then 6176. In 1772, although the slaves had increased to a maximum of 9834, the white population had dropped to 1314 and sugar production had actually declined due to decreasing soil fertility and the lack of new land. Already Montserrat's economy had become marginal to the imperial market system which had created it.

By the end of the eighteenth century, the characteristic West Indian racial hierarchy was firmly in place. The groups of whites, free persons of colour, and black slaves were rigidly structured legally, economically, socially and culturally. Whites were the owners, managers and overseers, professionals and clergy. Black slaves were the field and mill hands, the domestics and, occasionally, the artisans. Some free coloured persons were landholding slave-owners. Most were hucksters, small shopkeepers, clerks and artisans (Goveia, 1965, p. 228). The local legislature functioned largely to control the behaviour of the slaves. Sexual liaisons, rooted in social hierarchy and differential power, were probably the most prevalent form of inter-group relations. White men married white women but consorted freely with coloured and black women, both free and slave. Mating and family relationships were highly unstable among the slave population, largely because the men had no legal rights or obligations with regard to their mates or children and because families could be divided at the will of their owner.

The post-emancipation period

With the end of slavery in 1834, various new modes of attaching labour to the land were attempted and some alteration in the island's pattern of land tenure took place. Most important here is that emancipation initiated Montserrat's development as a 'migration-oriented' or 'migration-dependent' society (Philpott, 1968, 1973, 1977).

While ex-slaves in larger colonies such as Jamaica and British Guiana took to the uncultivated bush and mountain lands to start their own smallholdings, there was no unalienated land in Montserrat. Instead, as soon as the so-called 'apprenticeship system' which compelled ex-slaves to serve their former owners for 40.5 hours a week ended in 1838, newly-free Montserratians began leaving the island by the boatload for richer colonies such as Trinidad (Hall, 1971, p. 41). The estate owners, apprehensive about the disappearance of their labourers, pressured the British government into halting the practice of paying bounties for the removal of workers from one colony to another. While this measure slowed emigration, the exodus has never been halted.

Rather than allow the black labourers their own land, yet faced with mass emigration and a chronic shortage of capital, the estate owners adopted two measures in the 1840s to maintain agricultural production. Both of these measures – tenancy-at-will and share-cropping – were contentious political issues from their inception. Tenancy-at-will supposedly compensated for pitifully low wages by allowing ex-slaves to occupy a cottage on the estate and cultivate a small provision ground. Share-cropping involved the labourer performing all the sugar cultivation on estate land and receiving one-third or one-half of the resulting crop as his reward. Undoubtedly, the planters had little choice. In addition to the emigration of much of the labour force, a highly destructive earthquake in 1843 destroyed crops and buildings. Furthermore, the British government's reduction of the duty on foreign sugar in 1846 caused a severe depression in the West Indian market and the estate owners were no longer able to obtain loans from British merchants. The Montserrat planters also failed to introduce new technology. The island's agricultural system continued to be based almost entirely on the hoe and the cutlass until the mass exodus to Britain in the 1950s.

In 1849–50 the island suffered a severe drought and smallpox outbreak. Several of the estates changed hands and some of the least productive ones were divided into smallholdings which provided the basis for the development of a local 'peasantry'. Most of the island's arable land, however, remained as undivided estates even in the 1950s. The cultivation of limes began on two of the larger estates. By 1878, lime orchards totalled 120,000 trees and the juice became a very significant export commodity. The spread of lime cultivation and an improved market for sugar undoubtedly contributed to a reduction of emigration which took effect around 1857. From then until 1890, labour migration was relatively insignificant but by the end of the century it increased again. Once more the West Indian sugar industry was severely depressed, owing largely to inefficient production and competition from beet sugar, and Montserrat was particularly hard-hit.

Adding to the economic difficulties of the island, the older lime orchards were blighted by insects in 1892 and in 1896 considerable damage and some loss of life were caused by a flood and a series of earthquakes which continued intermittently until 1900. Finally, on 7 August 1899, a hurricane struck the island, wiping out the lime orchards, killing 100 people, injuring another 1000 and leaving 9000 homeless.

Like many small islands, Montserrat has been highly vulnerable to natural disasters – hurricanes, epidemics, earthquakes and (now) volcanic eruptions. Furthermore, the small scale of the island has intensified the social, economic and political problems rooted in monocrop plantation agriculture, racial hierarchy and the vagaries of world markets.

Twentieth-century migration

Thus Montserrat entered the twentieth century with the prospects for its two main agricultural export commodities – sugar and limes – virtually eliminated. The export of human labour assumed a renewed importance. When the United States government took over the building of the Panama Canal in 1904, British West Indian labour was in great demand. The *Montserrat Herald* of 12 May 1906 reported that 'scores of persons of the labouring and other classes have been leaving our shores by almost every steamer for Panama and Colon'. At the same time a new crop, Sea Island cotton, was introduced to revive the island's agricultural export trade. By 1903 some 700 acres were planted in cotton and by the 1930s more than 4000 acres were cultivated annually. Although production dropped radically in the 1950s, cotton remained the island's most important agricultural export in 1970.

Despite the completion of the Panama Canal, large-scale emigration continued into the 1920s. The natural increase in the island's population was more than counterbalanced by emigration to the United States, Canada, Cuba and San Domingo. This period was marked by the growth of Montserratian 'communities' in New York and Boston which continue to have social and economic significance for the island. But in 1924 the United States passed the first of a series of acts which restricted West Indian immigration. Once again, as one migratory outlet shrank, another expanded. The development of the oil refineries in the Dutch islands of Curaçao and Aruba in the late 1930s and 1940s created a demand for labour which was met by large numbers of Montserratians and other British West Indians, like the Nevisians described by Byron in the previous chapter.

In the long series of post-emancipation migrations that have been outlined, patterns of behaviour, values and expectations developed which have incorporated migration as an integral part of Montserrat's social system. This prepared the islanders for the largest migration of all, that to the United Kingdom.

Migration to Britain

The staggering scale of Montserratian migration to Britain was produced by a unique combination of economic and political events in the early 1950s. The island's cotton industry enjoyed a brief period of post-war prosperity during which the price nearly doubled. However, in 1952, competition from Sudanese cotton on the British market left most of the West Indian crop unsold. Montserrat planters reacted by lowering production.

In the same year, constitutional changes brought about the first election to the island's legislative council based on full adult suffrage. As in other West Indian territories, trade union activity became the pathway to political office. As the election approached, work stoppages on the estates were frequent and class conflict became so explicit and critical that a commission of enquiry was appointed. The union leaders complained to the enquiry about low wages paid to estate labourers and also that 'relics of immediate post-chattel slavery days' – share-cropping and tenancy-at-will – kept the workers in a state of servility and duress.

During this period, many of the Montserratians in Curaçao and Aruba began returning to the island or migrating directly to England. They contributed disproportionately to the exodus both through departing again themselves and financing the passage of others. Although only six Montserratians applied for passports to go to England in 1952, the number increased rapidly the following year when an Italian line began calling at the island on the return run from South America. In 1955 alone, 1145 Montserratians applied for passports. As the migration mounted, a Spanish line also began picking up passengers. The construction of an airstrip in 1956 also made air connections with Britain possible.

A few months before the election in early 1958, an island-wide strike paralysed the cotton industry. The workers returned only after being promised a second commission of enquiry. The commission attributed the sharp decline in cotton production to an extreme shortage of labour brought about by emigration to the United Kingdom and the reluctance of remaining people to work for low wages when remittances were arriving from migrants abroad. Furthermore, those who were willing to work during this period were less afraid to strike as many had some alternative income from outside the island.

While the commission recommended new wage rates of $1.30 (British West Indian) a day for men and $0.90 for women, compared with $0.90 and $0.60 respectively in 1953, this was not sufficient to stem the emigration of workers, which continued at a substantial level until curtailed by the Commonwealth Immigrants Act in 1962. Remittances became increasingly significant during this 'emigration decade' spanning the early 1950s to the early 1960s. In 1951, the value of the cotton crop was $620,000 while remittances amounted to $72,400. In 1960, the situation was virtually reversed; remittances totalled $617,000 while income from cotton dropped to $162,000. By 1962, primarily due to the shortage of labour, estate production ceased entirely. Share-cropping had been eliminated three years earlier.

Evidence based on census data, passport applications, and statistics collected by the Migrant Services Division of the West Indian federal government on migration to Britain suggest that between 4000 and 4500 Montserratians moved to Britain during the decade. Between the censuses of 1946 and 1960, there was a net emigration of 5399 people, leaving the island with a resident population of 12,167. Government figures show that between 1955 and 1961 alone 3835 Montserratians arrived in Britain.

Predictably, the vast majority (84.3 per cent) of Montserratian migrants were between the ages of 15 and 49. More surprising is that female migration was at least equal to male. There is a tendency in most migrations for men to send back for women; more important in the Montserratian case, however, are social and economic factors which stimulate women to leave in order to meet obligations owed to relatives left behind. They do not all migrate simply to join men who precede them.

From 1959, the considerable number of children applying for passports reflected a tendency to bring children to England after the adults became established. Other characteristics such as education, occupational skills, or land ownership, did not act as important selective factors in the migration to Britain. Demographically, the British migration took much of the working-age population out of the island; the old and the young were left behind.

The Incorporation of Migration into the Island's Social Fabric

The incorporation of migration into Montserratian society and culture, which had been ongoing over the preceding century, became a full-scale 'way of life' with the massive movement to Britain. The story is a similar one to that told by Margaret Byron for Nevis in Chapter 6. Some social institutions, notably family and household organization, were adaptive and may even have impelled migration. On the other hand, the island's class structure was altered somewhat by migration while the agrarian structure, especially plantation agriculture, was totally changed.

Household organization and migration

West Indian family and household organization is a controversial subject which has spawned a voluminous literature (see R.T. Smith, 1963; M.G. Smith, 1966). I have myself discussed Montserratian village-level family structure and domestic organization in the 1960s in considerable detail (Philpott, 1973). Here I will touch only on those aspects of family organization which directly impel or are impinged upon by migration.

Three types of marital union are practised in Montserrat: extra-residential unions, consensual cohabitation, and legal marriage. Throughout the course of their lives, couples may move from one type to another and most eventually marry legally with a religious service. The type of union has implications for migration.

Extra-residential unions are those in which both partners are normally resident members of different household groups. Such unions take place most commonly between a single man and a single woman; less frequently, between a married man and a single woman. The majority of children born in any given year are to parents in extra-residential unions. These children are normally incorporated into the mother's natal household with the mother's mother or her mother-surrogate acting as the 'social mother' while the biological mother is often treated, behaviourally and terminologically, more as an elder sibling or an aunt of the child. This practice contains a latent conflict. The daughter's motherhood is never entirely negated or unrecognized either in the household or the community. As motherhood is probably the most esteemed social role open to lower-class women, a daughter in this position may increasingly move to perform the role more fully. Furthermore, while both legal and social norms indicate that a father should continue to provide support for his children even after an extra-residential union has terminated, the sanctions are weak and male contributions in such cases are often sporadic or non-existent. Consequently, the need and desire of the biological mother to assume the full parental role, including its economic aspect, underlies the high rate of female migration.

Although *legal marriage* is not regarded as a prerequisite to child-bearing, it is connected with the attainment of a certain socio-economic standing, a demonstration of 'ambition'. A husband is expected to provide a house and support for his wife and children, and sometimes her children by other men that she may bring to the union with her. In addition, a marriage must be marked by relatively expensive ceremonies in the church and in the community. For most, as already suggested, migration is the main means of achieving the social and economic requisites of marriage. In addition, a considerable number of younger migrants have married in England. Unlike many West Indian societies, *consensual cohabitation* ('common-law marriage') is the least significant form of union in Montserrat.

Mobilization of passage money

Montserratian migration to distant labour centres calls for the use of commercial transport, very costly in terms of lower-class earning power. Consequently, accumulation of the necessary fare constitutes an important problem for prospective migrants. For example, the cheapest passage by boat to England in 1954 was $312; in the same year male estate workers in Montserrat were earning only $0.90 a day for a few months of the year. In the overwhelming majority of cases, passage money was mobilized on the basis of domestic group or kinship relations.

A quite typical case is that of a migrant who made it possible for 22 people – his children, siblings, sibling's children, and their actual or intended spouses – to migrate to London over a twelve-year period. The enduring quality of most of the relations involved was rooted in earlier shared membership in the same household group. For example, he brought over his three full sisters who had been reared in the same household with him; he did not help any of his father's 29 'outside' children who had been raised in other households.

The principal sources of passage money, in order of importance, were siblings, parents, parent's siblings, spouses, cousins reared in the same domestic group, intended spouses, and personal income (Philpott, 1973, pp. 132–3). It has already been suggested that one migration, to some degree, finances the next. A considerable number of initial passages to Britain, for example, were paid for by fathers or siblings absent in Curaçao and Aruba. Furthermore, on the basis of lasting sibling ties, it is not uncommon for Montserratians who migrated to the United States many years earlier to pay the passage and act as legal sponsors for siblings' children whom they have never seen.

Fostering of children

As many migrants have young children, the care of these dependents during their absence is another important domestic problem. Fostering of children, however, is not unusual anywhere in the West Indies. Due to the variable marriage careers of their parents and the changing contingencies of closely-related households within communities, children frequently are members of several household groups before they reach adolescence. The fostering of migrants' children, then, is an accentuation of prevailing practice.

During my earlier fieldwork, approximately one third of the children under 16 lived in household groups in which neither of their parents was a resident member (Philpott, 1968). Two-thirds of these children were left with maternal kin and most of the rest with paternal kin. Grandmothers were the single most important category of fosterers, mothers' mothers caring for approximately 47 per cent of the children and fathers' mothers for 18 per cent.

There is a transactional nature to such fostering arrangements. In the urban labour centres to which most migration takes place, a child represents an economic liability preventing the parents from realizing their full wage-earning potential. Consequently, most migrants leave their children in the island, believing it is a cheaper and healthier place to raise children. They do not usually pay their child's passage abroad until they are old enough to enter the urban work force. Some migrants have even sent their foreign-born children back to the island for similar reasons.

In terms of day-to-day household activities in Montserrat, on the other hand, children are indispensable. They care for the goats and sheep, gather firewood, carry water, climb trees for mangoes and breadfruit, and run numerous errands to the shops and post office. In addition, girls launder, cook meals and look after younger children. This fostering process is infused with the morality which stresses 'helping the family'. Older people are well aware that freeing a daughter or son to migrate by caring for their children provides the best assurance of financial security for all concerned.

The migrant, in turn, is expected to meet what I have termed 'remittance obligations' (Philpott, 1968).

Remittance obligations

The basic principles of these obligations are quite simple. The migrant is expected to send money ('breaks') and clothing to his mother or mother-substitute. To a much lesser extent, and depending on the nature and emotional content of the pre-existing relationship, the migrant is expected to do the same for his father. A migrant may also send occasional 'breaks' for close kin, mainly siblings reared in the same household, or attempt to finance their passage should they wish to migrate. Finally, migrants should send support for their children left on the island. As a female migrant usually leaves her children with her mother, she is under dual obligation to remit money to her mother's household. Consequently, some grandmothers, realizing their potential control over the migrant may be diminished, are reluctant to give up the grandchildren if and when the mother sends for them.

Children are implicitly taught these expectations in the home and the community through the praise of migrants who 'send a good break' and through the condemnation of the 'worthless-minded' kin who do not 'notice their families'. A migrant who reputedly returned from America with 32 trunkloads of gifts for distribution to his family and friends, and a woman who sent her brother a car from America in the days when the only other cars were owned by the wealthiest estate owners, have become near-legendary figures. Furthermore, when children collect the mail at the local post office, the excitement over receiving a registered letter containing money or the disappointment of receiving nothing continually reminds them of the ideal migrant behaviour.

Moreover, the significance of migration is given ceremonial expression at the community level. Aside from the *rites de passage* at birth, marriage and death, the only major ceremonial occasions are those connected with migration – namely feasts when leaving, when returning, or when marrying while away from the island. While such feasts have a religious aspect, they are clearly a form of status validation, sometimes locally referred to as 'signing the progress'. The feast at departure marks a major turning-point in the migrant's life.

When a migrant marries while away from the island it is his or her 'duty' to send home money to 'keep up the feeding'. In cases where both the bride and groom are Montserratian (and this is usually so), the money is sent to both the bride's family and the groom's family for separate feasts. This obligation appears to be generally met. In thus demonstrating that the home community is a group before which the change in status must be validated, the continuing nature of the migrant's ties to the community is also implicitly expressed. The biggest feasts are usually held when a migrant returns from a long period of migration. They mark, as it were, the migrants' reintegration into the community; their success and prestige.

The homeward flow of remittances and clothing is symbolically reciprocated by the people in Montserrat through the periodic sending of parcels of cassava bread, various weeds and bush leaves used to make tea, salt pork and occasionally rum. Ultimately, the continuing interrelationship between migrant and home community is expressed through the exchange of goods and services.

Migration and class structure

Up to the present day Montserrat manifests a modified version of the 'colour-class' system which emerged after emancipation. The *upper class* consists of the resident owners or managers of larger businesses, expatriate colonial officials, professionals, expatriate religious officials, bank managers and the larger merchants. Most of the people in this small stratum are white or light-skinned ('coloured'). The *middle class* is primarily made up of the civil servants who staff the post office, the public works department, the hospital, the courts and police, and the education department. Bank employees and some shopkeepers are also part of the middle class. Phenotypically, the middle class is entirely coloured or black. The *lower class*, primarily black, comprises approximately 90 per cent of the island's population. Most do not have a regular job, but tend to work at sporadic wage labour, subsistence agriculture and some at commercial fishing. Because of the small scale of Montserratian society and the high degree of personal interaction, 'class lines' are more fluid than in many Caribbean societies.

Lower-class black mobility into the middle class seems to be connected with the attrition of the island's coloured population through selective migration, particularly to the United States. The primary requisite for social mobility among lower-class Montserratian blacks was the attainment of secondary or higher education. Yet the cost of fees, uniforms, supplies and incidentals was prohibitive for lower-class students; for many years after the secondary school opened it was virtually the preserve of the middle and upper classes. However, remittances from migrants in Curaçao and Aruba, and scholarships established by a Montserratian migrants' organization in the United States, gradually made such education more accessible. Since the 1960s, most secondary school students have been drawn from the lower class and are often financed by remittances from Britain.

Although the number of middle-class jobs has increased over the past 25 years, these positions have been easily filled by secondary school graduates. The social expectations pertaining to the work an educated person can and should do channel most graduates into the civil service, teaching, banking and commerce. Thus, while secondary education has permitted upward mobility for some in the island's social hierarchy, for most it is a form of 'anticipatory socialization' that prepares people for migration and work outside the island.

Hence migration has had a significant impact on the socio-economic hierarchies of the island and has promoted a certain degree of upward mobility. At the same time, changes in political, economic and educational processes have stimulated further migration.

Statistically, recent migration has been largely a lower-class phenomenon; yet class has not acted as an important selective factor. Considerable migration has taken place at all social levels. Consequently, if class distinctions are to be made with regard to migration, they can only be based on *migrant ideology*, defined as 'the cognitive model which the migrant holds [regarding] the nature and goals of his migration' (Philpott, 1968, p. 474).

Lower-class workers perceive migration initially as a temporary phase, mainly to gain money and improve their positions *vis-à-vis* a local island group. Expectations about 'helping the family' with remittances and about the care of migrants' children are also encompassed. In the upper class, migration is more likely to be viewed as permanent, or very long-term, made on the basis of universalistic standards of professional or occupational advancement. No particular expectations about remittances or child care are held. Middle-class migrant ideology incorporates elements of both.

Obviously, in any particular case, this polarized abstraction is not meant to apply without qualification. Few upper-class Montserratian migrants fail to take some account of their attainment relative to the island society; nor do lower-class migrants totally ignore supra-local standards regarding adequate wage levels or working conditions. The abstraction, however, points up some rather important differences in class conceptions about migration.

There are two potential avenues of social and economic advance for the lower class: education and migration. Secondary education is expensive in Montserrat. Consequently, the most feasible and likely alternative for the lower-class islander is to migrate with the hope of ultimately returning to the island a richer person. While some lower-class migrants manage to acquire enough skill, education or money while abroad to gain entrance to the middle class upon their return to the island, this is rare. However, returned migrants are generally accorded higher status in their local communities than they enjoyed prior to migrating.

References to a man's migration – even when referring to middle-aged married men – are often couched in terms which imply the attainment of adulthood: 'He went out to make himself a man' or 'I was only half a man before I went out'. The main reason for returned migrants' enhanced social standing is their assumed affluence but they also enjoy greater esteem merely by virtue of having had a socially valued experience; i.e. having 'been out', particularly to England or the United States.

While migration experience is valued in itself, a permanent upward alteration of the returned migrant's standing in the community is based largely on the amount of money he brings home. Usually such wealth is manifested in new or improved houses, running water, flush toilets, indoor kitchens, refrigerators, radios, household furnishings and other acquisitions with which returned migrants have consistently altered the material culture of their home areas.

Yet, for the most part, returned migrants produce very little social or economic innovation and, indeed, the potential for such innovation seems very limited. Returned migrants cannot return to the ordinary labouring jobs which they held before migrating without loss of newly-enhanced status. Consequently a suitable capital investment is necessary to maintain a relatively independent position. Such investments tend to be culturally defined; notably rum shops, vans or trucks for use as buses, and cattle. As the demand for rum and bus services has been less than the supply in recent years and pasture land is very scarce, returned migrants, the would-be entrepreneurs, are increasingly forced to re-emigrate.

Agrarian Transformation and Residential Tourism

At the community level, the most immediate and obvious economic effects of migration to Britain were the total collapse of plantation (estate) agricultural production, a major decline in peasant (smallholder) market and subsistence agriculture, and an increased dependence on cash income, initially largely from remittances. The reduced emphasis on agriculture manifested a general attitude that 'working with the hoe' was hard, unprofitable, unprestigious, and to be avoided if at all possible. Cotton production was particularly unpredictable and, consequently, householders who received moderate remittances or other income were reluctant to risk losing it in expenses for wages, land rental, fertilizer and seed.

The difficulty of obtaining agricultural labour was, of course, another factor. While property development and government road work created some demand for male labour in the mid-1960s, many men were underemployed. Employment prospects were worse for women. Yet small producers complained that they could not hire people to help them with their crops. They contended that, since the migration to Britain, many of those remaining behind were 'just lazy'. Such views were exaggerated but intermittent remittances and sporadic wage labour made it possible for younger men in particular to shun agricultural labour.

Commoditization and the decline of reciprocal labour exchanges

One of the more crucial consequences of greatly reduced agricultural production and increased dependence on cash income was the related decline in *reciprocal labour exchanges*. Prior to the collapse of the estates, landless men and women earned part of their income as agricultural wage labourers and part as share-croppers. While the estate owners and managers obtained labour through asymmetric cash transactions, the share-croppers co-operated with fellow villagers in reciprocal labour exchanges, locally called 'maroons'.

In a maroon, groups of 5 to 30 men or women, depending on the size and nature of the task to be performed, gathered at the land of the holder to prepare the ground, weed, or harvest the crop. While the organizer of a maroon was expected to 'give back the day' in similar labour for each of those who participated, the maroon also had aspects of a fête. Most householders were directly involved in share-cropping approximately equal amounts of land and stood in a relatively equal socio-economic position *vis-à-vis* each other. As the men and women of most households were involved in different, and sometimes two or more, maroon groups, this institution created a considerable degree of economic interdependence between households within the community. Since the end of estate agriculture, maroons have almost disappeared. Inter-household dependence decreased as more and more households dropped out of agricultural production, depending instead on remittances or other sources of income. With the increasing emphasis on cash purchases, labourers would work only for wages.

Real estate development and 'residential tourism'

After the collapse of the estates and the loss of the public revenues they generated, the local government began to encourage their sale as housing plots for wealthy expatriates from the United Kingdom, the United States and Canada to build retirement or winter-time residences. The impetus from the government side came from the Chief Minister, W.H. Bramble, who had come to prominence in the 1950s as the populist trade union leader who had led the opposition to the estate owners. A deal was struck between the government and the development companies in 1960 and the Montserrat Real Estate Company, located in New York and the largest of the three participating companies, opened a local office in Plymouth in May 1961. Two estates on the western side of the island and one on the eastern side were subsequently subdivided into housing lots under this scheme.

Over the years, some 2000 lots were sold and approximately 400 houses built. While most of the lots were sold to expatriates, some were also bought by local middle-class people and some by migrants overseas. The housing development created a construction boom and a great demand for local carpenters, plumbers, electricians, other tradespeople, and even for unskilled labour. The construction of these houses also introduced a new segment into the local population, i.e. the expatriates, who came to reside on the island permanently or for several months each year. They could be called 'residential tourists'. As well as contributing cash income to the local economy, these expatriates would subsequently come to play a significant role in the post-volcano debates over identity, redevelopment and the future of the island. All of the housing estates have been evacuated, although only Spanish Pointe, on the eastern side, has suffered actual destruction of houses.

At the same time as this housing development was booming, there was also considerable economic retrenchment in Britain and, to a lesser extent, the United States and Canada. Hence, many migrants overseas began to build houses on the estates, or to build new or renovate older houses in the villages they had left behind. Ironically, the very retrenchment which was undermining the migratory system actually fuelled the local economy. In turn, other residents in the villages who had family members working in construction began to up-grade their village houses using the standards of the housing estates as a model. Thus, for the past 25 years, the main product of the island has been 'growing homes' and, despite a very precarious economic situation, an appearance of material prosperity came to pervade the island.

The impact of Hurricane Hugo

In September of 1989 another natural disaster – Hurricane Hugo – hit the island. More than 90 per cent of the houses were either destroyed or seriously damaged by the storm, although amazingly there were almost no fatalities. According to most local accounts, there was barely a leaf left on this normally lush green island. Agriculture was further devastated and international emergency aid programmes were instituted. However, within a period of months not only the vegetation began to return but the housing economy, given the further demand for repair and rebuilding, became hotter than ever. Insurance, relief and aid monies poured into the island. Construction workers were even brought in from other Caribbean islands. Residents came to complain that the narrow island roads were becoming clogged with newly-purchased Japanese cars. Based on the greed and selfishness manifested by a few islanders during Hurricane Hugo, some later saw the volcanic eruptions as divine retribution.

The aftermath of the volcano

For Montserratians at home and abroad, the aftermath of the recent major volcanic events has been marked with confusion over whether the island was to be redeveloped with a new 'capital' in the north or totally evacuated. Anger and frustration directed at the volcano shifted to dissatisfaction with the efforts of local leaders and, in particular, the British government.

In early August 1997, Britain's International Development Secretary, Clare Short, announced in London that the Montserratians who had left the island 'might well return when they know what our commitment is to the north'. However, by mid-August, a scientific report of a trend to 'more violent and hazardous' volcanic behaviour prompted the British government to announce the possibility of evacuating the remaining population. The destroyer HMS Liverpool was sent to stand by and a financial offer of £2500 per adult was made to relocate people to neighbouring islands or Britain. This offer was much less than what had been requested and, consequently, local Chief Minister Bertrand Osborne resigned in the ensuing outrage and another member of the island council, David Brandt, took his place. At the same time, Clare Short accused the Montserrat leaders of 'hysterical scaremongering' in their requests for more help from London and suggested they would be asking for 'golden elephants next'.

With the prospect of total evacuation looming, Montserrat's plight attracted worldwide media attention and a number of major newspapers and wire-services sent correspondents to the island for extended stays. Around the same time, international fund-raising efforts were undertaken, most notably a benefit concert in September at the Royal Albert Hall. This was organized by the Beatles' producer, Sir George Martin (who also operated a recording studio in Montserrat), and included music stars Elton John, Paul McCartney and Eric Clapton. When much of this high-profile attention, and a subsequent parliamentary report, focused on the shabby treatment Monserrat was receiving, the British Prime Minister, Tony Blair, undertook to deal with the Montserratian situation himself and essentially repudiated his own minister. An improved relocation scheme was worked out for Montserratians wishing to leave the island, and, at the same time, promises were made about monies for redevelopment. Furthermore, in February 1998, largely as a result of the Montserrat crisis, the British government proposed to offer full passport and citizenship rights to all inhabitants of the remaining 'dependent territories'.

Within the island, further controversies emerged after insurance companies cancelled existing policies or refused to issue new insurance. There were also some complaints of looting expressed on the Evergreen Network, an e-mail news group which was started by an expatriate American doctor to disseminate up-to-the-minute news of the volcano and other Montserrat matters. The insurance and looting issues were particularly pertinent to the expatriates with large homes on the housing estates, many of whom had invested their life savings into their properties. Local home-owners were also affected, although many of the houses were uninsured and totally lost.

The bulk (more than 4000) of the evacuees went to Antigua, St Kitts, Nevis, and the Virgin Islands. After the relocation package was announced, many more went on to England, usually to join relatives already there. This relocation of Montserratians to Britain included many older people for whom the shift to the cold climate, and often unfamiliar society, was extremely traumatic. A not untypical story is that of Thomas Fenton (pseudonym), the islander referred to earlier who had brought over 22 of his kin during the time he worked in a gasworks in London in the 1950s. Fenton had returned to the island in 1963 and was spending his retirement in the village home he had had improved while he was away in England, a retirement supported by his savings and by raising and selling some cattle. When the volcano erupted, his village was one of the first to be evacuated. He and his wife were moved to a shelter in a church in the north of the island where his wife died from the stress of the situation. A few months later, Fenton relocated to London along with his two sons and their families. He first went to live with his sister whom he had taken to England more than 30 years earlier. But she died while he was staying with her. 'Since then, I have been moved from pillar to post. I have changed addresses five times ... My friend, I can tell you life is not easy. I can say death is better than the problems I am going through.' While a friend is trying to get him into a home for the elderly, he is unable to cook and do his own washing; he is dependent on his sister and a grand-daughter to come and perform these domestic duties for him.

A similar sad story emerges from the death of an 86-year-old Montserratian woman, Mary Galloway (pseudonym), who had been evacuated to England in November 1997. Her last days were particularly difficult. In Montserrat, she had been involved in an accident and had one of her legs amputated. When she arrived in England, she was in need of constant nursing attention which her family was unable to offer her. In England she was entitled to free nursing care, but social services could not find a place for her near her family in London. Instead, she was placed in a home in Durham, 450 km away. Her children visited her there when they could, but the journey was time-consuming and expensive. Alone in a strange country, Mary went downhill rapidly. As another instance of Montserratians coming together, more than 200 people attended her funeral in London.

Global Culture: Island Identity

The political contestation in the wake of the volcanic crisis and the everyday struggles of so many Montserratians to rebuild their lives and reconstitute their social world have also brought issues of culture and identity to the fore. This heightened consciousness is manifested in various ways – but most explicitly in the e-mail and Internet groups which have emerged since the emergency began. Facing whether there would continue to be 'a Montserrat', in effect, led to asking: What does it mean to be Montserratian? What is Montserratian culture?

Social identity and the sense of place

Throughout the various ebbs and flows of migration, Montserratian society has progressively become more transnational and Montserratian culture has become increasingly globalized. However, as the discussion above shows, the island has remained a strong focal point for material and symbolic relations and the islanders, even in the diaspora, have maintained a strong sense of place. Until recently this sense of place has not generally been manifested in an explicit 'island identity' and certainly not in the form of 'identity politics'. For Montserratians, as with any human group, identity is and has been related to particular contexts, the principal of which I will discuss briefly here.

Intra-Caribbean identity

In the 1960s, during my research in Montserrat, any sense of Montserratian identity tended to be implicitly expressed in relation to or in opposition to other West Indian islanders. The positive features of Montserratian identity were unstated but were implied through often negative or stereotypical characterizations of other islanders, usually in situations of inter-island labour migration, temporary residence, or marriage. This construction of identity cognitively seemed to radiate out from Montserratians to Nevisians, who were seen as being most like Montserratians; then on to the Antiguans who, while close physically and culturally, nevertheless, were often said to be 'too fast' or 'too trixified' (too cunning, devious). Among English-speaking islanders, furthest from Montserratians were the Jamaicans who were often depicted quite negatively (Philpott, 1973, p. 171), based largely on stories from migrants overseas as contact within the Caribbean was extremely rare. The 'Big Islander/Small Islander' distinction, which bisects the West Indies, further reinforces mutual negative stereotyping.

Spanish-speaking or French-speaking islanders were beyond the cultural understanding or social experience of most village-level Montserratians. Any construction of identity *vis-à-vis* these 'foreign' islanders came largely out of migratory experience. For example, some older Montserratians who had worked as cane-cutters in Cuba in earlier days claimed Cubans, Puerto Ricans and other Spanish-speaking islanders were 'too quick with the knife'. These inter-island constructions of identity, however, were intersected by class and even local differentiation within the island itself as well as in the migrant context.

The international situation

In the context of international migration to Britain and the United States, the inter-island distinctions became much less important. In the 'host' societies to which Montserratians migrated, the islanders were not recognized in a particularistic way based on island origin. Rather, in official and popular discourses in these societies, Montserratians were usually lumped together, 'racialized', with other West Indians, Africans and even South Asians.

Thus, until recently, there has been almost no manifestation of an explicit (and political) Montserratian identity in the international context. True, in the early days of migration, named island-based voluntary associations were organized in the United States, Britain and later in Canada. However, as I have extensively documented elsewhere (Philpott, 1973 and 1977), the primary basis for the emergence of an explicit Montserratian identity lay in the continuing interaction and communication among Montserratians in the various host societies. Particularly in the British migration, Montserratian villagers largely reconstituted their former social world in the context of the new urban environment. For example, migrants from Tuitt's, a now-destroyed village on the eastern side of the island, who moved to and continue to live mainly in Hackney and Stoke Newington in London, often referred to Birmingham as 'Long Ground' – because that is where most of the migrants from a nearby village of that name in Montserrat (the first one to be inundated with mud from the volcano) lived.

Social relations based on island background, particularly in non-work contexts, were very important. Kin and friendship ties are continued through residential proximity and reinforced by a strong tendency towards island endogamy among migrants who married in England. This tendency both reflected the migrants' 'return ideology' and commitment to Montserrat and recognized the potential for disagreement, in an inter-island marriage, over which partner's island the couple would return to – if and when they returned to the West Indies. Montserratian migrants participate together in many church groups, in rotating credit associations ('boxes'), and in weddings, Christmas parties and other celebrations. They often perform personal services such as hairdressing, construction work and car repair for each other on the basis of reciprocity. These functional and adaptive social networks provided the potential underpinning for a transnational Montserratian society and an emergent Montserratian identity.

Discourses of the diaspora

Thus, until recently, there was no explicit invocation of Montserratian identity in the context of international migration. Rather, 'the fact that people were being blocked out of and refused an identity and identification within the majority nation' (Hall, 1991, p. 52) led many to the adoption of 'black' identity as an act of emergent political consciousness in racially structured societies such as Britain or the United States. The children of Montserratians born in Britain are often deeply involved with the diaspora culture currently being articulated in Britain (Gilroy, 1987, 1993). This culture is concerned with 'the struggle for different ways to be "British", ways to stay and be different, to be British and something else complexly related to Africa and the Americas, to shared histories of enslavement, racist subordination, cultural survival, hybridization, resistance, and political rebellion' (Clifford, 1994, p. 308).

In August 1997, a memorial service for victims of the Montserrat volcano was held in Toronto by Montserratians living there. A number of islanders spoke movingly of the dead, expressed anger at the havoc the volcano was wreaking on their 'dream' of return to the island, and unanimously affirmed their faith that they would retire there nevertheless. When a local politician, in the midst of urging Canadian government aid for Montserrat, referred explicitly to 'the Montserratian diaspora', many in the audience appeared somewhat mystified by this reference. Yet, since the volcanic crisis, Montserratians are clearly engaged in a 'diasporic discourse' (Clifford, 1994, p. 310). In such memorial services and other social gatherings held in Britain, the United States, Canada and many Caribbean islands, as well as in more overt political actions, there have been increasing articulations of Montserratian identity and attempts to define a distinctive Montserratian community in the context of disruption and displacement. This new articulation has been particularly prominent in 'cyberspace'. The e-mail news group has already been mentioned, several web sites focus on the volcanic activity, and, in early 1998, the Emerald Network, a web site for the 'international Montserratian community' was launched by a Montserratian in Boston.

External ly, the particularity of the volcanic disaster, the media and celebrity attention, and the necessity for the British government and the aid agencies to respond specifically to the 'Montserrat crisis', have promoted wider recognition of a Montserratian identity. For instance, one Montserratian who has worked in an auto parts factory in Toronto for many years, told me that because of all the television coverage, his fellow workers tell him they now know who he is: 'before they had never heard of Montserrat'. The increasing focus on Montserratian identity has, in turn, led to attempts to articulate Montserrat's 'culture'.

The crystallization of culture

The very practice which has contributed to the maintenance of the island as a viable society – migration – has also contributed to the transition of the distinctive local culture toward 'global culture, locally expressed' (Olwig, 1993). As already indicated, housing and other aspects of material culture tend to reflect various waves of migration. Furthermore, with the decline of local agriculture and increasing reliance on cash purchases, the local cuisine has increasingly shifted to the use of imported food such as frozen chicken. These food tastes are partly formed by the migrants who bring them back from the US and Britain, but also because of decreased availability of local foods and provisions.

Key rituals such as the traditional Montserrat wake and funeral have been influenced as well. Even in the 1960s, the village practice at the time of death was that all members of the immediate family were responsible for preparing the dead person for burial within 24 hours. Women had to sew special clothing for the body and prepare special foods for the wake before the burial and for a breakfast the following morning. Men were responsible for arranging to have a coffin built and to have the grave dug. During the wake throughout the night, hymn-singing (the 'sacred') was alternated with vulgar and ribald games (the 'profane'). By 1987, when I re-visited Montserrat, village funerals had changed dramatically. The body was kept refrigerated for a number of days. Professional undertakers were employed, the traditional wake was no longer held, and the type of food eaten after the funeral was no longer the specialized food of earlier times. These changes in funerary practice were mainly due to more general cultural trends but, with regard to migration, they made it possible for relatives spread throughout the western world to return home for the ceremony.

As the threat to an immediate end to the 'place' of Montserrat receded somewhat, concerns were raised (via the e-mail medium) that the 'culture' of Montserrat was also in danger. 'The culture of Montserrat is what make Montserrat what it is,' one correspondent wrote; 'We lose our culture, and we lose our identity' (cf. Olwig and Hastrup, 1997, p. 11). Interestingly, the perceived cultural loss cannot be attributed directly to the volcano but rather to globalizing or regionalizing trends which have been ongoing throughout the Caribbean for some time. The threatened culture tends to be seen, in an almost folkloric way, as an assemblage of disappearing or defunct cultural practices: Miss Goosie, a stilted masked figure in the New Year's Day parade; String Band, the 'old time' music; the Jumbie Table, the table set for the spirits on festive occasions; and, of course, Goat Water, the island's still highly-valued festive kid stew.

Another instance is the Jumbie Dance, a distinctive Montserratian spirit possession ritual similar to other syncretic Afro-Caribbean religious prac-tices such as voodoo or shango. In this dance, usually held at weddings or in times of serious illness and always accompanied by a fête, dancers entered trance and voiced messages from ancestral spirits which acted as protection or cures. In the 1960s, the religious authorities on the island believed this then-denigrated practice had disappeared but I was witness to its continuing vitality at the village level. By 1987, in one of my return visits to the island, I found the jumbie dance had been reduced to an occasional commercial performance for tourists at a local hotel. One day I visited one of the remaining master musicians of the jumbie dance in his small cabin in a village. I asked him about playing the old-time music and he rather grate-fully got out his big 'woo-woo' drum which had been used to call forth the spirits in the middle of the dance. As he played for me, his aged wife began to dance and then to enter trance. Precisely at this moment, a young man in the neighbouring house cranked up his 40-watt stereo playing a current reg-gae and, symbolically and actually, put an end to the jumbie dance.

Other correspondents viewed culture in a more dynamic way.

'Remember that culture is people not a piece of merchandise,' e-mailed one Montserratian; 'It will remain or disappear only if the people want or do not want it'. Furthermore, contemporary cultural production has been flourishing with new calypsos and socas by local musicians chronicling the volcanic events and their aftermath.

Whither Montserrat?

Today, in the wake of the volcanic disasters and disruptions, Montserratians are more dispersed than at any time in their long migratory history. For some who have lost everything, the emotional and financial costs may prove to be too high to hold on to their 'dream' of this little Caribbean island. For many – perhaps most – others, left in the island and scattered throughout the world, there is more explicit invocation of island identity and concern with the integrity of Montserratian 'culture' than ever before. At the very time when many anthropologists are abandoning anthropology's 'localizing strategies' and tendency to unequivocally asociate 'place' and 'culture' (cf. Gupta and Ferguson, 1992, 1997), Montserratians are beginning to articulate such associations.

This countervailing tendency, I suggest, emerges from the current political context which impels Montserratians, as citizens of a 'dependent territory' of Great Britain, to make political claims. As the volcanic activity and immediate crisis subside somewhat, questions of resettlement and redevelopment have already arisen. The British government has promised substantial sums of money for building a new 'capital' in the north of the island and for other undefined development projects. The question of who will benefit from future development looms large. Debate and potential conflict over the specific material and social needs of residents and expatriates are beginning to emerge in the political arena. In addition to abstract notions of 'citizenship', issues of race, cultural belonging, and self-identify may play into the legitimation of claims in this debate and in the way development will be negotiated by different interests.

The concern with culture and identity, however, is not simply a calculated political strategy. It is a powerful emotional force embedded in the personal relationships which have so long sustained islanders on the island and abroad who have been deeply threatened by the recent natural disaster. The island – perhaps all islands – serves as a potent symbolic focal point. The vigorous efforts of Montserratians to hold on to or reconstitute the place, culture and identity of Montserrat ensure that the island will continue to be a socially meaningful space in an increasingly globalized, deterritorialized world.

References

Akenson, D.H. (1997) *If the Irish Ran the World: Montserrat, 1630–1730*. Montreal: McGill-Queen's University Press.

Clifford, J. (1994) 'Diasporas', *Cultural Anthropology*, 9(3), pp. 302–38.

Fergus, H.A. (1994) *Montserrat: History of a Caribbean Colony*. Basingstoke: Macmillan.

Gilroy, P. (1987) *There Ain't No Black in the Union Jack: The Cultural Politics of Race and Nation*. London: Hutchinson.

Gilroy, P. (1993) *The Black Atlantic: Double Consciousness and Modernity*. Cambridge, MA: Harvard University Press.

Gipson, H.L. (1960) *The British Isles and the American Colonies: The Southern Plantations, 1748–1754*. New York: Knopf.

Goveia, E.V. (1965) *Slave Society in the British Leeward Islands at the End of the Eighteeenth Century*. New Haven: Yale University Press.

Gupta, A. and Ferguson, J. (1992) 'Beyond "culture": space, identity, and the politics of difference', *Cultural Anthropology*, 7(1), pp. 6–23.

Gupta, A. and Ferguson, J. (eds) (1997) *Anthropological Locations: Boundaries and Grounds of a Field Science*. Berkeley: University of California Press.

Hall, S. (1991) 'Old and new identities, old and new ethnicities', in A.D. King (ed.), *Culture, Globalization and the World System: Contemporary Conditions for the Representation of Identity*. Binghampton: State University of New York, pp. 41–68.

Olwig, K.F. (1993) *Global Culture, Island Identity: Continuity and Change in the Afro-Caribbean Community of Nevis*. Chur, Switzerland: Harwood.

Olwig, K.F. and Hastrup, K. (eds) (1997) *Siting Culture: The Shifting Anthropological Object*. London: Routledge.

Philpott, S.B. (1968) 'Remittance obligations, social networks, and choice among Montserratian migrants in Britain', *Man*, 1(4), pp. 465–76.

Philpott, S.B. (1973) *West Indian Migration: The Montserrat Case*. London: Athlone Press.

Philpott, S.B. (1977) 'The Montserratians: migration dependency and the maintenance of island ties in England', in J.L. Watson (ed.), *Between Two Cultures: Migrants and Minorities in Britain*. Oxford: Blackwell, pp. 90–119.

Smith, M.G. (1962) *West Indian Family Structure*. Seattle: University of Washington Press.

Smith, R.T. (1963) 'Culture and social structure in the Caribbean: some recent work on family and kinship studies', *Comparative Studies in Society and History*, 6(1), pp. 24–46.

8

Caribbean Migration: Motivation and Choice of Destination in West Indian Literature

Brian Hudson

'Why are you running away from your country?'
'Why do so many of your people come here?'
The Emigrants, George Lamming (1980)

Caribbean Migration Research and Creative Literature

Migration within, between and from the islands of the Caribbean has long been a topic of systematic investigation by social scientists, including sociologists and geographers (Duany, 1994; Richardson, 1992). These studies have been concerned mainly with a limited range of issues such as the characteristics of the migrants and of the places of origin and destination, and the political, social and economic factors that encourage migration. Much of this work has been at an 'objective' level, tending to rely largely on the abundant statistical data sources compiled by bureaucratic agencies. Some scholars, on the other hand, have sought to explore the Caribbean migration experience at the individual level. Among these are George Gmelch (1992) and John Western (1992), both of whom made studies of Barbadian migrants based on oral histories obtained by personal interviews. Western augmented his discourse by drawing on the creative works of Caribbean writers, including V.S. Naipaul, Austin Clarke and Derek Walcott, as well as those of other authors, notably Virginia Woolf and Salman Rushdie. In a study which seeks to 'explain' Caribbean migration, Elizabeth Thomas-Hope (1992) quotes extensively from Braithwaite's poem, *The Emigrants*, but these verses are confined to the Introduction and Conclusion of the book. Thomas-Hope draws her data from 'in-depth' interviews and surveys conducted in Jamaica, Barbados and St Vincent, but her book explores the topic of migration in theoretical terms rather than at the individual level.

More recently, some scholars have made more specific use of creative literature to examine the phenomenon of Caribbean migration. Claire Alexander (1995) focuses on two novels by the Barbadian writer George Lamming to explore the themes of exile, 'otherness' and arrival, while in another chapter of the same collection, Robert Aldrich (1995) analyses the works of francophone writers, particularly those of Martinique and Guadeloupe, to demonstrate the way in which the French Antillean literary tradition is an expression of the migrant experience. Much of the literature of the anglophone West Indies can be seen as a similar expression of the migrant experience, and it is because of this that there is more than sufficient material for a study of the kind presented in this essay. As Cobham (1979, p. 29) observes, 'In spite of their diversity, West Indian writers are held together by a background of trauma and promise which they share with the entire Caribbean'.

This chapter, like that by Aldrich (1995), is a contribution to the study of 'a full body of literature that arguably hangs together through a relationship with a migratory record or history' (White, 1995, p. 2), in this case focusing on writers from the anglophone Caribbean. The purpose, however, is more akin to that of Thomas-Hope's (1992) book, for it seeks to throw light on the migrants' motives for leaving home and the factors which influence their choice of destination. The method is one which may be seen to complement that of Thomas-Hope, being based on the belief that 'Literature, though it may also be many other things, is social evidence and testimony' (Coser, 1963, p. 2).

In her study, Thomas-Hope discusses several theoretical perspectives on Caribbean migration, including the economic-materialist, the societal and the political standpoints, but she emphasizes that it is necessary to focus less on specific events or negative factors in the objective environment and instead to seek explanation in the context of prevailing conditions of life and livelihood in the region and the localities from which the migrants come. She argues that historical-structural factors manifest in global inequalities, societal tensions between individuals and the state, as well as between source and destination countries '... are significant in the decision-making processes relating to migration through their impact on the nature of images which are formed in the minds of potential migrants' (Thomas-Hope, 1992, p. 22). In the words of three West Indian writers, the Trinidadian 'yearned after the outside world' (Naipaul, 1961, p. 186), the Barbadian had a 'desperate determination to leave the small-island country' (Clarke, 1980, p. 70), and for the Jamaican 'it became imperative to go to England, to bring to fulfilment these experiences which began at school' (Gladwell, 1969, p. 35).

The following discussion draws on the rich creative literature of the anglophone Caribbean to throw light on this theme, particularly as it relates to the lives of individuals. The material has been selected from novels, short stories and autobiographies by Caribbean authors, all of whom have shared in the migration experience, to illustrate and illuminate this phenomenon at the intra-island, intra-Caribbean and extra-Caribbean scales. The pieces discussed reflect various aspects of the migration process, including the influence of geographical proximity of destinations and changing opportunities as economic and political developments alter the regional and global context over time. The relationship between class and gender on the one hand, and migrant opportunity and motivation on the other, can also be discerned, while chance events, including vicissitudes in personal relationships, may influence individual decisions. What is clear is that the causes of migration are generally complex, and indeed often are 'not consciously understood by the actors themselves' (Thomas-Hope, 1992, p. 30).

'West Indians have been migration-prone almost from the very beginning of European colonization' (Boswell, 1989, p. 107), but it is with the twentieth century, especially the three decades from 1940 to 1970, that this chapter is mainly concerned. Not only was this a time when large numbers of West Indians emigrated to Europe, the USA and Canada, but it was also the period which saw the flowering of Caribbean literature, itself largely an expression of a newly found national and regional identity, and in no small measure stimulated by the migration experience (King, 1979). Indeed, much of the West Indian literature written in the post-war years was concerned with the experiences of Caribbean migrants, this theme sometimes being explicit in the titles of works such as George Lamming's novel *The Emigrants* (1980) and Wallace Collins's autobiography *Jamaican Migrant* (1965).

While post-war emigration is an important, even dominant, theme in much Caribbean literature, many books make reference to earlier migrations within the region and beyond, some of these movements continuing to this day. The reader gains the impression that migration within and between the islands and to and from more distant places is an established part of everyday West Indian life which over time has itself adjusted to the important influences of these human movements. Indeed, it will be seen that literary evidence strongly supports the view of empirical researchers in the field that migration can be regarded as an integral part of Caribbean culture (Richardson, 1992; Thomas-Hope, 1992).

This chapter examines from a geographical perspective a broad sample of Caribbean creative literature in English in order to understand and explain the migrant experience at the level of the individual. The geographical approach emphasizes the spatial dimension of this migratory phenomenon while, at the same time, exploring the theme of causality in a process of decision-making which has tended to generate distinct waves and patterns of movement over time. Among the major factors which have influenced decisions to migrate and the choice of destinations are the structural roles of colonialism and education which limit and distort people's perceptions of the world and its countries, and foster an economic and mental dependency on the 'mother country'. The account which follows examines migration at a range of spatial scales and oriented to several destinations: intra-island migration; migration within the Caribbean region; and emigration to Britain and North America. Special attention is given to the role of schools and the colonially-imprinted education system. For the precise location of the many islands and territories mentioned in my account, reference can be made to Figure 6.1 in Margaret Byron's Chapter 6 (see p. 116).

Intra-Island Migration

While internal migration, commonly in the form of movement from rural to urban areas, occurs throughout the Caribbean region, it is probably of greater significance in the larger islands such as Jamaica, where it has been treated by novelists from Herbert de Lisser in *Jane's Career* (1914) to Richard Thelwell in *The Harder They Come* (1980). In *Jane's Career* there are also references to migrations overseas to Costa Rica where banana plantations provided work for Jamaican labourers, as well as to Panama where canal construction offered labouring jobs.

Among the Jamaican writers whose works reflect the experience of rural–urban migration is Claude McKay (1889–1948). McKay's autobiographical *My Green Hills of Jamaica* (1979) describes his 'change from country child to small-town kid' in Montego Bay (McKay, 1979, p. 13) and his later move to the city of Kingston, as well as his eventual decision to emigrate to the USA where he became a prominent figure in the Harlem Renaissance. McKay's move from his rural village to the small town where his brother lived necessitated adjustments, including going about shod instead of barefoot, in an environment where people and places 'were kind of chic' and life 'less rude' and 'more refined' (McKay, 1979, p. 14). This, together with a better schooling and occasional trips by train to Kingston, no doubt prepared the country boy for his eventual move to the capital city. It seems likely, too, that the American influence to which, according to McKay, schools in the Montego Bay area were being exposed, was a factor contributing to his later decision to migrate to the USA. Although the young McKay returned to his rural family home before going to Kingston and then emigrating to New York, his life-course may be seen as a classic example of step-wise migration (Conway, 1980). In this case the movement had an hierarchical as well as a spatial component as the migrant moved from rural settlement to small town and then to the city. It also had an international dimension, the ultimate destination being one of the world's great metropolitan centres. In the Caribbean, even the early stages of this step-wise migration process often involve overseas travel as migrants from small islands move to larger and more developed ones (Conway, 1983). I shall return to this particular point later in this chapter.

Intra-Caribbean Migration

Among the smaller islands, inter-island migration for shorter or longer periods has long been commonplace, as Margaret Byron noted in Chapter 6. In the words of Elizabeth Thomas-Hope, 'One of the notable features of Caribbean migration has been its development of spatial fields which provide habitual, even institutionalized means of extending the limits of "small-island" opportunities to incorporate a wider world. Caribbean migrants assume the permeability of national boundaries...' (Thomas-Hope, 1992, p. 20). This feature of Caribbean life is reflected in the works of novelists such as Antigua's Jamaica Kincaid (1983) in *Annie John*, and Grenada's Jean Buffong (1992) in *Under the Silk Cotton Tree*. At the age of sixteen, after a quarrel with her father, Annie John's mother left her parental home in Dominica and sailed to Antigua where she married and settled down. The novel suggests that Annie's mother kept in touch with her mother in Dominica. Annie's father last saw his parents when he was a little boy, although they, too, kept in touch with him, sending presents from South America to which they had emigrated. It is significant that, when leaving Dominica, Annie's mother went to neither of the nearest island territories, Martinique and Guadeloupe, which were and remain French, and that on her stormy voyage she bypassed the tiny British island of Montserrat en route to the larger English-speaking Antigua (Kincaid, 1983). Although smaller in area than Dominica, Antigua has long been more developed economically than the rugged island to the south. A more detailed exploration of the work of Jamaica Kincaid is provided by Rachel Hughes in the next chapter of this book.

Even more than in Kincaid's *Annie John*, Jean Buffong's (1992) novel, *Under the Silk Cotton Tree*, set in the author's native Grenada, evokes a way of life in which migration between small Caribbean islands, and particularly from smaller to larger and more developed neighbours and nearby mainland countries, is commonplace. Apart from the adjacent and very similar St Vincent and the Grenadines, Grenada's nearest neighbour is Trinidad and Tobago, a relatively large twin-island country with one of the most developed economies in the region, one which benefited from the exploitation of the nearby Venezuelan oil-fields even before its own oil resources were developed. No wonder that, 'With people trafficking between the islands ... half of Trinidad people have family in Grenada' (Buffong, 1992, p. 89). For example, one of Buffong's Grenadian characters, Miss Sagoo, has a daughter in Trinidad. 'Apparently, the woman left Grenada and went overseas years ago, first to Venezuela, then somewhere in the Virgin Islands before she settled down in Trinidad. She only came back to Grenada once to see her mother, then she went back, saying Grenada too slow for her' (Buffong, 1992, p. 11).

Throughout Buffong's novel there are references to travel between Grenada and other Caribbean territories, especially Trinidad, and to sojourns of various lengths in those places where migrant people often maintained links with their Grenadian families. The father of Flora, the first-person narrator in Buffong's novel, had left her mother and gone to live in St Croix, in the US Virgin Islands, a colony where the economy was boosted by tourism and 'offshore' industry. Consequently the Virgin Islands, especially St Croix, are mentioned frequently in the book, but there are many other places to which the novel's Grenadian characters migrate, at least for a time. These include the Netherlands Antilles island of Aruba which, like Trinidad, had established refineries for Venezuelan oil, neighbouring St Vincent, and even 'all the way up in Jamaica' (Buffong, 1992, p. 110).

In Buffong's Grenada, migration is not all one-way. Not only do many Grenadians return home after periods abroad, but people come to the island from other countries, including one widely travelled St Vincent woman, formerly an inter-island trader, who had been in Grenada so long that the locals had stopped calling her 'Vincee' (Buffong, 1992, pp. 103–4). Another woman, residing on the Grenadian island of Petit Martinique, was 'really from Venezuela', having followed a man back to Grenada (p. 91).

The patterns and economic character of intra-Caribbean migration have changed over time, something reflected in the literature discussed above. Earlier movements from the islands to Central America for plantation work or canal constuction jobs were superseded by migrations to Venezuela, Trinidad and the Netherlands Antilles generated by the rise of the oil industry. More recently other economic developments have affected the patterns of movement, including tourism in places such as the US Virgin Islands, a territory frequently mentioned in Buffong's (1992) novel. As Buffong and Kincaid both observe, movement within the Caribbean continues, some of it related to small-scale trading, and throughout the region there are many people who live on islands which are not the ones on which they were born. Similarly, in North America, Europe and now in many other parts of the world, there are large numbers of Caribbean people, substantial West Indian communities even, who reflect the migration which began as a trickle before the Second World War, but which became a flood afterwards.

Migration from the Caribbean

One of the many West Indian novelists who left their homeland shortly after World War II is George Lamming, whose works, notably *The Emigrants* (1954) and *The Pleasures of Exile* (1960), reflect the Caribbean migration experience.[1] Lamming's first novel, *In the Castle of My Skin* (1953), portrays the childhood and growth into manhood of a Barbadian boy. The boy, 'G', has a grandmother who went to Panama when the canal opened, settling in the Canal Zone, and his friend, Trumper, leaves for the USA, returning to Barbados a few years later. 'G', like the author himself, goes off to Trinidad, and Lamming's second novel, *The Emigrants*, starts with the shipboard observations of a passenger who, after a four-year sojourn in Trinidad, is emigrating to England, again reflecting the author's own experience. Aboard the ship one West Indian migrant says to a group of England-bound fellow passengers, 'All you make a step forward ... There was a time when all you stowaway to Trinidad' (Lamming, 1980, p. 39). Another observes, 'Well they's small islan' people, you know ... an' they lucky to have a big island next door. Those o' we who know what life mean in a big town got to come far to get better' (p. 39). Clearly migration to the metropole is seen as the ultimate step in what is often a hierarchical step-wise movement which, as I noted before, may start in a rural area of a small island and progress through small town to large, often involving going to a bigger island in the region before leaving the Caribbean for the 'great world beyond'.

Aboard ship, Lamming's emigrants discuss their reasons for migrating to an overseas destination of which they are largely ignorant. Their plans are generally vague. Most hope to find work of some kind, or to further their education and gain qualifications which will improve their chances of employment there or back home. What they all seek is 'a better break', a phrase used repeatedly in the novel. Their attitude towards the islands they are leaving is ambivalent. While one says, 'Trinidad ain't no place for a man to live' (p. 37) and another, 'Me see worst in Jamaica' (p. 38), the same people are proud of their origins, and argue passionately about the merits of their homelands, be it the excellence of their beaches or the quality of their education. The migrant's dilemma is expressed by Tornado, who says, 'This blasted world ... is a hell of a place. Why the hell a man got to leave where he born when he ain't thief not'in, nor kill nobody, an to make it worse to go somewhere he don't like' (p. 42). Unlike most of the others, Tornado had already spent some time in Britain where he served in the Royal Air Force, an experience which led him to 'hate like poison ... that said England' (p. 43).

With limited opportunities at home, the prospect of material advancement is a powerful attraction which can overcome fears of the unknown and even a repulsion based on rumoured or firsthand knowledge of the migrants' destination. Even a country which is 'so cold that only white people does live there' can attract migrants from the sunny Caribbean when 'they say that it have more work and better pay in England' (Selvon, 1979, p. 15).

Some, however, may be moved by a sense of adventure or the desire for a change; as one of Lamming's emigrants explains,

> Well, 'tis simply because ah little tired. Ah sick, bored. Ah doan' care w'at ah do next, but ah can't stan' in Trinidad no more 'cause ah know w'at rum taste like, an' ah know w'at woman, taste like, an' if you know dose two you know

Trinidad. Is a good life, an' I ain't know any better for me or any of de others, power or no power. But it only goes to show dat too much of a good thing can get you sick. (Lamming, 1980, p. 64)

For others it is not so much a desire for change as an urge to prove something. 'All them want to prove to somebody dat dem doin' something or that they can do something' (Lamming, 1980, p. 66). For many that 'somebody' may be 'themself'.

Them want to prove that them is themself. That is what they want to prove. An' them got to feel an' search an' be afraid all the time 'cause them doan' know what them must prove. A man who feel him got to prove himself start wid de first disadvantage that him ain't know w'at him ought to prove. (Lamming, 1980, pp. 66–7)

The speaker then refers to a search for a West Indian sense of identity which was a response to the experience of European colonialism and the consequent transplantation of people from Africa, India, China and elsewhere that gave rise to the heterogeneous Caribbean population.

If most of the emigrants can put forward only vague notions about their motives for leaving the Caribbean, they do not seem even to question the choice of England as their destination. They are 'a sample of the men who are called her subjects and whose only certain knowledge said that to be in England was all that mattered' (Lamming, 1980, p. 107). On being asked, 'Why do so many of your people come here? ... Collis cannot find an answer' (p. 138). Part of the answer, we are told, was that, 'England was not only a place, but a heritage. Some of us might have expressed a certain hostility to that heritage, but it remained, nevertheless, a hostility to something that was already a part of us' (Lamming, 1980, p. 229). Centuries of political domination and economic subordination had developed a dependency which shaped the mental attitudes of West Indian people (Thomas-Hope, 1992). Even when the search for identity turned towards Africa, like Boca in Buffong's (1992) novel, migrants were more likely to end up in England than in the less familiar continent. On their arrival in England, despite some strange and puzzling features, there are many things with which the West Indians are already very familiar because of Britain's commercial and cultural links with her Caribbean colonies. It is with a feeling of recognition that arriving migrants see the bombed-out buildings of the recent war which had dominated the news back home. 'The War. It was fought here, and you read about it' (Lamming, 1980, p. 118). They are excited to see the factories where cosmetics, razor blades, suspenders, paint and other products familiar to them in their homelands are manufactured. 'Read partner. Look what they make. They make everything here on this side. All England like this. Everything we get back home they make here, ol' man' (pp. 119–20). England and all things English have been part of the West Indians' lives since birth. As the character known simply as 'the Jamaican' expresses it, '[F]rom the time we born, in school an' after school, we wus hearin' about them' (Lamming, 1980, p. 86).

This last quotation neatly opens up the next part of my analysis: the role of education. As Julia Crane (1971) observes in her study of the anglophone Netherlands Antillean island of Saba, young Sabians are 'educated to emigrate'. The role of schools in the Caribbean is of great importance in shaping migration propensities, particularly for those who have the best access to formal education.

The Role of Schools

Dependency is not only a consequence of political and economic ties; it is a state of mind deliberately fostered by the colonial powers (Thomas-Hope, 1992). Schools played a very important role in the cultural colonization of Caribbean minds, and were used by the British colonial authorities to promote the interests of imperialism. This was something that many West Indian writers recognized and portrayed in their books (Hudson, 1994). In the words of Amon Saba Saakana (1987, p. 102), 'A society which educates its people away from its own history and environment is a colonial society.' At Joyce Gladwell's (1969, p. 35) Jamaican boarding school, her 'mind and imagination were fed on English scenes and English thoughts'. In her autobiography, *Brown Face, Big Master*, she records, 'The curriculum was imported from England; the books and subject matter were English. Even the exercise books were sent from Foyles in London' (Gladwell, 1969, p. 32). This merely continued the conditioning process which had begun at her elementary school where she 'learned to read from books prepared for English schools, illustrated with pictures of rolling English wheat-fields and well-clad English children climbing over styles' (p. 32).

Another Jamaican, the actor and autobiographer Charles Hyatt (1989, p. 135), records similar memories of his schooling: 'For in them days readin' book and history book and geography book didn' have in anything 'bout Jamaica. Only 'bout England.' As a schoolboy he was taught to sing songs such as 'Rule Britannia', 'The British Grenadiers' and even 'On Ilkley Moor Baat 'At'! 'Them use to drill them songs inna wi head an heart' (Hyatt, 1989, p. 111).

Schooling appears to have been little different in Trinidad. There Merle Hodge's (1981) character, Tee, learns to read using a primer which recounts the adventures of two English children, recites English nursery rhymes, writes in a copy book which features the king's head on the cover, and sings songs such as 'Land of Hope and Glory', 'Loch Lomond' and 'Men of Harlech', as well as 'God Save the King'. From her reading, Tee comes to believe that 'Reality and Rightness ... were to be found Abroad' (Hodge, 1981, p. 61), and her education trains her to think of England as 'The Mother Country' (p. 30). Tee's class is taught to recite 'Children of the Empire Ye are

Brothers All' (p. 26).

Several Caribbean writers have described how imperial and royal occasions were celebrated at school. Among Charles Hyatt's most vivid memories is Coronation Day:

> Out came the little red, white an blue Union Jack flag dem, the balloon dem wid the King and Queen face pon dem, de enamel mugs with the flag an King and Queen, the cups and saucers, the drinking glasses, everything! Everything that had a surface had the king and queen face pon it. Exercise book, ruler, rubber, pencil, pencil box. (Hyatt, 1989, pp. 111–12)

In a more formal literary style, George Lamming's semi-autobiographical novel, *In the Castle of My Skin*, describes the celebration in Barbadian schools of imperial events such as the late Queen's birthday and Empire Day:

> The school wore a uniform of flags; doors, windows and partitions on all sides carried the colours of the school's king ... In every corner of the school the tri-colour Union Jack flew its message. The colours though three in number had by constant repetition produced something vast and terrible, a kind of pressure or presence of which everyone was part. (Lamming, 1970, pp. 31–2)

Evidence from West Indian literature attests that being part of the British Empire was something that was often stressed in the Caribbean classroom (Hudson, 1994). An amusing but disturbing illustration of this is given in the autobiographical work, *Growing up Stupid Under the Union Jack* (1980), by the Barbadian-born novelist Austin Clarke, who emigrated to Canada. In a class conducted by the headmaster, the boys are questioned about a wall map of the world on which the British Empire is prominently displayed in red. These young Barbadians were taught to regard it as *their* empire, to which they 'as free people belonged'. Indeed, they were educated to consider themselves English; 'We English, colonial, overseas and dominion to the spit and polish of our spine' (Clarke, 1980, p. 154).

Clarke was one of those West Indian writers who left the Caribbean to further their education – he studied at the University of Toronto – and who eventually settled overseas. While it is true that, 'During the immediate post-war years, West Indian literature was a reflection of growing nationalism, hopes of a regional federation, feelings of anti-colonialism and interest in local culture' (King, 1979, p. 3), the effectiveness of British colonial education is evident in the very fact that most writers from the anglophone Caribbean migrated overseas. Like the majority of their fellow islanders, they usually went to Britain, the USA (Britain's successor as the Caribbean's major imperialist power), or Canada, a large Commonwealth neighbour with considerable interests in the region. Significantly, in a survey of schoolchildren conducted in seven English-speaking Caribbean countries in 1980, Morrissey (1983) found that most of those questioned expressed a desire to leave their homeland; the United States, Canada and Britain were the preferred overseas destinations.

USA and Canada

As Thomas-Hope (1992, p. 6) observes, the main destinations of Caribbean migrants are 'metropolises of Europe and North America with which ties have been strongest and upon which dependency has been greatest ... those countries to which people look in the pursuit of realizing personal objectives and goals'. Not only did Thomas-Hope's survey of potential migrants in Jamaica, Barbados and St Vincent also show that the USA, Canada and Britain were the preferred destinations, but her study suggests that most people 'did not even think' (p. 87) of other countries. Her study indicated, however, that as a migrant destination, Britain was now generally regarded as less attractive than the USA or Canada, particularly among Jamaicans.

The powerful pull of Britain has already been discussed, and the attraction of North America can be partly explained by its relative proximity to the Caribbean, a factor also mentioned previously. Other reasons for choosing North America instead of Britain include disappointment and unpleasant experiences reported by earlier migrants to the 'mother country', and the tightening of restrictions on immigration into the United Kingdom, notably the Commonwealth Immigrants Act of 1962. For some West Indians, however, North America, despite its reputation for racism, had a special appeal that was related to its perceived greater opportunities and new ideas which made Europe, particularly Britain, dull and old-fashioned in comparison.

It was to America that Claude McKay chose to go after the successful publication in 1912 of his book of poems, *Songs of Jamaica*. His English friend, Mr Jekyll, was at first horrified at this decision, warning, 'Claude, do you know how negroes are treated in America?' McKay replied that he did know but that 'America was a great modern land' to which he wanted to go. (McKay, 1979, pp. 79–80)

McKay had given poetry readings to audiences which included British and American residents as well as light-skinned upper-class Jamaicans, and he met 'two distinguished coloured American artistes [who] were playing at a theatre in Kingston' (McKay, 1979, p. 80). One of them, Miss Henrietta Vinton Davis, later 'became one of the leading personages and one of the most brilliant speakers of the Marcus Garvey Black Star Line and the Back-to-Africa movement' (p. 81). Miss Davis told McKay that she was a friend of US black leader Booker T. Washington, founder of the Tuskegee Institute which she often visited and where many West Indians were students. In McKay's words, 'The fame of Mr Washington has spread all over the world and I wanted to be in his school' (p. 82).

> Going to America was the greatest event in the history of our hills; America was the land of education and opportunity. Even though Miss Davis had told us that it was not a good place for coloured people – we all believed in it. We thought of England, France, Germany and the rest of Europe – some of our children were

being educated there – but we thought of America in a different sense. It was the new land to which all people who had youth and a youthful mind turned. Surely there would be opportunity in this land even for a negro. (McKay, 1979, pp. 84-5)

Even in that most British of Caribbean islands, Barbados, many people were attracted by the lure of the USA. Lamming's character, Trumper, 'had always dreamt of going to America' (Lamming, 1970, p. 239), and his friend, 'G', saw him off when he departed with hundreds of other migrants to a country about which, 'There ain't much to say ... except that the United States is a place where a man-can-make pots-of-money' (p. 316). Despite this remark which Trumper makes on his return to Barbados, he does have a great deal to say about America, a place which makes one 'feel that where you been livin' before is a kind of cage' (p. 319). Trumper plans to return to the USA, but as an individual migrant: 'The emigration scheme might be alright for some people, but it don't give you the kind of freedom you want' (p. 317).

During the Second World War, US servicemen were stationed in the Caribbean, and their way of life impressed many West Indians, encouraging potential migrants to look towards their American neighbours rather than to Britain. 'The Yanks turn Trinidad upside down, an' when they finish they let we see who was who. They is a great people, those Yankee people' (Lamming, 1980, p. 117).

Less influential as a military and economic power, and without the media dominance enjoyed by the USA, Canada nevertheless has a strong presence in the anglophone Caribbean, particularly in the fields of trade and commerce. There were, too, the bonds of Empire, eventually transmuted into the Commonwealth, which maintained a variety of links, not least in the areas of education and culture in which the Caribbean and Canada shared a common heritage with Britain. This was one of the attractions which drew West Indians to Canada, among them writer Austin Clarke, a Barbadian author whose work was briefly introduced earlier.

It was higher education which attracted several of Clarke's characters in his collection of short stories, *Nine Men Who Laughed*. One character who, unlike others in the book, had not gone to Canada to enter university was, nevertheless, quite well educated and aspired to a good job. Like the others, 'In Barbados ... [he] was schooled and trained in the British system of values ... But we ... nine men [who] emigrated to Canada ... were also marked out as members of the colonial elite, persons whose futures were assured both in Barbados and elsewhere in the English-speaking world' (Clarke, 1986, p. 3). Indeed his proud father had objected strongly to his son's decision to leave Barbados for Canada: 'You come telling me you going to Canada as a' immigrant? To be a stranger? Where Canada is? What is Canada? ... Canada is no place for you, man. The son of a Barbadian plantation owner? This land was in our family before Canada was even discovered by the blasted Eskimos and the Red Indians' (Clarke, 1986, p. 48). Nevertheless, even with an upper-class background and a good education, West Indians in Canada, as elsewhere, often had to take menial jobs, particularly if they were working in the country illegally: 'In the eight years he had spent in the country, he had lain low for the first five, as a non-landed immigrant, in and out of low-paying jobs given specifically to non-landed immigrants, and all the time waiting for an amnesty' (p. 36). As one observes, 'Wessindians don't have much money, because they does-get the worst and lowest jobs in Toronto' (p. 61).

When migrants' hopes are not realized in one place, they often move on to another, sometimes going to a different country. Autobiographer Wallace Collins (1965), having migrated from Jamaica to Britain in 1954, left there for Canada in 1962, largely because of widespread English prejudice agaist black immigrants at a time when 'that infernal immigration bill was being discussed across the land' (Collins, 1965, p. 121). He wrote, 'My wife and I ... eventually agreed to migrate to Canada where we thought that we would at least be free from this prejudicial bellyache' (p. 122). Apparently this move did not improve things greatly, for Collins found that he missed England very much. 'After living and working in Canada for the past two years ... I am able to state ... that in Great Britain a man has more political and social rights under the law, than in any other country in the world' (Collins, 1965, p. 122).

Many migrants who are disappointed with their lives overseas return home. So do some who made a success of things abroad but respond to the pull of home. For the Jamaican Anthony Winkler, 'Living in America as an immigrant was ... like living in a vivid dream ... I felt no love for the land. It did not smell right. Even after thirteen years it still had the alien, unrecognizable spoor of a foreign place ... It did not sound right' (Winkler, 1995, p. 3). This is a quotation from Winkler's autobiographical account of a year which he spent back in Jamaica, published under the title *Going Home to Teach*. Nevertheless, in the notes about the author at the end of the book it says, 'Mr Winkler currently lives in Atlanta, Georgia, with his wife and two children'. Winkler's case illustrates the problems of identity, ambiguity and ambivalence typically experienced by the Caribbean migrant, indeed by migrants worldwide. On achieving the goal of emigration, the migrant feels alienated in the chosen foreign environment. Having arrived in their desired destination, many emigrants experience a longing for the familiar things left behind; yet, returning home, they feel compelled to go back to the foreign land which they do not love.

Conclusion

This study of selected works from the extensive field of Caribbean literature in English shows it to be a rich source of qualitative data on migration within and from the region. Its particular value is to portray the migration experience from the point of view of the individual, showing how a variety of push and pull factors, the role of image, and chance events contribute to decisions to move and influence the choice of destinations. So complex are the processes involved that even the migrants themselves rarely comprehend them, often making decisions without a full knowledge of their implications or of all the possible alternatives available.

Through the study of many personal experiences drawn from auto-biographies and semi-autobiographical fiction, we are able to analyse the migration phenomenon in a way which illustrates the interplay of roles performed by a wide variety of factors, notably geographical, economic and social elements within the structure of colonialism and, especially, colonial education. In this literary survey we have encountered male labourers and female traders, cane cutters, teachers and writers. Clearly, gender and class, matters noted but not systematically treated in this chapter, are important aspects of the topic of Caribbean migration which require closer attention. The behaviour of many of the characters in the literature surveyed illustrates how small-island Caribbean people tend to treat the entire region as their immediate world, largely ignoring political boundaries, and supports the idea that migration is an essential element of West Indian culture.

This chapter has focused largely on the aspirations of individuals who perceive greater opportunities for self-advancement overseas. Their choice of destination has generally been limited to places about which they have a certain knowledge, namely nearby islands and territories in the Caribbean (particularly where English is spoken), and metropolitan countries to which political and economic ties are strong. The important role of education, especially schooling, has been identified, as well as the changing opportunities for employment. Future literary research on Caribbean migration might focus usefully on specific aspects of the problem which has been treated only broadly in this study. Among themes which could be explored in greater depth are class, gender and age differences, colour, education, the search for identity, ambivalence, and return migration. Caribbean literature undoubtedly contains a wealth of material which will reward more detailed study.

Acknowledgement

The author gratefully acknowledges the useful comments on early drafts of this chapter by the editors, Russell King and John Connell.

Note

1. The dates given here are for first publication of these books. My own references to these (and other) books are often, perforce, to later editions. Fuller details of publication dates are given in the References.

References

Aldrich, R. (1995) 'From Francité to Créolité. French West Indian literature comes home', in R. King, J. Connell and P. White (eds), *Writing across Worlds: Literature and Migration*. London: Routledge, pp. 101–24.

Alexander, C. (1995) 'Rivers to cross. Exile and transformation in the Caribbean migration novels of George Lamming', in R. King, J. Connell and P. White (eds), *Writing across Worlds: Literature and Migration*. London: Routledge, pp. 57–69.

Boswell, T.D. (1989) 'Population and political geography of the present-day West Indies', in R.C. West and J.P. Augelli (eds), *Middle America: its Lands and Peoples*. Englewood Cliffs, NJ: Prentice Hall, pp. 103–27.

Buffong, J. (1992) *Under the Silk Cotton Tree*. London: Women's Press.

Clarke, A. (1980) *Growing up Stupid Under the Union Jack*. Toronto: McClelland & Stewart.

Clarke, A. (1986) *Nine Men Who Laughed*. Markham, Ontario: Penguin Books Canada.

Cobham, R. (1979) 'The background', in B. King (ed.), *West Indian Literature*. London: Macmillan, pp. 9–29.

Collins, W. (1965) *Jamaican Migrant*. London: Routledge & Kegan Paul.

Conway, D. (1980) 'Step-wise migration: toward a classification of the mechanism', *International Migration Review*, 14(1), pp. 3–14.

Conway, D. (1983) 'The commuter zone as a relocation choice of low income migrants moving in a step-wise pattern to Port of Spain, Trinidad', *Caribbean Geography*, 1(2), pp. 89–106.

Coser, L.A. (1963) 'Introduction', in L.A. Coser (ed.), *Sociology through Literature*. Englewood Cliffs, NJ: Prentice Hall, pp. 2–7.

Crane, J.G. (1971) *Educated to Emigrate: The Social Organization of Saba*. Assen: van Gorcum.

Duany, J. (1994) 'Beyond the safety valve: recent trends in Caribbean migration', *Social and Economic Studies*, 43(1), pp. 95–122.

Gladwell, J. (1969) *Brown Face, Big Master*. London: Inter-Varsity Press; first published 1954.

Gmelch, G. (1992) *Double Passage. The Lives of Caribbean Migrants Abroad and Back Home*. Ann Arbor: University of Michigan Press.

Hodge, M. (1981) *Crick Crack, Monkey*. Oxford: Heinemann; first published 1970.

Hudson, B.J. (1994) 'Geography in colonial schools: the classroom experience in West Indian literature', *Geography*, 79(4), pp. 322–9.

Hyatt, C. (1989) *When Me Was a Boy*. Kingston: Institute of Jamaica.

Kincaid, J. (1983) *Annie John*. New York: Farrar, Straus & Giroux.

King, B. (1979) 'Introduction', in B. King (ed.), *West Indian Literature*. London: Macmillan, pp. 1–8.

Lamming, G. (1960) *The Pleasures of Exile*. London: Michael Joseph.

Lamming, G. (1970) *In the Castle of My Skin*. New York: Collier Books; first published 1953.

Lamming, G. (1980) *The Emigrants*. London: Allison & Busby; first published 1954.

Lisser, H.G. de (1914) *Jane's Career*. London: Methuen.

McKay, C. (1979) *My Green Hills of Jamaica*. Kingston and Port of Spain: Heinemann Educational Books.

Morrissey, M. (1983) 'Country preferences of school children in seven Caribbean territories', *Caribbean Quarterly*, 29(3-4), pp. 1–19.

Naipaul, V.S. (1961) *A House for Mr Biswas*. London: André Deutsch.

Richardson, B.C. (1992) *The Caribbean in the Wider World 1492–1992*. Cambridge: Cambridge University Press.

Saakana, A.S. (1987) *The Colonial Legacy in Caribbean Literature*. London: Karnak House.

Selvon, S. (1979) *The Lonely Londoners*. London: Longman Drumbeat; first published 1956.

Thelwell, M. (1980) *The Harder They Come*. London: Pluto Press.

Thomas-Hope, E.M. (1992) *Explanation in Caribbean Migration: Perception and the Image: Jamaica, Barbados, St Vincent*. London: Macmillan Caribbean.

Western, J. (1992) *A Passage to England: Barbadian Londoners Speak of Home*. London: UCL Press.

White, P. (1995) 'Geography, literature and migration', in R. King, J. Connell and P. White (eds), *Writing Across Worlds: Literature and Migration*. London: Routledge, pp. 1–19.

Winkler, A.C. (1995) *Going Home to Teach*. Kingston: Kingston Publishers.

9

Between the Devil and a Warm Blue Sea: Islands and the Migration Experience in the Fiction of Jamaica Kincaid

Rachel Hughes

[T]he first thing [Dinah] said to me when Mariah introduced us was 'So you are from the islands?' I don't know why but the way she said it made a fury rise up in me. I was about to respond to her in this way 'Which islands exactly do you mean? The Hawaiian Islands, the islands that make up Indonesia, or what?' (Kincaid, 1994, p. 56)

This chapter is concerned with the experience of migration from an island periphery under colonial rule, to a large, western metropole. My aim is to examine Antigua (an island in the Eastern Caribbean chain) as 'home' and New York City as a 'destination' place, through the fiction of Jamaica Kincaid. I am specifically interested in Kincaid's semi-autobiographical novel *Lucy* published in 1991.[1] As such, this exploration is one which focuses on women's migration experiences, specifically those of Caribbean women. Through *Lucy* I examine the journey of a young Afro-Caribbean woman (the character Lucy Josephine Potter), her strategies of 'becoming' in a destination place, and her negotiation of an individual history. This literary analysis of *Lucy* is linked to Kincaid's more direct, spoken statements in interviews and incidental written pieces about her own migration, her writing, other thematic concerns and the world at large.

Migration research in the social sciences, while rich and diverse in its own right, often fails to capture the experience of what it is to be a migrant; and be, or not be, part of a community, a nation, a society (King *et al.*, 1995, p. x). Creative literatures offer powerful insights into the nature of the migration process and the experience of being a migrant. Moreover, to date, literary works by Caribbean authors seem to have captured the special nature of women's experience [including migration] in the region more successfully than have academic analyses, exploring the complex ensembles of goals and expectations that have developed over time (Mintz and Price, 1985, p. 8).

Migration literatures, especially those detailing movement from a colonial periphery to the colonial motherland or another metropole, draw out the crises and shifts of a migrant identity and the construction of a new sense of one's 'place' in the uneven constellation of Empire. In Kincaid's writing, the exploitation of one place and people by another place (people) is seen anew, making transparent a colonial childhood and making precarious that colonized identity. As literary and political projects, Kincaid's stories play out the post-colonial strategy described by geographers Alison Blunt and Gillian Rose whereby 'in thinking through the structures of power that underpin a (peripheralized) identity, and choosing a hybrid identity or location, it is possible to displace the (larger) distinction between centre and margin so necessary to the colonizing master subject' (Blunt and Rose, 1994, p. 17). That is to say, while Kincaid writes in *Lucy* of the intimacies of a colonized identity, her conclusions and strategies of the self speak to a radical and encompassing post-colonial politics of location.

Islands and Empire

Jamaica Kincaid was born in 1949 in Antigua, then a British colony, leaving Antigua around the age of sixteen for New York City. Living and working still in the United States of America, Kincaid has published extensively over the last two decades including a collection of early writings entitled *At the Bottom of the River* (1978), and five novels: *Annie John* (1985), *A Small Place* (1988), *Lucy* (1991), *The Autobiography of My Mother* (1996) and *My Brother* (1997). Both *At the Bottom of the River* and *Annie John* concern a girl's childhood in 1950s colonial Antigua and draw heavily on the experiences of Kincaid's own life. *Lucy* – also a semi-autobiographical work – is a migration story of a young Afro-Caribbean woman's first few years in New York and her experiences working as an *au pair* to a wealthy, white family. *Annie John* and *Lucy* can be read as two parts to the one narrative, the character Lucy emerging from the younger version of herself in *Annie John*. Kincaid's third book, *A Small Place*, is concerned with colonialism and (neo)colonialism, environmental degradation and tourism in present-day Antigua and the Caribbean. In her most recent novel, *The Autobiography of My Mother*, Kincaid returns again to the Caribbean to investigate the fictional life of Xuela, a Dominican woman. Kincaid has identified herself as a 'pro-female' voice.[2] Her writing focuses on women, on relationships between mothers and daughters, between white women and black women and the relationship between women and the masculine venture of colonialism.

Throughout Kincaid's writing the island re-appears as an instinctive landform. Kincaid's Caribbean islands are idylls – fertile and fragile – but are also small places of exceptional heat and drought, vulnerability and frightening isolation. In much of her writing Kincaid is interested in the way in which her island home, colonial Antigua, was fused – culturally, economically, imaginatively – to that very distant and dominant island, England:

> When I saw England for the first time I was a child in school sitting at a desk. The England I was looking at was laid out on a map gently, beautifully, delicately, a very special jewel ... England was to be our source of myth and the source from which we got our sense of reality, our sense of what was meaningful and our sense of what was meaningless – and much about our own lives and much about the very idea of us headed that last list (Kincaid, 1991, p. 13).

Kincaid has repeatedly spoken of her own writing as fleshing out the relationship between the colonizer and the colonized, the powerful and powerless which, with its violent history of slavery and plantation profit, is so much a part of the colonial history of the Caribbean (Ferguson, 1944, p. 171). In one interview, Kincaid remarked of her childhood 'Even the business of breakfast was "Made in England" like almost everything around us, the exceptions being the sea, the sky, the air we breathed' (Kincaid, 1991, p. 14). Kincaid's Antigua is 'neither a motherland, nor a fatherland' for descendants of the African people forcibly transplanted to the Caribbean. Colonized first in the seventeenth century, Antigua did not begin to move toward formal independence from Britain until the late 1960s, finally attaining independence in 1981. Kincaid maintains that Antigua was a place where – in the last two decades of colonial rule (her childhood) – 'England and its glory was at its most theatrical, its most oppressive' (quoted in Cudjoe, 1990, p. 217).

Kincaid's Antigua is 'a small place, a small island' and she remarks:

> in a small place, people cultivate small events. Every event is a domestic event, the people in a small place cannot see themselves in a larger picture... [cannot see themselves] in a more demanding relationship [with the world], a relationship in which they are not victims all the time of every bad idea that flits across the mind of the world (Kincaid, 1988, pp. 52–6).

Writing of the multiple discourses of islands in western literature and science, Gillian Beer links islands and Empire by arguing that:

> island stories [can] serve to justify or repudiate colonialism, [but also] the island is both total and local, [it] emphazises both inhabiting and observing [and yet] there is a persistent trouble about how to sustain the necessary roles of observing and participating [and in this] the island site has been of particular importance (Beer, 1989, pp. 10, 21–2).

I would argue that Kincaid's writing, and in it her very different 'discourse of the island', does repudiate and undermine colonialism but does so by magnifying that island 'trouble' of which Beer writes. That 'trouble' for Kincaid marks the point of self-consciousness, vulnerability and invalidity of the project of colonialism, or of those other and more contemporary incursions whereby islands are 'professionally delimited' for academic study, western aid, or tourism development. Kincaid's Caribbean islands trip up the encompassing western eye which 'sees where the edges of an island are, [and sees therefore] a tidier project'(Beer, 1989, p. 22), by illustrating the degree of violence enacted and contained by the island-as-dominion, and the specificity and complexity of identity imbricated in such 'a small place'. Kincaid's discourses of the island include: the island as (environmentally and politically) vulnerable; the island as a (childhood) paradise; and the island as a site of concentrated and syncretic cultural resistance to what Beer terms 'the intense focus of the observer's eye [coming to] lodge so firmly on this minaturized zone, claim[ing] simultaneously empathy and control' (Beer, 1989, p. 23).

Migration and the Caribbean

> More than many areas of the world, the societies of the Antilles have been created by migrations, both voluntary and forced ... Edouard Glissant's statement that the Caribbean is a land of *enracinement* (rooting) and *errance* (wandering, drifting) is appropriate to both the history and the literature of the West Indies. (Aldrich, 1995, p. 101)

As Robert Aldrich observes, the South American Caribs and Arawaks were the first migrant populations of the Caribbean, but were virtually wiped out by the European conquistadors and their invasive plantation economies in the seventeenth century. In *Lucy*, Lucy's own grandmother, a Carib Indian, provides a textual link to these other, older histories of migration. The other part of Lucy's racial heritage is Afro-Caribbean, from the African people forcibly removed to the Caribbean as slave labourers in the sugar (and other) plantation industries. The momentum of migration continued into the late nineteenth century and the twentieth century: from the European nations to their Caribbean dominions in commercial enterprise, employment and administrative capacities; and to a greater extent, and conversely, toward the urban metropoles. Migrants left the islands in search of employment, education and a greater socio-economic freedom. Again, evidence of this long history of migration is found in Kincaid's novel: Lucy's father's mother left for England when he himself was a child, and Lucy's father's father went off to build the Panama Canal when Lucy's father was seven: 'my father never saw his father again' (Kincaid, 1994, p. 125).

In Kincaid's writing there is an acknowledgement of the inevitability of

Caribbean out-migration; the eponymous characters of *Annie John* and *Lucy* both expect to leave Antigua to gain further education and employment. In *Lucy* this 'culture of migration' demands that Lucy send back a remittance income to her mother, especially after her father's death. Kincaid herself has spoken of leaving her family and

> not being the sort of dutiful daughter that was expected of me. All the friends I grew up with who had left and gone to England or Canada would send back their paychecks to their families and everybody's family was building a nice house and my family was sort of deteriorating because my father was sick and couldn't support the family anymore and I wouldn't get a proper job [in New York]. (Kincaid, quoted in Ferguson, 1994, p. 179)

In *Lucy* we are presented with the reasons why Lucy behaves dissimilarly to her peers. Lucy's narrating voice illustrates 'those individual and complex "identity shifts" that result, both on the individual level and at larger-group levels, from the migration experience' (White, 1995, p. 2). Lucy's identity changes in response to her initial experiences and continues to change as she finds a more permanent residence in New York. Her identity is, as Stuart Hall suggests, a matter of 'becoming' as well as of 'being'. The character of Lucy attests to 'cultural identity as being not something which already exists, transcending place, time, history and culture. Far from being eternally fixed in some essentialized past, [identities] are subject to the continuous "play" of history, culture and power' (Hall, 1992, pp. 221–3).

The creative literature of women migrants challenges the hegemony of views of gender roles in migration, a situation derived largely from male-dominated research (White, 1995, p. 11). White traces, amongst other emergent themes in migrant women's writing, an emphasis on the role of gender in migration decision-making and adjustment to the host society. Women's writing also articulates the migration experiences of encapsulation, abandonment, rebellion and a celebration of the strength of an individual woman to survive under the (often contradictory) continuing 'presence' of the home place as well as the social and cultural pressures of the new place (White, 1995, pp. 10–13). In a study of the settlement abroad of women migrants from the Caribbean island of Nevis, Olwig also touches upon this important paradox. The success of Nevisian women migrants is contingent upon their meeting the demands of mothers and kin left behind, but at the same time they must develop new kinship and familial networks and make their own life abroad (Olwig, 1993, p. 163; see also Byron's Chapter 6 in this book).

With such theories of Caribbean migration and identity in train, this chapter follows some of Kincaid's specific concerns: the expectations held by Lucy of her destination, New York; the assimilating identities constructed for Lucy by other people abroad; the multiple shifts in identity desired and enacted by Lucy herself; the role of gender in her experience of migration; and her changing perceptions of 'home'. *Lucy* and Kincaid's autobiographical *New Yorker* article 'Putting myself together' (1995) can be seen to sketch a unique yet illustrative account of insistent exile and powerful self-transformation.

New World Territory

Kincaid writes, through Lucy's eyes, of a New York winter landscape which is uncompromising, cold, bare and anonymous. Lucy's first days demand many re-visions or re-conceptions of knowledge that she has never before questioned: that a sunny day may not feel warm to the skin, that trees appearing dead are only dormant in the cold, and that houses must be shut tight against the cold (Kincaid, 1994, p. 10). Landscape, then, is a continuing metaphor for fundamental and personal experiences of marginalization. Cityscapes and remoter, rural landscapes, the 'new' grounds which Lucy stands upon, refuse to contain or nourish her. Even after a year of living in the city, Lucy sees anew

> how cold and hard and shut up and tight the ground was. I noticed this because I used to wish it would open up and take me in, I felt so bad. If I dropped dead from despair as I was crossing the street, I would have to lie there in the cold. The ground would refuse me. (Kincaid, 1994, pp. 140–1)

Lucy's experiences of the Northern American countryside link her to a specific invisibility: that of the [urban] migrant, especially a black woman, excluded from the picturesque-rural, 'national' landscape. Kincaid's writing about Lucy's experiences of the rural landscapes of Lake Michigan bears comparison to the photographic work of the black British artist Ingrid Pollard. Pollard is specifically interested in the exclusion of non-whites from the significant and signifying landscape of the English countryside. As in Kincaid's writing, distinctly geographical metaphors are employed: the representational strategy becomes one of 're-territorialization', 'to disturb and subvert the settled relations of identification and recognition [in landscape]' (see Kinsman, 1995). For example, on the overnight train-ride to the country with Mariah (the mother of the New York family for which Lucy is an *au pair*), Lucy recognizes in Mariah's beloved view of new-ploughed fields a history of violent subjugation:

> [Mariah] drew up my blind and when I saw mile after mile of up-turned earth, I said, a cruel tone to my voice, 'Well thank God I didn't have to do that.' (Kincaid, 1994, p. 33)

Lucy later 'hears' the landscape: 'the land did not say, "Welcome. So glad you could come." It was more, "I dare you to stay here."' (p. 34)

The Great Lakes, like the city, disappoint Lucy. Having learnt of the Lake(s) from school geography books, its origin and history: 'now to see it up close was odd, for it looked so ordinary, gray, dirty, unfriendly, not a lake to make up a song about' (p. 35).

While Lucy negates any attachment to the new place,[3] she paradoxically craves, if subliminally, a geographical and psychological *location*. On her first night in New York Lucy dreams she is wearing a strange nightgown; suddenly she is 'compelled' to know 'where [the] nightgown came from'. Location and orientation to place is something which is actively sought out and desired, being the first part of sense-making in a foreign place, a construction of a relationship to that new place. At the same time, definite location and exclusionary landscapes bring new vulnerabilities. When Lucy looks at a world map she is granted a cartographic representation of the 'physical' distance between herself and her home. By virtue of the significance of her newly-won autonomy, Lucy also realizes the one-way nature of her migration, of there being 'no going back'. She sees that 'an ocean stood between me and the place I came from', but asks rhetorically 'Would it have made a difference if [the ocean] had been a teacup of water?' Expectations of herself and of New York have set her in one direction, that of a continuing, if frightening, autonomy and momentum: 'I could not go back' (Kincaid, 1994, pp. 9–10).

Kincaid, writing (in *Lucy*) of the United States and (elsewhere) of her childhood 'mother-country' England, explores the forceful themes of disappointment and resentment engendered in the migrant by arrival in the colonial 'centre'. Such disappointment results from the failure of the larger, more powerful nation to meet the expectations of a person coming from a place considered to be peripheral, subordinate to, or a copy of the real, happier, more beautiful, original place in which they have now arrived. When Kincaid travels to England, she is a child of Empire arriving at the imperial centre, and her anger and disappointment become more acute. She writes in an autobiographical piece 'On seeing England for the first time',

> finally then I saw England, the real England, not a picture, not a painting, not a story in a book, but England, for the first time. In me, the space between the idea of it and its reality had become filled with hatred ... And a great feeling of rage and disappointment came over me as I looked at England. (Kincaid, 1991, p. 16)

Lucy also experiences anger and disappointment in response to the cityscape of New York, perceiving a tiredness, 'an ordinary, dirty, worn down' surface scratched by the dreams and hopes of many migrants and potential migrants.

Race and Gender

Importantly, part of Lucy's migration experience is a continuation of a child-hood in which she has repeatedly and variously been told that she is some-thing that she is not.

> I was always being told I should be something, and then my whole upbringing was something I was not: it was English. (Kincaid, quoted in Cudjoe, 1990, p. 219)

When Lucy locates herself by looking at the world map, she immediately rejects that cartographic representation of her location and translates her experience of distance into an experience of exile; she is unable to go back to Antigua because of the nature of that island-place – its inhospitable famil-iarity, its routine, its capacity to imprison her. But these aspects of Antigua, so long resented, are dangers also of the new place. The domestic sphere in which she is 'placed' is one which replicates the dynamics of power of the colonial system; of employer and employee, or as Lucy understands it, of master and servant. She is to nanny for a wealthy, white nuclear family, her new home is the home of Mariah and Lewis and their children, a spacious city apartment. In this domestic space, her room is

> just off the kitchen – the maid's room. I was used to a small room but this was a different sort of small room. The ceiling was very high and the walls went all the way up to the ceiling, enclosing the room like a box – a box in which cargo trav-eling a long way should be shipped. But I was not cargo. I was only an unhappy young woman living in a maid's room, and I was not even the maid. (Kincaid, 1994, p. 7)

Lucy's conclusions about her room reflect her being located by Mariah and Lewis as an outsider. Lucy's awareness of her own racial and cultural heritage becomes apparent in the image of (human) cargo in a lengthy, contained passage. Lucy underscores a connection between her present situation – as a black woman working in a white, patriarchal family – and the discourses of the European colonial slave trade: the transportation of African people to labour under the plantocracies of the colonial Caribbean. In the presence of the family, especially in her diametrical relation to Lewis, Lucy is soon referred to as the 'Visitor' because (in their minds) she

> seemed not to be a part of things, as if I didn't live in their house with them, as if they weren't just like a family to me, as if I was just passing through, just saying one long Hallo!, and soon would be saying a quick Goodbye! So long! … Lewis looked at me, concern on his face. He said, 'Poor Visitor, poor Visitor' over and over. (pp. 13–14)

Again Lucy finds herself being told she should be something she is not: that she should feel she is part of the family, but she is not and cannot be. Mariah and Lewis both unconsciously play out a re-colonization of Lucy, a strategy of assimilation and containment while imagining they have her best interests at heart. They attempt to tell and show her things, to place her within their realm of understanding and care – a location which Lucy consistently resists. By virtue of her race and gender Lucy is also exoticized, her colour and culture fetishised by people that she meets at various junctures throughout the novel. When Mariah suggests Lucy study the paintings of Paul Gauguin, Mariah naively 'places' Lucy as adjacent to the exotic 'island women' who people his canvasses (see Ferguson, 1993, pp. 246–7). But Lucy has a far more complex reaction to Gauguin's work than Mariah expects – exploring her own racial, cultural and gender identification. First, Lucy identifies with the intimate yearnings of the French painter: 'I understood finding the place you were born in an unbearable prison and wanting something completely different from what you are familiar with', but also sees that 'his life could be found in the pages of a book; I had just begun to notice that the lives of men always are' (Kincaid, 1994, p. 95). As she gazes at the paintings, she becomes disillusioned with Gauguin's 'rebellion and despair' and sees only a 'perfume' of heroism about him. Gauguin's paradise, those islands on the fringe (albeit a different fringe) of the world, is Lucy's point of departure. Her island home, that 'unbearable prison', affords her no pleasurable, sun-drenched freedom. She begins to gauge the difference between his and her experience of exile: 'I was not a man; I was a young woman from the fringes of the world, and when I left my home I had wrapped around my shoulders the mantle of a servant' (p. 95). While Gauguin is someone who accumulated servants, Lucy understands herself as having been forced to become one. Kincaid has also spoken of being Afro-Caribbean in a racist society, in one instance being refused employment, but of not ever seeing herself in such a way:

> How was I to know I was a black girl? I never pass myself in a corridor and say, I am a black girl. I never see myself coming toward me as I come round a bend and say, There is a black girl coming toward me. (Kincaid, 1995, p. 100)

Lucy is exoticized by her lover Paul (also a painter): 'he asked me where I was from; he touched my hair, and I could tell that the texture of it was new to him' (Kincaid, 1994, p. 100). For a time Paul is part of (and useful to) Lucy's exploration of her own sexuality. However, when one day he drives her out to the country to show her a mansion in ruins, and tells her that it once had been the estate of 'a man who had made a great deal of money in the part of the world that [she] was from' their relationship sours. When he attempts to speak of travel and freedom, the freedom of great wealth and men who 'cross the great seas' Lucy retaliates:

> I pointed out the dead animals [along the side of the road] ... I tried to put a light note in my voice as I said, 'On their way to freedom, some people find riches, some people find death,' but I did not succeed. (Kincaid, 1994, p. 129)

Lucy also comes to recognize a continuity between Antigua and New York of gendered (male to female) violence. At one significant point, blurring the distinction between violence in public and private spaces, Lucy recalls the 'Devil-possession' of an Antiguan schoolmate in the girl's own home within her more recent fear engendered by a report in the press of a story of a young immigrant girl who had her throat cut in the New York subway:

> [In Antigua] on the one hand there was a girl who was beaten by a man she could not see; on the other there was a girl getting her throat cut by a man she could see. In this great big world, why should my life be reduced to these two possibilities? (Kincaid, 1994, p. 21)

Studies of contemporary Caribbean migration support the notion that gender is an important differentiating element within the migration experience (Momsen, 1993). For Caribbean men migration is seen as a natural extension of their wide array of extra-domestic activities; 'freedom' implies the ability to leave with the least fuss, leaving the fewest obligations behind (Olwig, 1993, p. 157). Women, on the other hand, are usually more closely associated with the domestic sphere as mothers or vital members of the extended household. As well as it being more difficult for women to leave, once they migrate they bear a far greater responsibility abroad than their male counterparts of continuing to support the households or kin they have left behind. This support is both economic, via remittances, and expressed through kinship; Caribbean migrant women are expected to build up new networks of friends and family abroad so that other kin may find it easier to follow and successfully settle in the host country (Olwig, 1993).

In fact, Lucy refuses most of her familial responsibilities when she migrates to New York. Most marked is her desire to be free of her mother's surveillance, which is a surveillance heightened by the bounded and limiting form of the small island. In *Annie John*, Kincaid's protagonist dreams of migration and exile in Belgium – a faraway real place she reads of in a book that becomes a fantasy-place where 'my mother would find it difficult to travel to and so would have to write me letters addressed in this way: Miss Annie Victoria John, Somewhere, Belgium' (Kincaid, 1985, p. 92). Lucy, also desiring isolation from the mother, leaves unopened any letters she receives in New York addressed in her mother's handwriting. Finally she burns all her mother's letters. In her recent and extensive study of Caribbean migrant life-stories in England, Mary Chamberlain notes rebellion and escape as significant motivations for young women's migration, quoting one woman as saying:

I felt that I'd come somewhere and, you know, away from the mother, and the mother, and the mother, and all that one restricted area ... I was glad for the freedom. That was me ('Beryl' quoted in Chamberlain, 1997, p. 97).

In much of Kincaid's writing, a crisis of anger and loss centres on the mother. Not only is the rift between the mother and daughter central, but the agony of this division can also be seen as a metaphor for another primary theme – racial and cultural domination (Simmons, 1994, p. 23). The mother-figure is one of strength, beauty, intelligence and reproductive labour. Simultaneously, Kincaid presents this same mother as a woman of fractured and borrowed identity.[4] The piety of Christian servitude, British etiquette and endless domestic labour come to define her maternal role. In Lucy's experience of adolescence, the home of the house and yard and the island-home have become zones of conflict between the mother/Mother country and daughter in which the mother attempts to control the child who consistently avoids and subverts such colonizing control. In broader literatures, other research examining gender relations in the Caribbean confirms

a double paradox: of patriarchy within a system of matrifocal and matrilocal families; and of a domestic ideology coexisting with the economic independence of women. The roots of this contemporary paradoxical situation lie in colonialism. (Momsen, 1993, p. 1)

Much of *Lucy* traces the difficulty of Lucy's enacting a geographical and psychological separation from her mother. This separation is made all the more difficult as Lucy is reminded of her mother in the thoughts and actions of Mariah whose loving attempts to include Lucy in the immediate nuclear family, and later in the 'family' of middle-class feminist ideals, are as stifling and colonizing as the love of Lucy's own mother. Lucy's racial, cultural and class difference – as well as her experience of colonialism – are potentially silenced in this way.

Lucy continues to experience a denial of her ever 'belonging' in New York. But at the same time Americans deny the 'difference' of the place from which Lucy has come. When Lucy meets Mariah and Lewis' many friends who have taken holidays in Antigua – or just 'the Islands' – Lucy wishes she came from a place that was 'filled with slag and unexpectedly erupting volcanoes', because, she says: 'somehow it made me ashamed to come from a place where the only thing to be said about it was "I had fun when I was there"' (Kincaid, 1994, p. 65). The paradox of coming so far from such a small place, so infinitely consumable, so consumed by past colonial incursions and present island tourism, produces for Lucy a great feeling of anger and an intensification of feelings of displacement and placelessness. The view of the tourist is one which is personified by the character of Dinah when she asks of Lucy: 'So you are from the islands?' Lucy recognizes Dinah as someone who gazes from a larger, powerful nation and for whom small places are consumable, glossy and generic – islands mean tropical islands – exactly where and with what local specificities is not important. For powerful nations and their citizens, as Gillian Beer notes, the idea of the island is no longer one of a fortress, defended by the sea, rather remote islands are accessible and commodified as leisure packages, but the commodification itself depends on an unchanged survival of long-standing western literary and scientific ideals of 'the island' (Beer, 1989, p. 21). Through the character of Dinah, Kincaid also draws attention to something Colleen Cohen and Frances Mascia-Lees have described as 'the excessive and almost hysterical production of a mythic Other – not atypical of Western tourists' (Cohen and Mascia-Lees, 1993, p. 141). Dinah negates Lucy's specificity – her race, place of origin and her experience of colonial rule – when she infers she knows 'the Islands', positing as 'natural' that everyone has been to Antigua, and had fun there. It is a similar invasion of Lucy as Mariah's claim of (Native American) Indian ancestry, to which Lucy responds (to herself) 'How do you get to be the sort of victor who can claim to be the vanquished also?' (Kincaid, 1994, p. 41). Lucy's observation, curiosity, participation and photography are aspects of a strategy of reversal of this invasive pseudo-anthropological gaze. Lucy positions herself as a powerful onlooker gazing at the New York 'natives' – the family, especially Mariah, and others like Dinah – with whom she is in turn fascinated, amused, infuriated and humiliated. When Mariah claims 'possession' of the 'trophy' of Indian blood Lucy asks her 'How do you get to be that way?' and 'the anguish on her face almost broke my heart, but I would not bend. It was hollow, my triumph, I could feel that, but I held onto it all the same' (pp. 40–1).

Triumph and Syncretic Identities

By conceiving new identities for herself and by resisting those bounded identities or locations set out for her by others, Lucy does undergo positive, enabling shifts of identity as part of her migration to New York in *Lucy*. Similar in theme (and more openly autobiographic) is Kincaid's written piece called 'Putting myself together' – about her first experiences of New York. This title suggests either a prior decimation or dismemberment of the self being brought about by the migration experience, or the wider experience of colonized people, living and shattering within the oppressive institutions of the dominant culture. Of course, these two readings of the title could speak to the experiences and contexts of both Antigua and New York. In 'Putting myself together' and *Lucy*, Kincaid writes of mimicry and participation in family and social systems which is partial and transgressive. For instance, she details dressing in clothes and hats and shoes from welfare shops, undertaking a deliberate, impersonating play:

[My clothes] had been worn by people who were alive when I had not been; by people who were far more prosperous than I could imagine being. As a result it took me a long time to get dressed, for I could not easily decide what combination of people, inconceivably older and more prosperous than I was, I wished to impersonate that day. (Kincaid, 1995, p. 98)

Mary Chamberlain argues that 'superficial and ephemeral as they may seem', clothes play a central role in the performance-experience of many Caribbean women's migration. As Chamberlain sees it, 'women refer to the clothes they wore, for dressing well place[d] them in the centre, as subjects, as creators of illusion' (Chamberlain, 1997, p. 105). It is through such iconography that women signalled their individuality, their difference.

Desire for an individual history and a recognition of difference continues to dictate Lucy's experiences in the latter part of the novel. Refusing assimilation, Lucy watches from 'outside' as Lewis' extra-marital affair causes the perfect family to fall like Rome around her. By her use of this metaphor of the fall of Rome, Kincaid provides multiple allusions:

'Rome' recalls the proscription of paganism (involved in how people like Dinah regard Lucy), the riots of Thessalonica (incidents in which islanders, formerly slaves, are expected to 'misbehave') and the invasion of 'territory' in the name of civilisation. [The metaphor] encapsulates not only familial disintegration but its colonizing aspect and Lucy's role as the outsider-servant who refuses at many levels to mimic or participate in the system. (Ferguson, 1993, p. 246)

Even after Lucy frees herself of Mariah and Lewis and moves into her own place, she remains unconnected to her new life, shut out of the society she has come into. From the window of the new flat she shares with her friend Peggy she can see only roofs: 'Everything I could see looked unreal to me; everything I could see made me feel I would never be part of it, never penetrate to the inside, never be taken in'. (Kincaid, 1994, p. 154).

There are many references toward the closure of *Lucy* to a falseness of things: the false lemon scent of Peggy's hair, the falseness of her relationship with Paul; the falseness of the 'landmarks' she had once imagined for herself which had now eventuated but were empty of happiness and satisfaction: her own house of sorts, a man in her own bed. The photographs she had begun to take and develop herself also represented a falseness of expectation: 'I would try and try to make a print that made more beautiful the thing I thought I had seen, that would reveal to me some of the things I had not seen, but I did not succeed' (p. 160). At this time, after a second winter in America, Lucy is consolidating and reviewing all that she has, that which she has lost and envisioning that which she wants to be:

I understood that I was inventing myself, and that I was doing this more in the way of a painter than a scientist. I could not count on precision or calculation; I could only count on intuition. I did not have anything exactly in mind, but when the picture was complete I would know. I did not have position, I did not have money at my disposal. I had memory, I had anger, I had despair. (p. 134)

Throughout the novel, however, Kincaid does not show her protagonist attempting to choose between identities. But this does not indicate that Lucy is unaware of the dictates of essentialized sexual, geographic or cultural oppositions. Lucy was quite aware, in Antigua, of the two options most readily available to her: to remain in a place where she cannot really be at home or to journey to a place where she will be a vulnerable outsider (Simmons, 1994, p. 121). Having come so far, this dichotomy prevails in the same way for Lucy in New York as it did in Antigua. Now that she is a migrant, a very different person and a very different daughter, any return to Antigua would position her again as a 'vulnerable outsider'. Her identity exists between being somehow, in New York, the devil – a figure she holds on to,[5] an excluded person but a powerful one – and the 'warm blue sea' of her Caribbean, and there being a powerless and colonized person. This sub-strait metaphor of 'betweenness' embodies the multiple contradictions of the migration experience; the places of the 'devil' and the 'warm blue sea' both contain overlapping territories of empowerment and disempowerment. The warm blue sea also represents in *Lucy*, and in Kincaid's other writings, an amniotic and idyllic state of childhood and cultural origin. The devil – though powerful and large-bodied – is still an outcast, representing a painful and fundamental exile.

Lucy instead chooses to speak to herself, make sense of her hybrid origins: the 'foul deed' of Columbus's colonialism and her own place – the island of Antigua – of which she has little real knowledge. In New York, re-inventing herself, Lucy traces her own personal marginalization back to the beginnings of her resistance to and awareness of colonialism. In her thoughts she returns to the island. She remembers herself as a schoolgirl, refusing to sing Rule Britannia. She also remembers as a child having a pen-friend on a nearby French territory – an island which was within view of her own island. At that time Lucy wished she lived under the French rather than the British colonial power because 'the stamps on her [friend's] letter were always cancelled with the French words for liberty, equality and fraternity' while Lucy's own letters showed only 'an image of a stony-face sour-mouth woman' (Kincaid, 1994, p. 136). The irony of her distaste is not lost on the older Lucy: 'I understand now that in spite of the words my pen pal and I were in the same boat.' Yet she is still able to see a flexibility and a disjunc-ture between the experience of Empire and the 'face' of that imperial power (Ferguson, 1993, p. 251). It is in this disjuncture and flexibility of interpreta-tion that Lucy suggests an emancipatory potential. Between the signifier, the 'face' of Empire and the receiver, 'the colonial subject', is a whole world, an entire language of subjective and complex experience or association.

Interestingly, Antigua continues to be a source of Lucy's identity as a woman, folding her mother's voice and strength into a language of female-ness, as a 'fertile soil' for her own creative representation of herself. At the same time Lucy is almost violent in her construction of herself as a unique and autonomous person – the antithesis of her mother, especially her mother as an Englishwoman, a dutiful and proper wife, mother and moral-ist. Lucy feels that New York provides the space for such a construction or transformation of self to occur, and she feels she could not return to Antigua, except perhaps to die there. Yet Lucy does not ever feel completely a part of the new place. It continues to seem unreal and unwelcome; the only conti-nuity in her experience in the physical landscape is a feeling of exclusion. Lucy is alienated too within the family for which she works; she is a woman, a black woman and an employee, a hired help, an outsider, the 'Visitor'. In the domestic interaction between Lucy, Mariah and Lewis, Kincaid's fiction vividly portrays the racism described by Afro-American theorist and writer bell hooks as the situation whereby 'everywhere we go there is pressure to silence our voices, to co-opt and undermine them. Mostly, of course, we are not there [within the space of the dominant culture]. We "never arrive" or "can't stay".' (hooks, 1990, p. 148)

Bronwen Walter writes of Irish migrants: 'the denial of difference to Irish people [in Britain] is part of the denial of the true extent of British imperial-ism' (Walter, 1995, p. 47). The same may be said for Caribbean migrants in America and the denial of the extent of British and American imperialism and neo-imperialism in the Caribbean. In response to being made invisible and excluded, Lucy chooses to maintain many different social and personal 'locations' throughout the novel. In this manner, Kincaid illustrates Homi Bhabha's notion that 'difference is not simply a matter of tradition. It is inscribed and re-inscribed through lived experience and is a complex, on-going negotiation relating to the political conditions of the present' (cited in Walter, 1995, p. 47).

Conclusion

In this reading of Kincaid, especially her novel *Lucy*, I have sought to extend the current literature seeking to understand women's experience of migra-tion. In *Annie John* and *Lucy*, Kincaid's Caribbean contexts portray a culture of migration; however Lucy's migration takes unexpected turns. Instead of providing remittance support and operating as an arrival point for the migration of other family members, Lucy refuses almost all her economic and emotional responsibilities to her family, especially to her mother. Not only has she out-grown Antigua but, on leaving, she refuses to be tied to a household she sees as restrictively colonial and controlling. In these ways Lucy belies our ideas about migrant populations in destination societies and Caribbean women's migration experiences in particular. Lucy's migration stresses enormity and permanence, two qualities of Caribbean women's migration explored by Chamberlain (see Chamberlain, 1997, p. 101); but contrary to Chamberlain's observed norms, Lucy *does not* express a compulsion to be re-united with her family in Antigua and *does* narrate her experience in a more masculine vein of autonomy and heroism.

However, in shutting out contact with her home, Lucy still does not find complete freedom from Antigua. One of the ironies of Lucy's migration experience is a realization of the sameness of things, that: 'As each day unfolded before me, I could see the present take a shape – the shape of my past' (Kincaid, 1994, p. 90). In regard to relationship to landscape and the expectations of people she meets, Lucy, like most migrants, experiences a sense of exile, a feeling described by Caribbean writer George Lamming as one of 'inadequacy and irrelevance of function in a society whose past one can't alter, and whose future is always beyond reach' (Lamming, 1995, p. 12).

Debate around notions of 'centre', 'periphery' and 'exile' has long contributed to the Caribbean imagination and Caribbean literature. The idea of the Caribbean as a marginalized place has been replaced with a more self-assured post-colonial re-drawing of centre and periphery diagrammatics. Post-colonial authors like Kincaid write not from a cosy aspect of the centre (as accepted and canonized), nor from that place defined by the centre as the periphery, but from another 'place'. In *this* place is nurtured a strong 'third voice' of literature and cultural identity and the possibilities of greater and continuing self-design. Kincaid's writing of Antigua and England, and Antigua and America, emphasizes that

> [it has] always been difficult to distinguish the inside of a place from the outside; indeed, it is precisely in part the presence of the outside within which helps construct the specificity of the local place. (Massey, 1992, p. 13)

That is to say, in re-presenting place generally and Antigua specifically, Kincaid erodes the spatial distinction between 'local' and 'international'. The conventional map of Empire or The World is reversed, so that the (subjectively attributed) smallness of a small place – a small island – is made large. The exploitation of Antigua, like sadness, 'weighs worlds' (Kincaid, 1986, p. 84). Conversely, New York and England are shrunk to their particularly perverse size. In much of Kincaid's writing she regards England and the so-called 'universal' norm as the very local perspective of that small place – England – which, through the criminal activities of colonialism, has been able to control so much of the rest of the world (Tiffin, 1990, p. 37). The 'small place' is not, in the end, Antigua only – it is England as well. Real and perceived distances are brought squarely into question: 'I used to think that just a change in venue would banish forever from my life the things I most despised. But that was not to be so' (Kincaid, 1994, p. 90).

Broadly speaking, *Lucy* illustrates the way in which a place of origin, one's 'home' – through the expectations of others and one's own memories and fears – continues to interrupt the 'new' place and the new identities of that place. For Lucy and Kincaid, Antigua represents family and childhood, stagnation and repression within a patriarchal–colonial system, and a small-town familiarity which stifles individuality and creativity. Although Lucy is living in New York, Antigua remains the locus for her post-colonial awareness, her racial and cultural identification. Concurrently, the articulation of these identities and her long-sought emotional, sexual and economic independence are made possible by her being in New York. While Lucy has been critically politicized by an awareness of the colonial history of Antigua, and the contemporary 'place' of the small island of Antigua in the world, she wishes to write her *own* history: to 'be free to take everything just as it came and not see hundreds of years in every gesture, every spoken word, every face' (Kincaid, 1994, p. 31). For Lucy there is no real desire to return 'home'.

Notes

1. All page references are to the 1994 edition of *Lucy* published by Picador.

2. 'I have never had a male view. I don't have to say, make an effort to say "she". For me, it is *she*, understood. I identify myself as my sex, not my sexuality, my sex, I think that is the word.' (Kincaid quoted in Ferguson, 1994, p. 186)

3. Lucy states on leaving the Lakes: 'I would not miss the lake, I would not miss anything, for long ago I had decided not to miss anything. I sang songs: they were all about no pot of gold at the end of the rainbow.' (Kincaid, 1994, p. 81)

4. 'It became clear to me that I was writing about the mother [and] that the mother I was writing about was really Mother Country. It's like an egg; it's a perfect whole. It's all fused in some way or other.' (Kincaid quoted in Ferguson, 1994, pp. 176–7)

5. Of Lucy's name Lucy's mother – frustrated and angry – had once told her: 'I named you after Satan himself. Lucy short for Lucifer.' At the time Lucy 'went from feeling burdened and old and tired to feeling light, new, clean. I was transformed from failure to triumph.' The reason for this Lucy gives as being: '[from reading the Bible, Milton's *Paradise Lost* and Shakespeare] the fallen were well known to me, but I had not known that my own situation could even distantly be related to them' (Kincaid, 1994, p. 152). Lucy appropriates the centrality of Lucifer (central to the systems of belief and [dis]order into which she has been educated and has come to know so well) and finds herself liberated by such an identity. She finds herself not annihilated or extraneous but gloriously and dangerously exiled.

References

Aldrich, R. (1995) 'From Francité to Créolité: French West Indian literature comes home', in R. King, J. Connell and P. White (eds), *Writing Across Worlds: Literature and Migration*. London: Routledge, pp. 101–24.

Beer, G. (1989) 'Discourses of the island', in F. Amrine (ed.), *Literature and Science as Modes of Expression*. Boston: Kluwer Academic Publishers, pp. 1–27.

Blunt, A. and Rose, G. (1994) *Writing Women and Space: Colonial and Post-colonial Geographies*. New York: Guilford Press.

Chamberlain, M. (1997) 'Gender and the narratives of migration', *History Workshop Journal*, 43, pp. 87–108.

Cohen, C.B. and Mascia-Lees, F.E. (1993) 'The British Virgin Islands as nation and destination: representing and siting identity in a post-colonial Caribbean', *Social Analysis*, 33, pp. 130–51.

Cudjoe, S.R. (1990) 'Jamaica Kincaid and the modernist project: an interview', in S.R. Cudjoe (ed.), *Caribbean Women Writers: Essays from the First International Conference*. Massachusetts: Calaloux Publications, pp. 215–32.

Ferguson, M. (1993) *Colonialism and Gender Relations from Mary Wollstonecraft to Jamaica Kincaid: East Caribbean Connections*. New York: Columbia University Press.

Ferguson, M. (1994) 'A lot of memory: an interview with Jamaica Kincaid', *The Kenyon Review*, 16(1), pp. 163-88.

Hall, S. (1992) 'Cultural identity and cinematic representation', in M. Chan (ed.), *Exiles: Essays in Caribbean Cinema*. Trenton, NJ: Africa World Press, pp. 220–36.

hooks, b. (1990) *Yearning: Race, Gender and Cultural Politics*. Boston: South End Press.

Kincaid, J. (1985) *Annie John*. New York: Plume.

Kincaid, J. (1988) *A Small Place*. London: Virago Press.

Kincaid, J. (1991) 'On seeing England for the first time', *Harper's Magazine*, August, pp. 13–17.

Kincaid, J. (1992) *At the Bottom of the River*. New York: Plume; originally published 1978.

Kincaid, J. (1994) *Lucy*. London: Picador; originally published 1991.

Kincaid, J. (1995) 'Putting myself together', *New Yorker*, 20 February, pp. 93–101.

Kincaid, J. (1996) *The Autobiography of my Mother*. New York: Farrar, Straus & Giroux.

King, R., Connell, J. and White, P. (eds) (1995) *Writing Across Worlds: Literature and Migration*. London: Routledge.

Kinsman, P. (1995) 'Landscape, race and national identity: the photography of Ingrid Pollard', *Area*, 27(4), pp. 300–10.

Lamming, G. (1995) 'The occasion for speaking', in B. Ashcroft, G. Griffiths and H. Tiffin (eds), *The Post-colonial Studies Reader*. London: Routledge, pp. 12–17; piece originally written in 1960.

Massey, D. (1992) 'A place called home?', *New Formations*, 17, pp. 3–15.

Mintz, S.W. and Price, S. (1985) *Caribbean Contours*. Baltimore: John Hopkins Press.

Momsen, J.H. (ed.) (1993) *Women and Change in the Caribbean*. London: James Curry.

Olwig, K.F. (1993) 'The migration experience: Nevisian women at home and abroad', in J.H. Momsen (ed.), *Women and Change in the Caribbean*. London: James Curry, pp. 150–66.

Simmons, D. (1994) *Jamaica Kincaid*. New York: Twayne Publishers.

Tiffin, H. (1990) 'Decolonialization and audience: Erna Brodber's *Myal* and Jamaica Kincaid's *A Small Place*', *Span*, 30, pp. 27–38.

Walter, B. (1995) 'Irishness, gender and place', *Society and Space*, 13(1), pp. 35–50.

White, P. (1995) 'Geography, literature and migration', in R. King, J. Connell and P. White (eds), *Writing Across Worlds: Literature and Migration*. London: Routledge, pp. 1–19.

10

'My Island Home': The Politics and Poetics of the Torres Strait

John Connell

Six years I've been in the city
And every night I dream of the sea
They say home is where you find it
Will this place ever satisfy me?
For I come from the saltwater people
We always lived by the sea
Now I'm down here living in the city
And my island home is waiting for me
(*My Island Home*, Neil Murray, adapted by Christine Anu)

Like so many other island realms, the Torres Strait – between Australia and New Guinea – has increasingly become characterized by migration. Islanders have moved south to Australia, adopted new lifestyles but remained involved in the various ramifications of island social life. This chapter examines the structure and context of that migration, and traces the manner in which two migrants from the islands, Eddie Mabo and Christine Anu, in quite different ways – through politics and music – have both transformed island life and changed wider perceptions of the geography of indigenous Australia.

Some Historical and Geographical Background

The shallow Torres Strait, the 150 kilometre-wide gap between the northern tip of Australia, Cape York, and the south coast of Papua New Guinea, is studded with islands, coral reefs and sandbanks (Figure 10.1). Seventeen are occupied by indigenous Melanesians, whose culture and ethnicity are closer

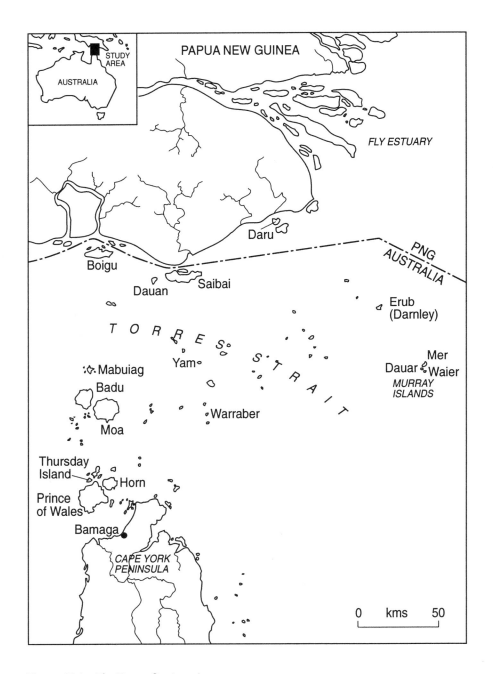

Figure 10.1 The Torres Strait region

to the Papuans to the north rather than to the Aboriginal population of Australia, with which the islands are politically integrated. Culturally and linguistically the indigenous population of the Torres Strait region consists of two main groups: the Meriam Mir- (Mer-) speaking people who reside on the high islands on the eastern side of the Torres Strait, and Kala Lagaw Ya speakers in the low-lying central and western islands. Prior to European contact, the inhabitants of the region were organized in small communities, with alliances based primarily on intermarriage, raiding and the reciprocal exchange of food and other resources; mixed horticulture, hunting and fishing economies prevailed. Despite cultural and economic variations across the Strait, contemporary Torres Strait Islanders are officially recognized as a distinct, indigenous minority within Australia. More importantly, they see themselves as distinct and unlike either Aborigines or Papua New Guineans.

In 1606, the Spanish explorer, Luis de Torres, became the first European to pass through the Strait. Despite intermittent European contacts, a new phase of history only began with the arrival of the London Missionary Society on Darnley Island in 1871, ushering in the 'coming of the light' and the era of 'pearlers, pastors and protectors' (Beckett, 1987). The discovery of pearlshell transformed the economy of the islands – creating wage employment – and the colony of Queensland quickly annexed the islands in 1879. For almost a century the exploitation of marine resources – either pearls or bêches-de-mer – was the basis of a colonial or neo-colonial economy. In the post-war years, however, the pearl-shelling industry, the only significant productive sector in the Strait, experienced rapid decline, and had disappeared by the 1970s. Government intervention became an ever more important element of the region's history. The public sector grew steadily, providing some new jobs, but never sufficient to meet local needs. By the 1970s a significant proportion of the islanders' incomes came from unemployment benefits and other social service allowances (Duncan, 1975, p. 58), a situation that has subsequently continued.

At the time of mission contact the islander population was probably between 4000 and 5000 but, like some other Melanesian populations, declined in subsequent decades. At the first official count in 1913 there were a mere 2368 islanders. When reasonably reliable census data became available in the 1950s the population was still no more than about 6000, almost all resident in the islands, but was growing extremely rapidly (Beckett, 1987, pp. 26, 38, 69–70). Even allowing for doubts over the reliability of census data, subsequent growth was exceptionally rapid; in 1981 the number of Torres Strait Islanders had increased to 15,230 and by 1996 to 28,740, at which time less than 4750 remained in the islands, with most of the others elsewhere in Queensland.

Out-migration during the early post-war decades was extraordinarily rapid. Economic change and migration restructured the islands of the Torres

Strait, resulting in three local divisions: the Prince of Wales group, including Thursday Island which is the urban administrative centre (with nearby airport, retail sector and a significant European and 'mixed race' population); the 'outer islands', which have more subsistence-oriented economies; and the mainland 'cape communities', composed of islanders resettled on Cape York, mainly at Bamaga. An increasingly significant population lived further south in Queensland.

Government funding levels throughout the Strait increased from the 1970s, producing a situation of 'welfare colonialism' (Beckett, 1987) at exactly the time that both state and federal governments were seeking to reduce the extent of dependency on the public purse. The effects of high levels of unemployment were cushioned by the Australian welfare state, and some attractions of island residence were accentuated (not least to Papua New Guineans who became seasonal and long-term migrants into the islands). The Strait is now highly dependent on government grants and expenditures, and on the welfare sector, which contributes at least 42 per cent of the regional economy's turnover (Arthur, 1992, pp. 22–3). It remains economically depressed, since marine activities, mining, tourism and other ventures have only flourished briefly. Moreover the impact of substantial welfare spending has contributed to 'a reluctance to garden or to increase subsistence fishing' despite the fact that subsistence activities are of real and symbolic value since local 'resources have a high social and gastronomic significance which makes them central to the reproduction of social relations' (Davis, 1995, pp. 9–10, 25). On several islands, especially those most distant from Cape York, almost all employment is government-sponsored, and private sector jobs are non-existent. On Saibai, for example, there are a small number of jobs as councillors, council staff and government department employees, which generate about a quarter of local income. The other three-quarters come from welfare payments, which provide income security. Indeed, because of increased welfare payments, there was a fivefold increase in Saibai's real income between 1973 and 1993 and 'an almost total underwriting of the cash economy by various levels of government' (Davis, 1995, pp. 10, 12, 24). At the centre, on Thursday Island (TI) especially, a combination of racism, paternalism and government subsidies had produced critical social problems: a quarter of the population were diabetic, infrastructure and services were lacking and high unemployment levels prevailed (Kehoe-Forutan, 1991). TI remained quite different from the other islands: 'TI was European, a transitory stopover and not actually anyone's home' (Kehoe-Forutan, 1988, p. 14). A place traditionally forbidden to the other islanders, it represented all that was wrong with policies of segregation and government control. To some extent, throughout the Strait, a 'deluge of money' (Singe, 1989, p. 226) and the emergence of 'government islands' had taken their toll: a curious condition of 'affluent poverty' (Scott, 1990, p. 392) had been put in place.

By the 1970s, the transition from economic exploitation to welfare colonialism had emphasized a new dependency; the productive potential of the Torres Strait had stagnated and declined, leaving the islands, like so many others in the Pacific, with MIRAB economies – dependent on MIgration (and sometimes the resultant Remittances), government Aid and hence with a Bureaucratic economy (Connell, 1988, pp. 81–2). Some island communities were 'in danger of becoming parasitic societies' (Singe, 1989, p. 221). The transition from subsistence to subsidy (Duncan, 1975) was well entrenched, a rapidly growing proportion of islanders lived on the mainland and there were long-standing fears over the erosion of local cultures, first expressed by Haddon's Cambridge Anthropological Expedition in 1898 which saw itself as being 'just in time to record the memory of a vanished past' (quoted by Sharp, 1993, p. xii). The future viability of island life appeared highly uncertain.

Islands of Emigration

Until the end of the Second World War, and to some extent much later, Torres Strait Islanders were restricted by law and administrative arrangements to living on their home islands, a similar situation to that elsewhere in Melanesia. Though some islanders travelled beyond the Strait as shipping crew, there was no evidence of distant residence. The demise of pearling dramatically transformed that situation. Indeed the right to leave the islands coincided with the collapse of that industry (Sharp, 1993, p. 35), stimulating rapid migration in search of employment. By 1986 over 80 per cent of Torres Strait Islanders were resident on the Australian mainland (Taylor and Arthur, 1993). 'Islands of emigration' had been produced in no more than a quarter of a century.

Migration was a response to the limited economic base of the islands, a fall in local labour demand, a growing labour force and a greater recognition that personal freedom, living standards, wage levels and employment opportunities were all poorer in the islands than on the mainland. A new settlement was established in 1948 at Bamaga on Cape York to resettle flood victims from Saibai; this settlement also attracted people from other islands to work in agricultural and sawmilling projects (Beckett, 1987, p. 70; Singe, 1989, pp. 184–5). After an early phase of cane-cutting employment, the majority of islanders found work on the Queensland railways – occupations that were unattractive to white Australians because of low wages and poor working conditions (Beckett, 1987, pp. 72, 202; Fisk *et al.*, 1974, p. 49). The employment history of Eddie Koiki Mabo (Loos and Mabo, 1996) was typical of many others who preceded and followed him. Chain migration was typical.

An economic rationale was not the only reason for migration. Social factors were important too, including migration in response to family feuding.

Young, single men perceived themselves as 'repressed and exploited by their elders' (Beckett, 1987, p. 127) – the very stuff of small island life. Overall, migration to the mainland was a safety valve for growing unemployment, social and political discontent and, increasingly, for growing population pressures (Beckett, 1987, pp. 67, 127).

In 1960 there were no more than about 500 islanders working on the mainland, mainly single men from the Eastern Islands (Beckett, 1987, p. 72), though some island councils were still discouraging and blocking migration. From then onwards, as island employment contracted, the balance rapidly tilted towards mainland Australia. It also tilted towards urban Queensland, where employment opportunities, with wage levels three times those in the islands, slowly opened up (Fisk *et al.*, 1974). By 1971 there were 3000 islanders on the Queensland mainland, with 71 per cent in the three northern coastal towns of Cairns, Townsville and Mackay (and a further 11 per cent in Brisbane). At that time some 78 per cent of all islanders were in Queensland (Beckett, 1987, p. 180), but others had migrated to Western Australian pearling towns, urban New South Wales and elsewhere. Dispersion subsequently became more rapid, so that by 1991 little more than half (54 per cent) of all islanders were actually in Queensland. By 1996 some 5670 islanders were enumerated in the Torres Strait, though 930 of them lived on Cape York and a further 1440 in TI (where they barely made up the majority on the island), out of a total of 28,740 islanders. Away from TI and Cape York, island populations were tiny; Badu had much the largest population with 506 islanders, followed by Murray Island (Mer) with 391.[1] By contrast there were 2650 in Cairns and more than 1000 in each of Brisbane, Mackay and Townsville. Islanders had become dispersed, the islands thinly populated and an urban future seemingly evident.

It was concluded in the 1970s that though some migrants might make return visits, it was unlikely that many would resettle permanently: 'after a few years in the south there is little likelihood that they will return to the Torres Strait with its poor housing, chronic unemployment, low wages, extravagant prices, restrictive atmosphere and limited future' (Singe, 1989, p. 219; see also Duncan, 1974, p. 101; Fisk *et al.*, 1974, pp. 27–42). Duncan observed that though 'some of the older folk remaining on the outer islands were quite optimistic about their migrated offspring returning to their home island', in practice such optimism was 'difficult to accept'. Hence she 'interpreted the notion of remigration (i.e. return migration) as an expression of sentiment' (1975, p. 63) rather than a realistic intent. From the 1970s to the 1990s migrant islanders intimated that they would return only if employment and education improved (Fisk *et al.*, 1974, p. 31; Taylor and Arthur, 1993, p. 30). Since these barely changed, return migration was minimal. Migrants appeared to have gone for good.

Despite fears of permanent emigration – and even the complete depopulation of the smallest islands – by the 1980s the population imbalance was changing. The islands began to experience the first trickle of return migration and the most rapid phase of population decline ended. For the first time in more than a quarter of a century, the period 1991–6 witnessed an increase in the islander population of the Torres Strait. Return migration was stimulated by the retirement of many of the early migrants, the ability to receive the full range of social security benefits in the islands, some access to income from rock lobster and trochus fisheries and public sector employment (Taylor and Arthur, 1993, p. 30), improved transport (at relatively low cost) within and beyond the Strait and, most recently, a growing pride in Torres Strait identity. The ethos of the islands was changing.

Eddie Koiki Mabo, Land and Independence

Migration from the Strait minimized political friction, whilst new welfare payments reduced the impact of unemployment; hence although island administration remained paternalistic, racist and often distant, there was little demand for change in political status. There was some interest in greater autonomy in the early 1970s, stimulated by border disputes with Papua New Guinea, but the independence of PNG in 1975 came as a shock to many islanders, who had long regarded themselves as economically superior (Beckett, 1987, p. 210). At a conference on the border issue in the following year, there was minority support for independence, with a migrant islander, Eddie Mabo, commenting:

> I think independence should be accepted as a long-term policy. We are a people of unique identity and we should work towards an ultimate goal of independence. We want to be recognized separately from our Papuan brothers and from our Australian brothers. We *are* the Islanders. (Mabo, 1976, p. 35)

In the same year there were calls for independence from other mainland islanders. They had formed a Torres United Party and proposed 'The Free Nation of Torres Strait', to be financed from leasing oil and gas prospecting rights, postage stamps, fisheries and electricity generation from underwater turbines. The proposal had little overt support in the Strait.

In the mid-1980s islanders, this time within the islands, again began to call for sovereignty over the Strait. This move should be seen primarily as an expression of discontent and dissatisfaction with continuing government control and inadequate service provision, and as a means of demonstrating islander distinctiveness from Aborigines (Beckett, 1987, p. 196), rather than as a move to achieve full political independence (Kehoe-Forutan, 1988). Pressure for independence accelerated as Australia moved towards the bicentennial celebration of settlement in 1988; this pressure was spearheaded by the Island Coordinating Council which represented every island. The ICC had created in 1987 an Independence Working Party Committee led by Jim Akee, a mainland islander (and former Secretary of the Torres United Party) who had returned to live on TI after many years away. Getano Lui, the Deputy Chairman of the ICC, argued that the Strait was 'being invaded from the north and south', with increased European exploitation of resources (mineral and marine) and Papuan migration and fishing incursions. Many mainland islanders supported independence, because of their own loss of identity through inclusion with the Aboriginal population (formally in the Aboriginal and Torres Strait Islander Commission, ATSIC). Demand for independence came to a head in January 1988 when a conference of islanders decided to press the case for secession with the United Nations' Special Committee on Decolonization, despite local opposition, especially on outer islands (where loss of social security benefits was perceived to be a problem). The move for independence petered out after islanders were promised a new regional council and the rest of the world was uninterested in their claims.

The first Torres Strait Regional Development Plan was produced in 1990, and emphasized five development goals. The first and last of these were 'Regional Identity and Greater Self-Reliance' and 'Torres Strait Islander Ways', an affirmation of the need and desire to break with welfare colonialism (Lea *et al.*, 1990; Lea, 1994). The quest for separate identity strengthened in 1992, led by Getano Lui, now ICC Chairman, as islanders sought to achieve administrative autonomy, with a local parliament, more control over service delivery, separation from ATSIC and the right to determine policies on natural resource development (Lui, 1994). Unlike the situation in 1988, when calls for secession were largely calls for assistance, the 1992 demands were realistic attempts to acquire self-government by 2001, with a distinct constitution similar to that in Australia's island territories – Norfolk Island, Christmas Island and the Cocos (Keeling) Islands. Negotiations eventually focused on regional autonomy and were given impetus by the landmark Mabo land case, to be described in more detail shortly. In 1994 a new Torres Strait Regional Authority was established, with greater autonomy than the Council it replaced, a separate budget and the authority to draw up and implement development plans, but still dependent upon ATSIC funding. A 1997 Parliamentary Inquiry recommended a more powerful Torres Strait Regional Assembly, with full independence from ATSIC, and noted that there was no longer any desire for independence, but there were aspirations to territory status (Parliament of Australia, 1997, pp. 38–9).

At the core of the Torres United Party's demand for independence were land issues. In the 1970s the Party's Chairman, Carlemo Wacando, a Darnley Islander resident in Townsville, had successfully challenged in the High Court the validity of the 1879 annexation of the islands by the Queensland government, and this laid the basis for sustained arguments over indigenous control of lands and resources. That islanders had no secure land title to their islands was increasingly evident in discussions of the Australia–PNG border issue, a ruling that island ex-servicemen were not entitled to war service home loans because of their lack of land tenure, and an attempt by the Queensland Premier to turn islander (and Aboriginal) reserves into fifty-year leases: all decisions that were taken with some degree of insensitivity to islanders and their aspirations (Kehoe-Forutan, 1988, pp. 11–12). Islanders, especially those on the mainland, began to challenge these distant authorities.

Land issues culminated in the 'Crown versus Mabo' case, finally decided in the Australian High Court in 1992, which not only changed land tenure in the Torres Strait, but transformed the contemporary history of Australia in its removal of the doctrine of *terra nullius*: the proposition that land was vacant and not under any ownership regime when white settlement first occurred. The case had begun a decade earlier when Eddie Koiki Mabo, a Mer (Murray) Islander long resident in Townsville, and four other Murray Islanders, began legal proceedings to establish their traditional ownership of land on Murray Island. They claimed continuous enjoyment of their land rights and argued that these rights had not been extinguished by the annexation of the islands in 1879. The case took ten years, during which time Mabo and three other plaintiffs died. The court declared that the Meriam people were entitled 'as against the whole world to possession, occupation, use and enjoyment' of their traditional lands. The decision represented a significant change in the relationship between indigenous peoples and non-indigenous Australians (and became particularly important for Aboriginal Australians, because of the automatic extension of the decision throughout the country). The Prime Minister, Paul Keating, commented soon afterwards that: 'Mabo is an historic decision – we can make it an historic turning point, the basis of a new relationship between indigenous and non-Aboriginal Australia' (quoted in Crough, 1993, p. 5). The Australian government responded to the High Court decision by recognizing the existence of native title and establishing a National Native Title Tribunal to determine title claims (since, in practice, only the Meriam were given direct recognition). Above all it removed one of the most substantial barriers to the process of reconciliation between indigenous people and other Australians, and reshaped notions of Australian identity.

The Mabo judgement did not just confirm the inalienable right of Murray Islanders to their own land, but recognized their spiritual attachment to it and validated all that the islanders had believed in terms of their authority over their island. It restored and revitalized the crucial link between land and identity. Murray Islanders were recognized as the moral equals of other Australians, and the narrow conception of land as a tradeable commodity was abandoned (Sharp, 1996, p. 16). It also transformed Eddie Mabo into a national icon and a local hero: a man whom his lawyer regarded as 'a fighter for equal rights, a rebel, a free-thinker, a restless spirit, a reformer who saw far into the future and far into the past' (Loos and Mabo, 1996). Like many others of his generation, Mabo had left the Strait in the 1960s in search of employment, adventure and freedom from government controls but, unlike most others, he also moved outside islander circles, marrying a South Sea Islander (a descendant of nineteenth-century labour migrants from Vanuatu) and meeting radical unionists and academics with interests in indigenous matters. Such contacts had resulted in the Queensland authorities recommending to the Murray Island Council that he be banned from visiting, or returning to, Mer. The Council banned him but after his father died in 1975, he returned illegally but briefly to the island.

Although Mabo was the first plaintiff, he did not proceed alone, since his migratory status had produced marginality. Indeed to do so would have been problematic since the Murray Island Council had earlier resolved that those who left the island forfeited their land. Though the Council did not adhere to that position in the case of those who returned, the attitude lingered and migrants were even referred to as 'ex-Islanders' (Beckett, 1994, p. 11), a not uncommon attitude in Melanesia where only the village was the 'true' home. Yet migrants like Eddie Mabo were far from being ex-islanders; though Mabo had an ambivalent relationship to resident islanders, and some Mer people made him pay a heavy price for political activism, 'emotionally and intellectually he never left Mer, even though he lived most of his life on the mainland' (Loos and Mabo, 1996, p. 4). In becoming a cosmopolitan, a man of several worlds, he had restored and strengthened Torres Strait culture and identity. This was doubly ironic in that by moving away from Mer, and absorbing himself in Torres Strait culture through reading European writers' accounts (notably those of Haddon's Cambridge expedition of 1898), he lost credibility in the eyes of the High Court judge.

> Justice Moynihan did not regard Eddie Mabo as a real Meriam, but rather as an urbanized political activist, who seeing the main chance made up for his lack of roots by reading books. Compared with the oral tradition such knowledge was unauthentic ... It is ironic that while anthropologists become credible expert witnesses by writing, 'natives' render themselves unauthentic by reading: tainted with literacy it seems they can't go home again. (Beckett, 1994, pp. 19, 22)

In the end Mabo did return, but to a home that was forever changed.

Eddie Mabo himself did not live 'to hear the heroic music of the ultimate finding nor to partake of the events which became the foreground of a post-Mabo landscape, a time when his name became a household word' (Sharp, 1996, p.11). He died of cancer in January 1992, and was buried in Townsville, five months before the historic High Court judgment. After three years of mourning, the wooden cross that marked the grave was replaced by a marble headstone – a ceremony familiar throughout the Strait – but only a few hours later the grave was desecrated. Mabo's body was exhumed, and in September

1995 flown to Mer, where he was buried – accompanied by a Malo dance, to honour a 'great man' and stress the links that he had restored to the era before colonial and mission influence. For the wrong reasons, he had finally returned to the land of his ancestors. More than half the island population ignored the ceremony – a mark of intense island factionalism, and a further reminder of the significance of the social divisions at a different scale that Mabo had sought to overcome. Nonetheless one migrant had made his mark.

Saltwater People

Music has long played a central role in Torres Strait society. It has functioned as a vehicle for participation, a means by which islanders assert their position in island society and an expression of the nature of that society, reflecting and marking social boundaries and expressing kinship ties. Torres Strait songs are performed by choral groups that reinforce the importance of the group; as a Boigu island teacher has stressed 'Music is of intrinsic importance in our lives. As such, it is a social issue .... Music is a social event, a way we celebrate our culture together' (quoted in York, 1995, p. 28). The arrival of contemporary popular music has not fundamentally changed this.

With immigration from other Pacific islands, there was a long 'period of flirtation' with various South Sea styles of singing and dancing, mainly incorporating Polynesian songs and rhythms, until the inter-war years when islanders began to develop their own synthetic style that became dominant throughout the Strait (Beckett, 1981). That style became encapsulated in, and popularized by, the music of the Mills Sisters during the 1970s and 1980s. They combined traditional songs in the local languages, Creole and English, in forms which rhapsodized the islands. Lyrical and idyllic, they mainly sung in the Hawaiian 'pearly shells' tradition, incorporating calypso and country and western influences, with ukeleles and steel guitars. They sung of moons and stars and beaches, of a 'frangipani land', a 'tropical Queensland across the sea' where TI was an idyllic home, and problems were non-existent. They were enormously popular in the Torres Strait and the only islanders known to a wider world.

In 1995 a young black singer, Christine Anu, soared to the top of the popular music charts in Australia with her single 'My Island Home'. It was the first time that a Torres Strait Islander had received national acclaim in the music industry; her ethnicity was central to the themes and styles of her music, which drew much wider attention to the existence of Torres Strait Islanders.

Christine Anu was born in Cairns, where she lived until she was ten years old, one of seven children of a mother from Saibai and a father from Mabuaig. Her first language was Creole. Her family then returned to Mabuaig; her father was injured and unable to work, and living in Cairns was impossible on an invalid's pension. The Anu family became one of several to combine welfare and subsistence livelihoods in the islands. On Mabuaig electricity was absent, supplies arrived monthly from TI and television was replaced by singing, dancing and telling stories. Christine Anu recalled that 'It was bliss ... it's a child's playground to grow up on an island.'[2] She migrated again in order to finish her high school education in southern Queensland, then worked with the National Aboriginal/Islander Dance Theatre in Sydney and subsequently became a member of the national indigenous Bangarra Dance Theatre.

Her musical career evolved from dance, and in 1994 she recorded the album *Stylin' Up*, which included 'My Island Home'. The album was produced by David Bridie, who had previously worked with the mixed Papua New-Guinean-Australian group Not Drowning Waving, which had incorporated traditional Melanesian sounds into popular music. Bridie spoke Pidgin English, had some understanding of Melanesian culture, and travelled with Anu to TI where they collected samples of island sounds, local choirs, children singing and people talking Pidgin English, fragments of which were incorporated into the album.

Stylin' Up was widely acclaimed. *Juice Magazine's* critic observed that it was 'one of the most exquisite cross-cultural hybrids to be produced in this country .... Christine Anu has created one of the most sublime and stylish reflections of Australia's cultural melting-pot ever to be recorded' (Klein, 1995). At the same time *Vogue Australia* noted that 'Christine Anu unites two worlds in her superb debut album' whilst *Australian Style* spoke of her as being 'this country's first indigenous female pop music superstar'. The industry gave her awards and the album went 'gold' by the end of the year. A star had been made.

Christine Anu intended to produce an album that traced her heritage: 'I'm trying to sprinkle little pieces of my life into music, no-one has had my experiences'. In a musical sense, this came from combining the samples recorded in the islands with music produced in the studio using Torres Strait instruments, such as the *warup* (a narrow log covered with a goanna or snakeskin and played as a drum) and the *kolap* bean rattle (played by slapping against the body), alongside more familiar modern instruments. Traditional music styles were linked to contemporary forms like hiphop, reggae, funk and soul, each of which had already found a place in the Strait. Language too linked different realms; the album, if primarily in English, incorporated Anu's Kala Lagaw Ya language and Torres Strait Creole. As she has said, 'language, as much as our traditions, helps to give us identity and our sense of belonging'; moreover 'I believe language is the core of culture ... my traditional language and culture have been handed down through generations. I have a responsibility to see it is passed on ... in that sense I want to be a teacher, to write songs about my culture.' She brought the music of the Mills Sisters, many of whose concerts she had attended as a child, both forward into the modern world and more deeply into the past.

The most famous song on the album, 'My Island Home', opens with the sounds of water lapping against the hull of an aluminium dinghy – and throughout the song the water and sea are ever present: the island world of the Torres Strait, a place for nostalgia and yearning, evoked through a slow dance rhythm and layers of orchestral atmosphere. It brought an idyllic vision of the islands to an enormous public. The song had been first recorded by the Aboriginal Warumpi Band from the Northern Territory, and was reworked by Anu, so that 'west of Alice of Springs' became 'living in the city'. This adaptation and transformation emphasized some pan-indigenous expressions of identity, and the manner in which Torres Strait identity related to engagement with institutions and cultural traditions beyond the local region. 'My Island Home' also updated the music and maintained the spirit of the Mills Sisters, and other songs on *Stylin' Up* took similar themes. 'Wanem time?' promised that TI time was different

You're rushing through the day
The city makes you pay
Don't waste your life away
There's an island in the sun
Where the children they have fun
Don't forget where you belong
Even when you're far from home.

In 'SanE Wireless', inspired by a Creole song learned by Anu at primary school, she was also in a nostalgic mood: 'I've been thinking of home lately, I send my thoughts back ...'.

Other songs were simply evocative of diverse aspects of life in the islands: 'TamaOma', composed initially by a Murray Islander, had no particular meaning but was well known since a local TI band had previously taken it to number one position on the local radio station. 'Dive' triumphed oceans and reefs: 'I can hear the ocean's mighty thunder sound under the Milky Way'. 'Sik O' records a nineteenth-century event when island canoes repulsed a British warship; here Anu sought to represent the spirit of Torres Strait Islanders, their respect for the sea and the elements and their spirit of resistance to alien authority. 'The Monkey and the Turtle' combined Creole, English and Kala Lagaw Ya in an adaptation of an old Torres Strait nursery rhyme, a song known to every islander, and very much a 'uniting song'. 'Come On' provides a more challenging moment in describing the problem of alcohol abuse in islander communities, and the social breakdown that follows.

You wake up in the morning
Your mouth is arid, dry

Red blazing eyes a light
Like the flame inside
Your wife and children gone
Through tears you swallow pride.

Anu's father was an alcoholic and her own family suffered from the effects of alcohol abuse. Lighter and brighter songs remove the sombre tones, offer a 'light at the end of the tunnel', and disperse the more serious tone. In the majority of the album a triumphant old order was remembered, the songs and rhymes of the islands given prominence, and the old was fused with the new.

Perhaps the most poignant song on the album is 'Photograph', both a lament for the changes wreaked by missions and colonialism and a promise of a future that would draw its inspiration from the past. It was inspired by photographs of nineteenth-century islanders that Anu saw at Cambridge University on a visit with the Bangarra Dance Theatre. As Anu has said: 'if we acknowledge the way our ancestors lived, if we acknowledge the way that we live today, then tomorrow is a better place for our kids'. History was reclaimed.

Whose clothes are you wearing?
I can see they're from the mission world.
Why aren't you smiling?

In 'Photograph' Anu paid her own tribute to Mabo: 'I have a message to tell. Eddie Koiki Mabo is someone to look up to; he became a symbol, like a door that was opened. His is a very strong spirit that will remain for generations to come, because they will be walking through that door that he opened.' Hence in the song

You're with me every step I take
You're like the roots in the ground
Passing love to our family
You're the knowledge
You're the strength that I've found.

That strength enabled Anu to be the respected voice of a new generation, leaving her 'in the prime situation where I can pass down what has been our oral history for generations'.

But in all this the lure of the city could never be forgotten: a place of action and excitement, but where community life was still to be found.

Tidda [indigenous] girl has got her eyes up off the ground
Keeping time to the rhythm of the city sounds
Don't hold back, let it out, that's what community's about
Come along to my house, bring friends and family around.

The city was a place to 'Party', a place for 'Stylin' Up', representing the most obviously commercial element of the album, and a contemporary counterpoint to the past and the future in the Torres Strait. Displacement was an inescapable element of Torres Strait life; indeed it was Anu's own life.

Popular music here, as elsewhere, has become a mechanism of expression and empowerment, above and beyond more conventional political channels, intersecting with wider Australian institutions and audiences in diverse ways (Gibson, 1998, p. 165). The social concerns of the songs, all but the most 'traditional' being partly written by Anu, symbolically inscribe markers of indigeneity on the landscape and re-claim the islands as Melanesian places. The success of *Stylin' Up* boosted local pride in Torres Strait identity, language and culture, and enabled the Torres Strait to be heard and better understood in mainland Australia. A second migrant, Christine Anu, had added a new dimension to the processes of revitalization that Eddie Mabo had begun.

Stylin' Up

Torres Strait Islanders have long been a tiny and largely unknown minority within Australia, often categorized with Aborigines (as in ATSIC) and experiencing colonialism from its most rapacious resource-grabbing form through to a more contemporary welfare variant, whilst struggling to maintain a sense of culture and identity and control of local land and politics. That struggle was made more difficult through immigration from both north and south. Migration of Torres Strait Islanders contributed to a form of disintegration of island societies, removing kin and workers, often permanently, yet ironically it was two members of migrant communities, in quite different ways, who enabled the restoration and revitalization of island life. Migration both demoralized and denied island life but eventually contributed to a resurgence of identity and, in turn, migration itself declined.

The introduction of a single administration, a marine commodity economy and the conversion of islanders to Christianity, all made more explicit the existence of 'island custom', as opposed to the local variants that existed within the Strait. This was paralleled by the rigidification of geographical and social boundaries, including the growth of a pan-islander Creole language and, much later, during the Second World War when islanders experienced new mobility and opportunities, foreshadowing 'the emergence of

Torres Strait Islanders as one people with an awareness of a territory in common' (Sharp, 1996, pp. 169, 226). That process was accentuated, rather than disrupted, by emigration to the mainland, which emphasized the emergence of a new ethnic identity – 'the men of Torres Strait' – which was superimposed on island identities and had persuasive power within the Islands. In various ways migration, and participation in a wider system, created Torres Strait identity. To be aware of that regional identity it had to be perceived from beyond.

In the Torres Strait itself the rise of welfare colonialism ironically resulted in a situation where being an islander 'shed its penalties and attract[ed] rewards' – both social and economic – so that some mixed-race residents of Thursday Island began to identify as islanders (Beckett, 1987, p. 207). In the Strait, custom acquired a new relevance and, as this occurred, music and dancing became increasingly symbolic of islander identity. Simultaneously, in Queensland, migrant islanders, having failed to secure independence from government regulation, turned towards notions of island identity: 'The homes that many have not seen for many years become once again important, if not as places to return to, then as points of reference' (Beckett, 1987, p. 209). It had become advantageous for islanders, whether in the Strait or on the mainland, 'to deal with the state as members of an indigenous minority rather than as citizens' (Beckett, 1987, p. 12). Economic dependence and marginalisation further accentuated Torres Strait identity.

The Mabo decision on native title produced an authentic island hero, in death – as a national icon – even more powerful than in life, and further emphasized Torres Strait culture and identity. More than that, it finally placed the Torres Strait in national ideology. Eddie Mabo's land claim, in large part, grew out of the exile's longing to return home, to the island that he had effectively left in his 'teens, and that long afterwards he remembered as the place where he spent 'the best time of my life' (Loos and Mabo, 1996, p. 29). For him, like many others,

> for those on the mainland, the dream of modernity had faded and many had begun to think of going home, or if that was not a possibility, to worry about their land. To be an islander – even on the mainland – one must have an island, and to have an island, one must own a piece of it. Once again they grasped their past, as it flashed by. (Beckett, 1994, p. 23)

· Christine Anu, a second-generation migrant, who saw herself as a 'contemporary traditionalist [who wanted] to prove our culture is contemporary just as it is old' (quoted in Harbison, 1995, p. 6), popularized the anthemic 'My Island Home' to depict 'a place we all dream about, a place in our minds and in our hearts where we want to eventually go home, to die there or be buried there – we have sentiments to this place'. Both Christine Anu and Eddie Mabo, in quite different ways, held both nostalgic images and futuristic visions of an island realm that once again could become an acceptable, viable and culturally significant island home.

Acknowledgements

I am indebted to Jeremy Beckett, Chloe Flutter and Chris Gibson for their assistance. Lyrics from *My Island Home* by Neil Murray are used by permission of Rondor Music (Australia) Pty Ltd.

Notes

1. There are considerable doubts over the reliability of the census data on Torres Strait Islanders, especially in places distant from the Strait (see Taylor and Arthur, 1993), because of problems attached to self-identification. It is probable that numbers in the southern states are somewhat inflated.

2. Unless otherwise indicated, quotations from Christine Anu are from undated promotional material from the Australian Council for the Arts or from the ABC documentary 'Saltwater Soul' (19 May 1996).

References

Arthur, W. (1992) 'Culture and economy in border regions: the Torres Strait case', *Australian Aboriginal Studies*, 2, pp. 15–33.
Beckett, J. (1981) *Modern Music of Torres Strait*. Canberra: Australian Institute of Aboriginal and Torres Strait Islander Studies.
Beckett, J. (1987) *Torres Strait Islanders: Custom and Colonialism*. Cambridge: Cambridge University Press.
Beckett, J. (1994) 'The Murray Island land case and the problem of cultural continuity', in W. Sanders (ed.), *Mabo and Native Title: Origins and Institutional Implications*, Monograph No. 7. Canberra: Australian National University Centre for Aboriginal Economic Policy Research, pp. 2–24.
Connell, J. (1988) *Sovereignty and Survival: Island Microstates in the Third World*, Research Monograph No. 3. Sydney: University of Sydney, Department of Geography.
Crough, G. (1993) 'Implications of the High Court's Mabo decision', *Journal of Australian Political Economy*, 32(1), pp. 5–22.
Davis, R. (1995) *Looking Beyond the Borderline: Development Performance and Prospects of Saibai Island, Torres Strait*. Discussion Paper No. 80. Canberra: Australian National University Centre for Aboriginal Economic Policy Research.
Duncan, H. (1974) *Socio-Economic Conditions in the Torres Strait. A Survey of Four Reserve Islands*. Canberra: ANU.
Duncan, H. (1975) 'Life in the Torres Strait, II', *New Guinea*, 10, pp. 54–63.
Fisk, E.K., Duncan, H. and Kehl, A. (1974) *The Islander Population in the Cairns and Townsville Area*. Canberra: ANU.
Gibson, C. (1998) '"We sing our home, We dance our land": indigenous self-determination and contemporary geopolitics in Australian popular music', *Society and Space*, 161, pp. 163–84.

Harbison, M. (1995) 'Island way brought to life', *Koori Mail*, 17 May, p. 6.

Kehoe-Forutan, S. (1988) *Torres Strait Independence: A Chronicle of Events*, Research Report No. 1. Brisbane: University of Queensland, Department of Geographical Sciences.

Kehoe-Forutan, S. (1991) 'Self-management and the bureaucracy: the example of Thursday Island', in D. Lawrence and T. Cansfield-Smith (eds), *Sustainable Development for Traditional Inhabitants of the Torres Strait Region*. Townsville: Great Barrier Reef Marine Park Authority, pp. 421–6.

Klein, A. (1995) 'Stylin' Up', *Juice Magazine*, April, p. 63.

Lea, J.P. (1994) 'Regional development planning in the Torres Strait. New directions in Melanesian Australia', *Third World Planning Review*, 163, pp. 375–94.

Lea, J.P., Stanley, O.G. and Phibbs, P. (1990) *Torres Strait Regional Development Plan*. Sydney: University of Sydney, Department of Urban and Regional Planning.

Loos, N. and Mabo, K. (1996) *Edward Koiki Mabo: His Life and Struggle for Land Rights*. Brisbane: University of Queensland Press.

Lui, G. (1994) 'Torres Strait: towards 2001', *Race and Class*, 35, pp. 11–20.

Mabo, E.K. (1976) 'A perspective from Torres Strait', in J. Griffin (ed.), *The Torres Strait Border Issue: Consolidation, Conflict or Compromise*. Townsville: Townsville College of Advanced Education, pp. 34–5.

Parliament of the Commonwealth of Australia (1997) *Torres Strait Islanders: A New Deal*. Canberra: House of Representatives Standing Committee on Aboriginal and Torres Strait Islander Affairs.

Scott, J.M. (1990) 'Torres Strait independence. Issues in island development', in R. Babbage (ed.), *The Strategic Significance of Torres Strait*. Canberra: ANU Strategic and Defence Studies Centre, pp. 383–403.

Sharp, N. (1993) *Stars of Tagai: The Torres Strait Islanders*. Canberra: Aboriginal Studies Press.

Sharp, N. (1996) *No Ordinary Judgement: Mabo: the Murray Islanders' Land Case*. Canberra: Aboriginal Studies Press.

Singe, J. (1989) *The Torres Strait: People and History*. Brisbane: University of Queensland Press (2nd edn).

Taylor, J. and Arthur, W. (1993) 'Spatial redistribution of the Torres Strait Islander population: a preliminary analysis', *Australian Geographer*, 241, pp. 26–38.

York, F.A. (1995) 'Island song and musical growth: toward culturally based school music in the Torres Strait islands', *Research Studies in Music Education*, 4, pp. 28–38.

11

Speaking of Norfolk Island: From Dystopia to Utopia?

Peter Mühlhäusler and Elaine Stratford

Introduction

This chapter brings together a number of strands that its authors have worked on either singly or jointly in recent years: utopia, place-making, islands, ecolinguistics, artificial languages, and accommodation (the study of behavioural and linguistic convergence). These strands are important in migration research, illuminating how people modify language and their ideas of place through migration.

We also share an interest in a specific methodological issue: that small islands with small populations and shallow histories are ideal test cases for examining the complex relationships among imperialism, migration, language, and 'the good place'. Links among these relationships are important for new understandings of culture in human geography (see Stratford, 1999) or in relation to a novel subdiscipline such as ecolinguistics, which has a great deal of catching up to do in terms of developing its methodology and identifying an adequate data base. Interdisciplinary research is critical to such a project. We are also united by an assertion that – in significant ways – we speak into existence the worlds we inhabit.

After first demonstrating the importance of utopia to the work, we provide an historical account of Norfolk Island, linking the imperial project of utopia (and its 'under-belly', dystopia) to successive waves of migration to the island. Second, we examine how utopian languages are constructed, focusing initially on fictional and metalinguistic ideas about a putative Pacific utopia, and then on actual languages such as Pitcairn and Norfolkese. The issue of language planning – as an important component in making Pacific languages conform to some utopian ideal – is also noted.

Third, we investigate the role of ecolinguistic analysis in elucidating the cultural aspects of migration and the question of how people accommodate or adapt, achieving some kind of 'fit' between their language and the environments they encounter and appropriate. Finally, we explore the language of Norfolk tourism as an expression of a utopia created by a migrant people who nevertheless identify as indigenes, and who have maintained a constant struggle for independence from Australia since at least 1901.

The Study of Utopia and the Case of Norfolk Island

Scholars such as Barbara Goodwin and Keith Taylor (1982) or Ruth Levitas (1990) remind us that the study of utopia is not bounded within disciplines. Rather, utopia 'forms the concrete expression of a moment of possibility, which is however annihilated in the very process of being enunciated' (Bann, 1993, p. 1). For example, many students of utopia are quick to point out the pun which derives from Thomas More's 1516 rendering of *Utopia* – a word which can mean *ou*topia (no place) or *eu*topia (good place). This *double entendre* leads to the belief that the good place is unattainable except through representation, the operation of language, and ideology.

What can be said about utopia in general terms? Summarizing a considerable body of work about utopias, Ruth Levitas (1990) proposes that such places appear across cultures; are embedded in myths of origin and destination; tend to stem from religious and/or political fervour; vary in form and structure; are usually prescriptive; often involve the participation of marginal groups; and are both imagined and implemented in material ways.

The choice to study Norfolk Island as the principal object of a discussion about such matters is motivated by two factors. First, Norfolk Island is a place which has been perceived, represented, and experienced as both utopian and dystopian. This paradox is culturally and linguistically significant because of how language contributes to the changing symbolic and material interactions we have with place. Second, Norfolk Island has been the site of activity for at least three culturally and linguistically distinct groups of migrants: Anglo-Celtic explorers, colonists and convicts; Pitcairn Islanders; and English and Melanesian missionaries.

This chapter is part of ongoing individual and joint projects on the roles of language in perceptions of the environment; many of its conclusions thus remain tentative. However, what we have concluded here does not contradict our findings for other small island environments, and confirms the general usefulness of selecting island environments for study. It is their susceptibility to accelerated change, and the distinct ways of speaking about the environment in each of them, that make it possible to document historical causes of the present situation.

Norfolk Island, Migration and the Imperial Project of Utopia

Norfolk Island lies on the Norfolk Ridge in the South Pacific (Figure 11.1). It has a current population around 1500 people, though seasonal fluctuations can inflate that number; tourism, and a reputation as a tax-haven may have something to do with this (Treadgold, 1988). Somewhere near 45 per cent of the population are island-born Pitcairnese–Norfolkese, and most other residents are expatriate Australians and New Zealanders. The issue of identity

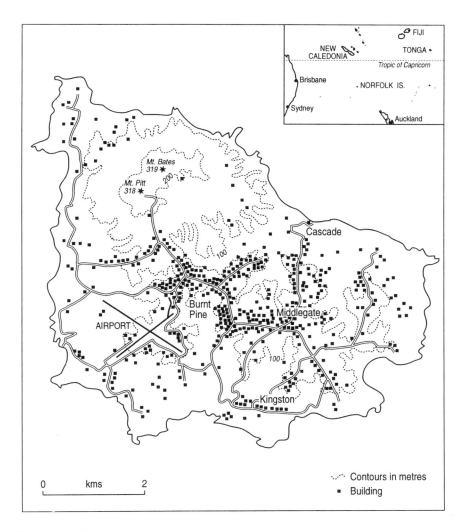

Figure 11.1   Norfolk Island: location and topography

according to ethnicity is problematic though, with intermarriage contributing to hybridity.

The island was 'discovered' by Captain James Cook on 10 October 1774 (Clune, 1967; Hoare, 1982). He named the island in honour of the noble Norfolk family whom he numbered among his patrons. After circumnavigating the island, Cook landed two small boats on 11 October, finding traces of Polynesian use, but no inhabitants (Hoare, 1988). Noting an abundance of pine and flax, both of which were in great demand in British manufacturing, and also the many species of flora and fauna, Cook is alleged to have recorded in his journal that the island was a 'paradise' (see Clarke, 1986, p. 9). His entries for 11 and 12 October 1774, while containing many positive references to the 'products' of Norfolk, do not contain this epithet.

Cook, of course, was among the most intrepid of European explorers whose activities were sponsored by imperial governments and 'great' families set on global expansion of markets, territories and geopolitical influence. Inevitably, Cook and later British settlers also mapped the island, thus constituting it as a Place for the British. Maps represent and make present; they reproduce the cultures that bring them into being. So maps *of* Norfolk Island, and maps *to* Norfolk Island, later used to facilitate successive waves of migration, say much about European ideas *about Europeans*.

Maps are thus ideological and idealistic, serving particular interests, in this case the interests of British colonial expansion. Maps of Norfolk construct the island. The language of cartography 'speaks' the island into existence. It legitimizes the subsequent use (and abuse) of the island as a place for new inhabitants and their cultural and linguistic practices. Nevertheless, this place is contested space: between convicts and soldiers; between settlers and distant administrators successively located in Van Diemen's Land (now Tasmania), New South Wales, and Canberra; between notions of indigenous and migrant cultures and languages; between ideas of the good place and its opposite.

Processes of exploring, mapping, naming, possessing, migrating, and administering are also implicated in a wider imperial politics of place in which some sites are rendered utopian and others dystopian. The act of forced migration, the transportation of people marked as criminal, is symptomatic of a much wider tendency of empire. This tendency to acquire, subordinate, and settle space (Said, 1989), and to create depositories for the home country's less desirable members, is part of imperial and colonial projects which characterize the cultural and linguistic history of Norfolk Island. An examination of representations of this Norfolk history will provide evidence for the shifting status of the island from Pacific paradise to 'Ocean Hell' (Murray, 1860, p. 366) and thence to tourist mini-Mecca, from a terrain dominated by cruelty to one suitable for the Christian home of Pitcairnese and birthplace of a people.

The first penal settlement: not entirely dystopian

Norfolk Island remained uninhabited for fourteen years after Cook's visit. Then, on 6 March 1788, under instruction from George III, Lieutenant Philip Gidley King landed a party of free settlers and convicts detached from the penal settlement established at Botany Bay by Captain Arthur Phillip. Chief among the reasons for this occupation, which numbered over 1100 people at its peak, was the desire to secure a resource base for naval supplies in the Pacific, a manoeuvre based on the abundance of flax and Norfolk Pine reported by Cook in 1774 (Clarke, 1988). While government officials and a large number of the settlers and convicts were speakers of English, other languages including Celtic Irish, Aboriginal languages, and Maori were represented as well (see Wright, 1986).

A great deal was achieved during King's two terms of office, from 1788 to 1790 and from 1791 to 1796 (Hoare, 1974). Despite clearing, cultivating, building, and other signs of progress, there were many difficulties in settling the island. Originally dominated by Norfolk Pine and dense subtropical rain forests full of waterfalls and teeming with birds, the small island had been hard to clear and cultivate. Plagues of caterpillars and grubs, along with rats and 'parroquets', made matters worse (Hoare, 1982, p. 6). Moreover, the 'sub-standard' quality of the pine and flax meant that these materials were found unsuitable for use in mast and sail-making as initially intended.

When it was clear that some of the early economic motivations for settlement would go unfulfilled, and especially from 1800, the penal colony once considered 'endurable and at times perhaps enjoyable' (Rigg, 1988, p. 97) became a place of considerable hardship and cruelty. After four years under threat of evacuation from the colonial government, the island was finally abandoned in February 1814. Its mostly wooden buildings were either burned or demolished to discourage occupation by migrants of other European nations (Rickard, 1995).

The second penal settlement: the nightmare decades

In 1825, the colonial government of Van Diemen's Land (then administering Norfolk Island) instigated a new wave of forced migration to the island, the maximum number of convicts being about 1200 (on the wider context of penal life in the Antipodes see Hughes, 1987). Those responsible for the return of convicts to the island incarcerated there the most 'hardened' convicts transported to Australia.

Subsequently, it became 'a place where Satan never sleeps' (Clune, 1967, p. 259) or, as Governor Darling was reported to say, a location for the 'extremest punishment short of Death' (Hoare, 1982, p. 35). Further, Darling apparently required that 'the felon who is sent there is forever excluded from all hope of return' (Clune, 1967, p. 113). Indeed, six free women who had accompanied their husbands to Norfolk were returned to New South Wales because of the violence on the island, and the exclusively male settlement came to be described as 'a cesspool of sodomy, massacre, and exploitation' (Christian, 1983, p. 12).

Norfolk Island's second penal settlement was disestablished in 1855, and the last prisoners were moved to Van Diemen's Land in 1856. During the last months of the second penal settlement, the Bishop of New Zealand visited Norfolk Island. As if to lift the pall cast by decades of cruelty, he suggested that the anticipated arrival of the Pitcairn Islanders from further east in the Pacific would render Norfolk Island what nature had intended it to be, an earthly paradise: Utopia (Murray, 1860).

The Pitcairn migration

In 1790, nine mutineers of the *Bounty* led by Fletcher Christian landed on Pitcairn with six Tahitian men, twelve Tahitian woman, and an infant girl (Hoare, 1982). What appeared to be paradise soon turned into a violent society. When Pitcairn was rediscovered by a visiting American whaler in 1808, only nine Tahitian women, 25 'mixed race' children, and one male mutineer, Alexander Smith (also known as John Adams), had survived.

In 1856, these people were relocated to Norfolk Island, Pitcairn having proved unable to support the community's food and water needs (Emery, 1985). This migration was sponsored by the Pitcairn Fund Committee, founded by prominent clerics, military personnel and government officials in England. It was not easy to leave Pitcairn, which the people had called *Fenua Maitai* or the good land (Emery, 1985). However, by the time they departed Pitcairn, the community had become deeply religious and concerned for their children's numeracy and literacy in English.

From 1856, activities on Norfolk Island shifted to whaling, trade, and other means for the support of the 194 Pitcairn Islanders. They had been guaranteed various rights to land on Norfolk, and this promise proved an important drawcard in their migration. Despite early assurances from the colonial administrator that they would be given free reign over the lands of Norfolk Island, the Pitcairn Islanders reportedly gained title to as little as one quarter of the total land area of Norfolk, and the administration of the land was undertaken from New South Wales. Disgruntled, some 16 Pitcairnese went back to their place of origin in two episodes of return migration (Clune, 1967; Emery, 1985). That island's population reached its peak of 233 in 1937, and since then the population has declined to about 50 as more and more Pitcairn Islanders move to New Zealand and other English-speaking countries.

Around the time that the Pitcairn Islanders migrated to Norfolk Island, the Imperial Parliament severed the island's administrative ties to Van Diemen's Land.[1] In leaving the Pitcairnese to their fate in 1856 the British representative, Captain Denham, noted that they would require no charity, being in possession of 'a most fertile land' (Hoare, 1982, p. 72). From that time until 1867, only one other small influx of migrants occurred, when a schoolmaster, a wheelwright, a blacksmith and a millwright left from England to join the community, their passage paid for by the Pitcairn Island Committee Fund in order to secure some professional assistance on the island (Hoare, 1982, p. 79). Three of these migrants were to leave within a few years.

Importantly, with the arrival of the Pitcairnese on Norfolk Island, a tradition of bilingualism began, with both English and Pitcairn–Norfolk being integral to islander speech repertoires. Symptomatic of a push for independence, partly provoked by disagreements with Governor Denison over the maintenance of the island's infrastructure and agricultural activity, this new speech was to prove equivocal over several decades. In 1914, for example, a report on the Pitcairn–Norfolk language by the Secretary for the Australian Department of External Affairs noted:

> Its use contributes to maintain a spirit of exclusiveness among these folks and for this reason, as well as because it has no merit to justify its continued existence, it is hoped that its employment may be discouraged in every possible way. (Hoare, 1982, p. 90)[2]

Mission migration

In 1867, the Melanesian Mission training school was transferred from Auckland to Norfolk Island (Clune, 1967). A free grant of 40 hectares of land, with an additional 378 hectares priced at £4 per hectare was provided to the Mission (Hoare, 1982). Apparently this alarmed the Pitcairnese, who had been thwarted in their access to all the lands on the island.

St Barnabas Chapel and the Mission training college were built away from the capital of Kingston, but there were regular contacts between the two. The Mission, numbering some 210 scholars by 1899, was built along the lines of a British public school, serving the goal of converting Melanesians to Christianity and a British way of life (Hoare, 1982). At its peak, the college consisted of six dormitories each holding 30 boys, while a much smaller number of girls were divided among the households of married missionaries. The common language of the college was a mission *lingua franca* known as *Mota*, but in communicating with outsiders English and Melanesian Pidgin English appear to have been used.

The Mission College was closed down in 1920, in part because a new Bishop to replace the presiding Bishop Wilson wished to establish a mission in the Solomons, in part because mission members had found it increasingly difficult to stay out of Norfolk affairs and the push by the Norfolk Islanders for independence from Australia. According to Hoare (1974), Bishop Wilson had decided that Norfolk had become less desirable as a mission site because of these political difficulties between the islanders and the colonial administration. Throughout the life of the Melanesian Mission, official contacts with non-mission outsiders were strictly regimented and were also unlikely to have led to major linguistic cross-fertilization. However, the story of unofficial contacts remains to be written and may reveal closer links between Melanesians and Norfolk Islanders.

Migrants or indigenes? Twentieth-century struggles for independence

Merval Hoare (1982) provides a lot of detail about Norfolk Island during this century, as does a comprehensive Internet site at http://www.ozemail.com.au/~jbp/pd. This website is interesting for its popular representations of Norfolk Island, and for assertively promulgating a pro-independence discourse in relation to the island. The site also provides a potted history of modern Norfolk Island. In 1901, for example, when the Australian colonies federated, there was no mention made of the status of Norfolk Island, then administered from New South Wales. In 1914, the Norfolk Island Act was passed, and the island became a territory under the Commonwealth's authority; it remains a dependency of Australia to this day. Between 1926 and 1935, Norfolk Islanders lobbied the Commonwealth Government to honour those initial promises about land and autonomy dating from 1855. In 1935, the Commonwealth passed the Norfolk Island Printers and Newspapers Ordinance, which remained in effect until its repeal in 1964, and which banned the publication of any pamphlet, newspaper or newsletter on Norfolk Island without the express permission of the Australian Administrator. The drive for independence had become the target for imperial discipline.

In 1946, when signing the United Nations Charter, the Australian Government nominated the Cocos Islands and Papua as non-self-governing dependencies; Norfolk Island was neither mentioned nor considered. In 1948, the Norfolk Islanders were given the right to apply for Australian passports as citizens, reinforcing the discourse of dependency. In 1955, 583 islanders unsuccessfully petitioned Her Majesty Queen Elizabeth II to grant self-government to the island. Then, in 1972, the Australian Government signed the International Covenant on Economic, Social and Cultural Rights. Notwithstanding this international obligation, by 1996 the government had still failed to apply the Articles of the Covenant to Norfolk Island, including those declaring that all peoples have the right to self-determination. Between 1972 and 1978, Norfolk Islanders continued to seek independence from Australia, calling for a referendum to decide the issue; this referendum was refused in 1978.

However, the Norfolk Island Act of 1979 did grant some measure of self-government to the islanders, who elected a Legislative Assembly of nine members in August of that year, gaining an independent flag in 1980. Yet in the November of that year, as an additional signal of its suzerainty, the Australian Government declared that the 200-mile exclusion zone around Norfolk Island was part of the Australian fishing zone, and this remains the case today.

Since 1983 political and cultural manoeuvring has continued on both sides. At one point, in 1991, the Australian Government suggested Norfolk Islanders vote within the electorate of Canberra, a suggestion soundly defeated in an island referendum. Between 1993 and 1995, the Norfolk Islanders unambiguously declared their position as the indigenous people of both Pitcairn and Norfolk Islands, further politicizing their status as 'authentic'. This declaration was based on the ways in which they construct themselves as a whole people, with distinct language, culture, customs, traditions and genetic make-up.

John Connell (1990) suggests that a sense of 'homelessness' is one of the corollaries of migration, itself part of the transformations that people have experienced through the operations of modernity. Perhaps the islanders' declaration of belonging to a home, a place, is a way to counteract discontents of modernity such as progress, ideology, the power of the map and the dulling homogeneity of cultural imperialisms. Either way, Norfolk Island has a fascinating history and a challenging future. Migration has played a central part in this history, with successive waves of Anglo-Celtic, Pitcairnese, Melanesian and Australian influences on the cultural and linguistic complexion of the island and its people. Most recently tourism has become a major factor in the constitution of the island and its people, and no doubt the language of tourism will have implications for Norfolk Island's culture and language, to which we now turn.

Constructing the Languages of Utopia

Part of constructing the Pacific and its peoples is the construction of their languages, which takes four main forms. First is the construction of fictional utopian languages which are meant to illustrate the character of the inhabitants of the Pacific Utopia (Laycock, 1987). Second, there is the vast array of metalinguistic views on how noble savages would speak; these views have acted as powerful filters in the description of Pacific languages and have prevented linguists from gaining access to authentic emic understanding (Rensch, 1991). Third are the languages used by the inhabitants of utopian speech communities such as those of Pitcairn, Bonin, Pohnpei or Palmerston (see contributions to Wurm *et al.*, 1996). Finally, there is the emergence of language planning as an important force to make Pacific languages conform to utopian ideas.

Regarding the first point, we are lucky to be able to refer to Laycock's excellent summary of a vast body of languages designed for literary and social utopias, many of the former located in imagined island settings: Artus' (1774) Hermaphrodite Island, Huxley's (1962) *Island*, or Lichtenberger's (1923) Cuffycoat's Island. The languages designed or chosen vary greatly for the three works just mentioned: respectively Latin; an invented language, Pala; and English and primitive English mixed with seal-like barkings. Languages invented from scratch (*a priori* languages) dominated seventeenth- and eighteenth-century literary utopias but real languages, either in their original form or modified (*a posteriori* languages), are more frequent in later writings. As Laycock (1987, p. 144) notes, there are dozens of examples of English being *the* language of both imagined and real utopias, and this point deserves some passing comment. Hilton's (1933) observation on Shangri-la – 'the atmosphere is highly reminiscent of Oxford University' – means that Oxford English is perceived to be the yardstick with which the achievement of utopian ideals is often measured.

The imperial ideal behind these civilized orderly tongues has dominated European views of Pacific utopia – and indeed of dystopia. It is interesting, for example, to observe how comments about the disintegration of society coincide with comments on the disintegration of the language of its members. The post-imperial language of the most chaotic beach communities of the Pacific has been – by necessity – a form of Pidgin English. The crudeness, violence and lawlessness of Pohnpei, or the early days of settlement of Bonin and Pitcairn, were mirrored by the lawless and reduced English spoken in such places. Indeed, very considerable effort was made by missionaries, the reformed leaders of these communities, and later state educators to restore high English in these communities. The greatest praise from such reformers is reserved for those communities where an excellent English is spoken, although it is sometimes acknowledged – as in the case of Pitcairn – that 'it will be a difficult task to eradicate' informal varieties of English (Neill, 1938, p. 16).

We are fortunate also in having a number of comprehensive surveys of the changing views of European explorers and colonists about the languages of the noble savages spoken in the 'Pacific paradise' (Rensch, 1991; Schütz, 1994). One example perhaps reflects the influence of such philosophers as Jean-Jacques Rousseau. In an item published in the *Mercure de France* of November 1769, Philibert de Commerson comments on the Tahitian language:

> A very sonorous, very harmonious language, composed of about four or five hundred words lacking in declension or conjugation, that is completely without syntax, is sufficient for them to render all their ideas and to express their every need. It is characterised by a noble simplicity which, excluding neither tonal modifications nor emotional pantomime, protects it from the arrogant tautology which we call richness of language, which make us lose incenses of perception and speed of judgement in a maze of words. The Tahitian, to the contrary, names his object as soon as he perceives it. The tone in which he has pronounced the name of the

object has already indicated how he is affected by it. Paucity of words makes for rapid conversation. The operation of the soul, the movements of the heart, are isochronous with the movements of the lips. The speaker and the listener are always in unison. (Quoted in Rensch, 1991)

Rensch (1991, p. 404) notes that these views of the Tahitian language 'are representative of many of the reports that we have of early visitors to the Society Islands' and it is probably not accidental that present-day speakers of Pitcairn–Norfolk point with pride to the roots of their language in this noble 'ancient Tahitian'.

As European involvement with the South Seas and its peoples shifted from scientific and philosophical curiosity to trade, exploitation of the region's natural resources and colonization or missionization, their views on the character of the Pacific languages changed. Such languages were then portrayed as barbarian, primitive and degenerate (for details see Mühlhäusler, 1996). A reassessment of this negative characterization has occurred only recently with the revival of pride in indigenous languages and culture.

The Language of Pitcairn

The Pitcairn language is well described (see Holm, 1989; Källgård, 1998; Laycock, 1989 and 1990; Ross and Moverly, 1964). In the past, though, it has often been mischaracterized. Most common is the error (repeated in Holm, 1989) that Pitcairn is Creole, a nativized former Pidgin English. This assessment is in conflict with Laycock's point that at no time was English absent from the speech repertoire of the Pitcairn Islanders. It also contradicts Harrison's comments on 'the early community's conscious desire to identify itself with the English side of [its] lineage' (1985, p. 135). Indeed, Laycock (1989 and 1990) suggests that Tahitian English was moribund by the second generation of Pitcairn, when their first experience of the outside world, a five-month stay in Tahiti in 1830, revived it as a linguistic means of signalling a separate identity. Laycock argues that the Pitcairn language is a cant deliberately created from the remnants of Tahitian English by adults.

Examples of communities creating new forms of speech (new dialects, new varieties) – mainly by modifying existing forms – are not easily unearthed ... but there is no doubt that small communities can carry out what I call *'naive linguistic engineering'*. (Laycock, 1990, p. 622)

If Laycock is correct (and a number of linguistic indicators support his view) then the Pitcairn–Norfolk situation comes close to that described in Huxley's novel *Island*, where English and an invented language, Pala, exist side by side.

Language Planning and Pitcairn–Norfolk

The roots of language planning are several, ranging from the desire to recreate the language of paradise or unmake the confusion of the Babel situation, to having a language whose contours are a perfect fit with the contours of a perceived single external reality. Laycock (1987, p. 150) observes that the makers of utopian languages often share this view that ideal language should be iconic, or directly reflective of nature. Most recently, reviving extinct languages to reclaim lost ethnic identity has been added as a motif.

Modern language planning manifests in two forms. The first of these is *status planning*, which assigns differential roles (for example, national language, language of education, official language, heritage language) to the languages spoken within modern nation states. The second is *corpus planning*, which involves the development of writing systems, standardization of grammar and modernization of the lexicon.

Until very recently, status planning in the Pacific has been designed to establish metropolitan languages such as English and French to replace the region's indigenous ones, in no small measure because moral, political and economic progress was thought to depend on the replacement of uncivilized tongues. As an illustration of this bias, Gunther (1958, p. 59) declared:

> only Christianity can replace the myriad philosophies, legends, pagan practices and supernatural fears that 510 tongues [of Papua New Guinea] have engendered. It is only by removal of these 510 tongues and the acceptance of a common language that the end of many unnatural behaviours can be achieved.

The efforts of the Melanesian Mission on Norfolk to make Mota the mission language can be interpreted in this way. By replacing Melanesian languages with a single language – whose suitability to church lifestyle was carefully controlled by the missionaries – a second aim was also achieved: that of keeping a symbolic and hierarchical distance between Melanesians and Europeans, even as migration had pushed them together. Proposals to develop a 'vulgar Papuan' as the language of wider communication in German New Guinea and the Dutch creation of administrative Malay (Mühlhäusler, 1996, p. 175) are similar examples of streamlining linguistic diversity in the colonial Pacific. Status planning is also seen in the attempts of colonial educators to eradicate Pitcairn and Norfolk (see Holm, 1989, p. 549; Ross and Moverly, 1964, p. 116), thus making the island's people full members of the English-speaking empire, dissolving the physicality of migration and linking all under one imperial banner, one speech, one 'place'.

In those instances where local languages were given official status, language planners operated from the hypothesis that such languages suffered from serious shortcomings and needed to be developed. St Barnabas

Mission on Norfolk, for instance, was modelled on English public schools by old Etonians, evoking the kind of images Laycock (1987, p. 144) portrayed as characteristic of literary utopias. The Mission's Mota language was developed to be a civilized inhabitant of this utopian ecology. It was given a writing system, a standard grammar and doctrinal terminology, and was purged of paganism and obscenities – as indeed were many other mission languages (Mühlhäusler, 1996).

In contrast to these standardising tendencies of planned languages, non-standard forms of English are deliberate markers of identity and a means to preserve one's authenticity against the onslaught of outsiders. In the case of the Pitcairn Islanders, there are indications that they increasingly resorted to the Pitcairn language as they became more exposed to outside influences. Wiltshire's report (1939) states that the tendency to use Pitcairn 'dialect' is pronounced in the home and school, and that the older generations speak a purer English than the newer.

There are, however, significant variations and contradictory trends. Harrison (1985, p. 133) reports that only 'pure' English was spoken in several households in both Pitcairn and Norfolk, and she comments that:

> It is notable that those families in which English speech had a prominent place took unusual pride in the quality of their English; also, their speech exhibited a great propensity to code variation towards Nf E [Norfolk English] in a context in which B Nf [broad Norfolk English – the most un-English and 'purest' level of Norfolk] was the norm.

She also notes a significant revaluation of Norfolk in the more recent past (p. 137):

> It is not surprising that most Norfolk Islanders should have regarded English favourably in comparison with Nf English as a language with a long and respectable history, confirmed in status by its use in all 'important' and public affairs. By contrast, Nf was seen as a form which came into existence as an improvised mixture of Tahitian and English. Norfolk Islanders tended to look upon it as a haphazardly constructed medium, originally adopted for talk among people who were either unable or too lazy to speak English properly; they were doubtful about whether it might be justifiably called a language with a proper grammar ... It is only in the last thirty years that such languages have been increasingly recognised as proper means of expression in their own right, and as respectable subjects of study.

This revaluation occurred when social changes, in particular the loosening of previously tightly-knit family networks, had progressed too far to provide a stable ecology for Norfolk. Hence, positive feelings about the language are not consistently matched by linguistic practice (Harrison, 1985, pp. 146–7). Some have vowed to speak Norfolk to their children because they want the language to be carried on, but fail to follow their intentions because Norfolk English fits more easily into their particular domestic set-up. Some voice their admiration for those families whose members are responsible for 'keeping Norfolk going', yet speak only Norfolk English in their own homes. Some speak Broad Norfolk consistently to their peers, and ostensibly are in favour of the language, but cannot bring themselves to use Norfolk to their children and grandchildren. It is interesting to observe, however, that the attitude of some Norfolk Islanders has changed over the years; for example, some deliberately speak Norfolk to their grandchildren though they avoided the use of it to their own children. Some feel such distress at hearing others speaking Norfolk 'incorrectly' (that is, not according to the old system of grammar) that they say they would prefer to see the language die out altogether than be spoken incorrectly.

The preservation of Norfolk today is seen as a task for corpus planning. In particular, there is the development of a standard written form of the language undertaken by the Norfolk Islander, Alice Buffett, and the Australian linguist, Donald Laycock (Buffett and Laycock, 1988). In their booklet they emphasize that 'Norfolk Islanders now have a written language ... and a readily accessible description of the grammar' (p. v), which 'is a great benefit ... for Norfolk Islanders ... who may be encouraged to write in Norfolk more often and explore the written potential of their language' (p. vi). It remains to be seen what impact such 'empowerment' of Norfolk will have in the longer term; it appears more likely that the traditional language will be replaced by a slightly modified English. As Laycock (1990, p. 625) observes:

> However, there are pressures on Norfolk Island for members of the in-group – essentially, the Pitcairn descendants – or semi-outsiders with close connections with the in-group (by marriage or other association) to speak Pn [Pitcairn]. Thus, the variety described is not 'Modified Norfolk' but 'Modified English', or ... 'Instant Norfolk'. It is the variety spoken by those members of the in-group who have imperfectly acquired Pn (mostly as a consequence of having grown up in households where Pn was not spoken in their childhood) but who wish to assert their right of membership of the group by the use of the group language, to the extent of which they are capable.

> The rules for this kind of 'Instant Norfolk' are absurdly simple. First, only two phonological modifications are essential: the replacement of the diphthongs [ei] and [ou] (English *gate* and *home* respectively) by [eÓ] / [iÓ] and [œÓ] / [uÓ] (Norfolk *giet* and *hoem*, in the Laycock/Buffett orthography). Second, characteristic lexicon (such as *salan* 'person', *naawwi* 'swim', *ama'ula* 'clumsy') needs to be used with high frequency, along with conversational tags such as *daas et!* 'that's it!' Thirdly, the speech requires some sprinkling of a few features of Pn syntax, such as the unusual (and probably artificial) benefactive construction that inserts a beneficiary between quantifiers and nouns: *giw wan ai glaas a biya* 'give me a glass of beer', and also the Pn pronouns – of which by far the most important is *aklan* 'us' (that is, the in-group).

Thus, naive language planning, rather than scientific planning, appears to determine what future the language will have.

Ecolinguistics of Norfolk Island

Central to the concern of both ecolinguistics and those interested in the cultural aspects of migration is the question of adaptation – how do people achieve a fit between the contours of their language and the contours of the environment? (Alexander, 1995). Mühlhäusler (1996) has argued that the study of small islands with a shallow history of human population offers the best opportunity for receiving answers to this question. It may also be argued that a very different question – why do so many utopias fail? – will receive additional and hitherto rarely considered explanations.

The ecolinguistic history of the Pitcairn Islanders prior to their resettlement on Norfolk deserves a few comments. The languages spoken initially on Pitcairn were English, Tahitian, and Tahitian (Pidgin) English (Laycock, 1989). The environment of the island was quite similar to Tahiti and knowledge about how to cope with it existed among the Tahitian men and women who accompanied the mutineers. There are indeed a significant number of Tahitian-derived lexical items descriptive of local life forms. Källgård (1998) notes that Pitcairnese contains some words of Tahitian origin, including:

> *ape* – giant karo
> *fanfaia* – painted coral bass
> *fetuwe* – large sea urchin
> *himanu* – pandamus flower
> *mawloo* – white-tailed tropical bird

Language, of course, migrates. But equally striking are the very large proportion of words in the same domains adapted from English or coined by the Pitcairn Islanders. Quite a few of these words bear witness to the processes of adapting to a new environment:

> *bity-bity* – shellfish with razor-like protuberance
> *devil-fish* – kind of ray
> *tough cod* – fish with tough flesh
> *dream fish* – fish inducing hallucinations

Languages also perish or change through the processes of migration. An important and as yet unstudied feature of the ecolinguistic connection between Pitcairn and Norfolk Islands is the large-scale loss of Tahitian names. A comparison of Harrison's (1979) Norfolk vocabulary with Källgård's (1998) Pitcairn one shows that many Pitcairn names are no longer found on Norfolk, and the resettlement of the Pitcairn Islanders to the very different environment of Norfolk required renewed adaptation. For the Pitcairn Islanders, too, Norfolk initially was full of objects for which they lacked language. Clarke (1986, p. 146) reports that:

> Everything about their new home astonished the Pitcairners: the massive stone buildings were veritable castles, the cattle and horses were the first they had ever seen, as were gardens of English flowers, and exotic new fruits and vegetables.

Conversely, many life forms to which they had become accustomed on Pitcairn were absent on Norfolk.

None of the main groups of settlers (English, Pitcairn Islanders, Melanesians) was native to Norfolk, and all of them found this environment very different from what they had been used to. In effect, migration and encounters with novel landscapes have resulted in the modification of language and the development of neologisms, many of them indicative of adaptation to new places or of tendencies to impose (inappropriately) the knowledge of familiar places on these new lands.

An early illustration of this latter inclination relates to perceptions of the native flax. 'It took Lieutenant King some time after he first arrived at Norfolk to recognize the native iris which abounded on the island as the flax plant described by Captain Cook [and] as one of the assets of the place' (Wright, 1986, p. 26). Indeed, it is interesting that he realized that English did not offer the linguistic means to talk about its cultivation and processing. It was argued that such knowledge could be gained in two ways; by kidnapping two New Zealand Maoris, Tooki and Woodo, who were brought to Norfolk in 1793, and by learning Maori. Tooki and Woodo were kept in the Lieutenant Governor's residence and their language was studied, particularly in order to learn their methods of processing New Zealand flax. While the first objective was relatively straightforward, and knowledge of Maori culture was gained, the second objective was not met, flax processing being the work of women.

The two periods of penal settlement on Norfolk Island resulted in a massive change in the environmental circumstances there. On the one hand, large areas of land were cleared to sustain European crops and animals; on the other hand, the indigenous flora and fauna were 'mined' with few thoughts given to preserving indigenous life forms. The naming practices of Pitcairn/Norfolk lend support to the view that utilitarian considerations rather than perceptual prominence drive folk-naming practice.

Moreover, the new settlers did not come to accommodate or adapt to an existing paradise but to improve their material well-being. 'Nature's liberality'

(Christian, 1983, p. 158) was taken for granted and a number of ideological factors conspired to trigger off rapid environmental deterioration, first on Pitcairn and subsequently on Norfolk. Moverley (1960) lists the detrimental impact of introduced pests, plants and animals (particularly goats) on Pitcairn Island, a process which was made worse by the strong vegetarian bias of the islanders combined with a prejudice against manuring; hence large areas of land had to be under cultivation. Massive soil erosion has reduced several parts of Pitcairn to the state of a semi-desert. As Moverly observes, 'the religious philosophy of the Islanders does not encourage them to look beyond the morrow' (1960, p. 63).

On their later arrival on Norfolk, the Pitcairn Islanders found that their new home did

> possess a much larger area of fertile land, an equable climate, a good rainfall, and excellent fishing prospects, but also there was a substantial physical capital stock left behind from the convict era ... All this was made freely available to the Pitcairners together with fodder, seeds, plants, and food stores to help maintain them until their first harvest ... Nevertheless, since it was an aid package that for the most part had not been tailored to their specific needs and mode of living, but was rather the accidental by-product of the decision to abandon the penal settlement on Norfolk Island, it also led to early problems of adjustment and difficulties in absorbing all the aid effectively. (Treadgold, 1984, p. 37)

A consequence was that land was cleared but was often neglected. It became weedy and scrubby. Livestock, left to stray, bred indiscriminately, while pasture conditions deteriorated. Items of fixed capital rotted or rusted where they lay.

Language of Tourism

The role of tourism in creating images of paradise or utopia is well-known and documented, with the emphasis shifting to images of unspoilt nature and eco-paradise in recent times. Current tourist brochures of Norfolk Island bear out this trend. For instance, statements about an unspoilt island's natural attractions awaiting the anticipating tourist can only take in those who are entirely ignorant of the massive environmental impact that recent settlement has had on this island. As Nobbs (1991, p. 158) remarks:

> Wood Jones, an eminent naturalist, once commented that man's [sic] attempts to improve on nature through species introductions resulted in inconspicuous failures of conspicuous disaster. Nowhere is this more evident than on oceanic islands like Norfolk. The more spectacular impacts of settlement on Norfolk and its nearby islands have their origin in introduced animals and, to a lesser extent, plants.

One prominent claim that Norfolk Island has no pollution is another example of hyperbole. It has been stated that 'tourist accommodation has such a high water consumption that it can severely stress the water resource capacity of an island' (Abell, 1993), and that there has been a fair amount of pollution in shallow unconfined ground water as a result of intensive cattle farming. At the same time 'little is known about quality degradation in the deep ground water system' (Abell, 1993). A waste management study has identified several other difficulties including an inadequate provision for disposal of hazardous chemicals and plastics, and an over-reliance on incineration (UniMelb, 1994). Another claim that harsh sounds of noise pollution heard on the mainland are replaced by sounds of nature[3] is not shared by at least some residents, who have made submissions on the proposal to upgrade Norfolk Island Airport,[4] and the presence of television, compact disc players, telephones etc. in tourist accommodation would seem to cast doubt on this statement.

Until quite recently, tourism to Norfolk Island has lacked consistent management and forward planning, with visitor numbers falling. This trend has led to a government marketing strategy (Unity 2005) which recognizes the island's dependence on tourism and its fragile ecosystem. In the Unity 2005 policy the symbolic value of the Norfolk language is also recognized. Thus, signage promoting scenic or heritage attractions should use both internationally recognized symbols and, where appropriate, Norfolkese.

The official 1997 Norfolk Island Government Tourist Board brochure, while reassuring tourists that the language spoken on Norfolk is English, adds 'but you'll hear Islanders speaking Norfolk, a unique language derived from the speech of the Bounty mutineers and their Tahitian wives and companions who settled Pitcairn Island in 1790'. The same brochure refers to Norfolk Islanders as 'indigenous people', a clear departure from previous practice and one perhaps again echoing desires for an identity independent from Australian neocolonialism.[5] The Norfolk language is colourfully described as 'English ... spoken with a West country burr and a dash of Tahitian, reflecting the enduring influence of the original Pitcairners',[6] or as a curious mixture of Tahitian and old English.[7] It will be interesting to see the extent to which the language will become a resource of tourist revenue, (e.g. by being used on phonecards or tea towels) and how it can become part of attempts to develop an image in the market place that is largely unassailable, as desired in the Unity 2005 policy.[8]

Conclusion

Utopia is a vision to be pursued, to be fashioned. Somehow, it seems much less likely to be realized than dystopia, for rarely are we so satisfied with life that we cannot see and lament its drawbacks. Both utopia and dystopia are also place-bound, often organized around the construction of particular places and reflecting certain ideologies. Implicit in these processes of place-making is language, and the ability of language to accommodate changes in culture.

Whether voluntary or forced, acts of migration tend to be rendered as utopian – the pursuit of the good, or as dystopian – the creation of hell on earth. Again, language and culture are central to the ways in which migration is undertaken and recalled; they reflect and simultaneously constitute the place-making that migration precedes and demands.

Islands are interesting test cases by which to examine the effects of migration, language change and accommodation (including ecolinguistics), and the formalities of language planning on place. They are both isolated and isolating, and yet are linked into imperial projects of colonial expansion. Norfolk Island was discovered during one period of imperial prowess in the history of the British Isles, and was colonized in several waves as prison, refuge and mission. It is the indigenous place of the descendants of the migrant Pitcairnese, themselves descended from supposedly ignoble mutineers and noble Tahitian 'savages'. And it is a resistant outpost for an Australian Government reluctant to loose the strings of a colonial past.

Finally, through each of these phases in its history – phases based on migration tied to imperial visions – Norfolk Island as a place has been conceived as both dystopian and utopian. This paradoxical conception has been reflected in the languages which have been used to speak the island into existence and in those which have helped to maintain it as an entity separate from those from which it derives. English, Celtic, Maori, Tahitian, Pitcairn-Norfolk, Mota, Australian English, and the new eco-speak of tourism – each has contributed to the formation of the place that is Norfolk Island. There remains much to be done to investigate the cultural and linguistic past of this island and its migration history, and the possible effects on it of language planning and the globalizing tendencies of metropolitan languages.

Acknowledgements

We are grateful for a small ARC grant for 1997 which enabled us to carry out a pilot study on the ecolinguistics of islands. Thanks to Julia Winefield (University of Adelaide), Julie Kesby (ADFA) and Asha Bowen and Anna Griggs (University of Tasmania) for research assistance; and Darrel Tryon (ANU), David Hilliard (Flinders University) and John Connell (University of Sydney) for discussing our ideas and generously giving us access to their Norfolk materials.

Notes

1. According to Hoare (1982), Norfolk was again to become a dependency of New South Wales from 1896; in 1914 it formally became a territory of the Commonwealth of Australia.

2. Parenthetically, we note that this attack on languages other than English coincides with the beginning of the First World War. It was paralleled in mainland Australia by similar attacks on German and indigenous languages, and was a corollary of 'English only' assimilation policies which persisted until 1972.

3. Norfolk Island Government Tourist Board brochure, 1997.

4. See the Department of Transport's report *Upgrading of Norfolk Island Airport to Medium Jet Standard: Final Environmental Impact Statement*. Canberra: Australian Government Printing Service, 1980.

5. These quotes are from the brochure referred to in note 3, above, pp. 8 and 35.

6. Norfolk Island Government Tourist Bureau Fact Sheet, 1997.

7. In a tourist brochure entitled *Jason's Passport to Norfolk Island*, 1997, p. 4.

8. See *Unity 2005 – Norfolk Island Government Tourist Bureau Marketing Strategy*, 1995.

References

Abell, R. (1993) 'Aquifer vulnerability on small volcanic islands in the southwest Pacific region: an example from Norfolk Island', *AGSO Journal of Australian Geology and Geophysics*, 14(2–3), pp. 123–33.

Alexander, R. (1995) 'Integrating the ecological perspective: some linguistic self-reflections', in A. Fill (ed.), *Sprachökologie und Ökolinguistik*. Tübingen: Narr, pp. 131–48.

Artus, T. (1774) *Description de l'Isle des Hermaphrodites*. Brussels and Cologne.

Bann, S. (1993) 'Introduction', in K. Kumar and S. Bann (eds), *Utopians and the Millennium*. London: Reaktion, pp. 1–6.

Buffett, A. and Laycock, D. (1988) *Speak Norfolk Today*. Norfolk Island: Himii.

Christian, G. (1983) *Fragile Paradise: The Discovery of Fletcher Christian, Bounty Mutineer*. London: Book Club Associates.

Clarke, P. (1986) *Hell and Paradise*. Melbourne: Viking.

Clarke, F. (1988) 'The reasons for the settlement of Norfolk Island 1788', in R. Nobbs (ed.), *Norfolk Island and Its First Settlement 1788–1814*. Sydney: Library of Australian History, pp. 28–36.

Clune, F. (1967) *The Norfolk Island Story*. London: Angus & Robertson.

Connell, J. (1990) 'Modernity and its Discontents', in J. Connell (ed.), *Migration and Development in the South Pacific*, Pacific Research Monograph No. 24. Canberra: Australian National University, National Centre for Development Studies, pp. 1–28.

Emery, J. (1985) 'Norfolk Island: a place of extremist punishment, and *Fenua Maitai*', *Heritage Australia*, 4, pp. 5–11.

Goodwin, B. and Taylor, K. (1982) *The Politics of Utopia: A Study in Theory and Practice*. London: Hutchinson.

Gunther, J.T. (1958) 'The People', in J. Wilkes (ed.), *New Guinea and Australia*. Sydney: Angus & Robertson, pp. 46–74.

Harrison, J.F.C. (1979) *Robert Owen and the Owenites in Britain and America*. London: Routledge & Kegan Paul.

Harrison, S. (1985) 'The social setting of Norfolk', *English Worldwide*, 6(1), pp. 131–53.

Hilton, J. (1933) *Lost Horizons*. London: Macmillan.

Hoare, M. (1974) *Rambler's Guide to Norfolk Island*. Sydney: Pacific Publications.

Hoare, M. (1982) *Norfolk Island: An Outline of its History 1774–1981*. St Lucia: University of Queensland Press (3rd edn).

Hoare, M. (1988) 'The island's earliest visitors', in R. Nobbs (ed.), *Norfolk Island and Its First Settlement 1788–1814*. Sydney: Library of Australian History, pp. 18–22.

Holm, J. (1989) *Pidgins and Creoles, Volume II Reference Survey*. Cambridge: Cambridge University Press.

Hughes, R. (1987) *The Fatal Shore*. London: Collins Harvill.

Huxley, A. (1962) *Island*. London: Chatto & Windus.

Källgård, A. (1998) '"Fut yoli noo bin laane aklen?": a Pitcairn Island word list', in *Papers in Pidgin and Creole Linguistics 5*. Canberra: Pacific Linguistics.

Laycock, D. (1987) 'The languages of utopia', in E. Kamenka (ed.), *Utopia*. Melbourne: Oxford University Press, pp. 144–78.

Laycock, D. (1989) 'The interpretation of variation in Pitcairn, Norfolk', in U. Ammon (ed.), *Status and Function of Languages*. Berlin: de Gruyter, pp. 608–29.

Laycock, D. (1990) 'The interpretation of variation in Pitcairn, Norfolk', in J. Edmonson, C. Feagin and P. Mühlhäusler (eds), *Development and Diversity: Linguistic Variation Across Space and Time*. Arlington: Summer Institute of Linguistics and University of Texas, pp. 621–8.

Levitas, R. (1990) *The Concept of Utopia*. New York: Philip Adam.

Lichtenberger, A. (1923) *Pickles; or Récite à la mode anglaise*. Paris.

Moverley, A.V. (1960) 'Pitcairn Island: an economic survey', *Transactions and Proceedings of the Fiji Society*, 4(3), pp. 61–7.

Mühlhäusler, P. (1996) *Linguistic Ecology*. London: Routledge.

Murray, Rev. T.B. (1860) *Pitcairn: the Island, the People and the Pastor, to Which is Added a Short Notice of the Original Settlement and Present Condition of Norfolk Island*. London: Society for Promoting Christian Knowledge.

Neill, J.S. (1938) *Pitcairn Island: General Administrative Report*, Colonial Report No. 155, London: His Majesty's Stationery Office.

Nobbs, R. (1991) *Norfolk Island and Its First Settlement, 1788–1814*. Sydney: Library of Australian History.

Rensch, K. (1991) 'The language of the noble savage: early European perceptions of Tahitian', *Canberra Pacific Linguistics*. Canberra: Australian National University, C117, pp. 403–14.

Rickard, J. (1995) 'Historical sites: Norfolk Island', *Australian Historical Studies*, 26, pp. 480–4.

Rigg, V. (1988) 'Convict life: a "Tolerable Degree of Comfort"?' in R. Nobbs (ed.), *Norfolk Island and Its First Settlement 1788–1814*. Sydney: Library of Australian History, pp. 97–111.

Ross, A.S.C. and Moverley, A.W. (1964) *The Pitcairnese Language*. London: André Deutsch.

Said, E. (1989) 'Representing the colonized: anthropology's interlocutors', *Critical Inquiry*, 15, pp. 205–25.

Schütz, A. (1994) *The Voices of Eden: Australia History of Hawaiian Language Studies*. Honolulu: University of Hawaii Press.

Stratford, E. (ed.) (1999) *Australian Cultural Geographies*. Melbourne: Oxford University Press.

Treadgold, M. (1984) 'Growth, structural change and distribution in a very small economy: a case study of Norfolk Island', *The Journal of Developing Areas*, 19(1), pp. 35–58.

UniMelb (1994) *Norfolk Island Waste Management Study*. Melbourne: UniMelb Ltd.

Wiltshire, A.R.L. (1939) 'The local dialects of Norfolk and Pitcairn Island', *Journal of the Royal Australian Historical Society*, 25(4), pp. 331–7.

Wright, R. (1986) *The Forgotten Generation of Norfolk Island and Van Diemen's Land*. Sydney: Library of Australian History.

Wurm, S., Mühlhäusler, P. and Tryon, D. (eds) (1996) *Atlas of Languages of Intercultural Communication in the Pacific, Asia and the Americas*. Berlin and New York: de Gruyter.

12

Islanders in Space: Tongans Online

Helen Morton

Pacific Islanders, long renowned for their skills in navigating the vast Pacific Ocean, are now beginning to navigate cyberspace, using the latest communications technology to traverse their scattered communities. When the kingdom of Tonga established its first direct link with the Internet, the site was named Kalianet, after the giant double-hulled sea-going canoes formerly used by Tongans to navigate the Pacific. The introductory blurb on the site proclaims, 'The navigational skills required by these early travellers are again required by the modern Tongans using computers to navigate this satellite connection to the InterNET.' Through sites such as Kalianet, Pacific Islanders are continuing 'the contemporary process of what may be called world enlargement' (Hau'ofa, 1994, p. 151). As Epeli Hau'ofa notes, this process includes 'establishing new resource bases and expanded networks for communication' both within and beyond Oceania's 'sea of islands' (1994, p. 156).

Of all the Pacific nations, it is Tonga that has embraced most eagerly the new means of communications via satellites and computer technology, and this chapter examines the impact of this on Tongan migrants and their relationships with each other and with their homeland. After briefly outlining the history and nature of Tongan migration, Tonga's ventures into the world of satellite communications and the Internet will be described. Using as a case study the predominantly Tongan Internet message board, the Kava Bowl, the remainder of the chapter will explore the ways in which the Kava Bowl extends the international Tongan community and provides a forum for a range of social interactions to occur. Frequent discussions take place on the Kava Bowl in which issues of concern to the participants are raised and debated. Some examples of such discussions are drawn upon to demonstrate the ways in which participants share their experiences of migration and how Tongan cultural identity emerges as the most pivotal and strongly contested concern of the forum.

Development of the Tongan Diaspora

In some parts of the Pacific, migration to Western nations has occurred on such a large scale and in such a manner that, as John Connell comments, 'it is the new diaspora that extraordinarily rapidly has come to characterize the contemporary South Pacific' (1987, p. 399). The Tongan diaspora is extensive and characterized by both concentration, in cities such as Sydney, Auckland and Salt Lake City, and dispersal, with Tongans scattered throughout many countries of the world. The outflow of Tongans from their homeland has markedly increased since the late 1960s, such that the population of this archipelago of tiny, scattered islands has remained fairly stable, at around 100,000, for some years now despite a relatively high birth rate. Even on the most conservative estimates there are over 30,000 Tongans living overseas; others estimate that as many Tongans live abroad as in Tonga (Marcus, 1993, p. 27).

Tonga is a constitutional monarchy, and today's king is the fourth ruler in a dynasty established in the mid-nineteenth century by a warrior chief who united the Tongan islands and became King George Tupou I. The constitution promulgated in 1875 by Tupou I transformed some of Tonga's *hou'eiki* (chiefs) into *nopele* (nobles), who became the land-holding aristocracy of the nation. Although a British protectorate from 1900 to 1970, Tonga was never formally colonized and from the earliest legal codes prohibited the sale of land to foreigners. Instead, every male over sixteen years is legally entitled to apply to lease land from his noble, which becomes inheritable by the eldest son in perpetuity (James, 1993). As the population of Tonga grew, however, it was increasingly difficult for eligible males to access land and this became a significant factor stimulating emigration.

Tonga remained relatively isolated until the present king commenced his reign in 1965. A process of 'internationalizing' then began in which foreign capital was sought to boost development and, particularly in the 1970s, many Tongans emigrated (Marcus, 1993). This 'internationalization' has also led to the increasing use of land for commercial production, most recently the farming of pumpkin squash for the Japanese market (James, 1993), although a significant amount of subsistence farming still occurs (Van der Grijp, 1993).

Internal migration has occurred in Tonga since at least the mid-nineteenth century, as rural villagers moved to the semi-urban centres in order to place their children in the better schools and to seek paid employment. The 1996 census showed that the main island, Tongatapu, is now home to 68 per cent of the population, and the population of the capital, Nuku'alofa, continues to increase despite the difficulty of obtaining either land or employment there. Nuku'alofa often acts as a stepping-stone for rural dwellers who go on to migrate overseas.

Tongans have been drawn to 'developed' nations in search of better opportunities for their children and better wages with which to support both immediate and extended family members. Their paths to migration have been varied, with many initially relocating as individuals, to study or to work, then establishing themselves in their new country and sponsoring other family members to join them. Others have migrated as family groups, sometimes through 'step' migration in which, for example, families migrate to Australia after a period in New Zealand, or migrate to North America through American Samoa. Many Tongans overseas are overstayers who initially travelled on temporary visas; amnesties have given some the opportunity to formalize their status as residents.

Large-scale migration has had a profound impact in Tonga, with few areas of Tongan life left untouched by this process (Gailey, 1992; James, 1991). One of the most striking effects of migration has been the growing reliance of the Tongan economy on the remittances migrants send home (Ahlburg, 1991; Faeamani, 1995). While there seems to be a gradual shift from monetary to in-kind remittances and trading partnerships (Brown and Connell, 1993; James, 1991), there is nevertheless a continuing commitment by many immigrant Tongans to assist family members remaining in the islands (Vete, 1995).

The need for such assistance and the impetus for more Tongans to leave the islands are driven by the increasing difficulty of accessing land and employment, as well as the perceived material and educational benefits of relocating to more affluent nations (Cowling, 1990; Gailey, 1992). Most Tongans living overseas maintain close ties to kin they leave behind, by phone, letters, and by exchanging videotapes of important events. Tongan foodstuffs and goods may also be sent from Tonga to the diaspora, as gifts or for sale by family members. There is a constant movement of people between Tonga and the populations of Tongans overseas: young people leave Tonga temporarily to study or work, children are sent to or from Tonga to live with different family members, and older people move to stay for periods with their married children. Other ties to the homeland are also maintained, through such groups as ex-students' associations which raise funds for various causes in Tonga, and visits to Tonga to see kin and to attend events such as church conferences and the annual Heilala Festival.

Tongan migrants also establish and maintain connections with one another in their host countries, primarily through networks of kin and attendance at churches with Tongan congregations. They also draw on a range of connections according to their village of origin, school attended in Tonga, membership of kava-drinking groups, shared workplaces, and areas of residence (Cowling, 1990, p. 192). As the number of Tongans living in some areas has increased, their sense of community has also grown, as has a more self-conscious awareness of themselves as part of the international Tongan community.

Despite the overall advantages that migration has had for Tonga's economy, it remains a 'Third World' nation, with some islands still without electricity or piped water and largely reliant on subsistence production. Tonga's king, Tāufa'āhau Tupou IV, has for many years sought ways to boost development and escape the poverty trap of aid-dependent nations. As George Marcus has pointed out, Tonga's venture into the field of satellite communications can be seen in this context, with Tonga's sovereign status being used as an economic resource in the hope of bringing prosperity and security to this tiny nation (Marcus, 1993, pp. 26–7).

'A Genuine Space-Faring Nation'

In May 1994 the Crown Prince of Tonga, Tupouto'a, witnessed the launching of a Russian-built satellite from Kazakhstan into an orbital position ('slot') registered by Tonga. Afterwards, he boasted that Tonga was now 'a genuine space-faring nation' and elaborated on plans for Tonga to become 'the second largest commercial satellite nation in the world'.[1] Although ambitious, the Prince's vision was not entirely far-fetched, as by 1991 Tonga had already become the world's sixth largest user of orbital rights, after the USA, UK, Soviet Union, China and Japan.[2]

Tonga's foray into the world of satellite communications began with the formation of Friendly Islands Satellite Communications Ltd, known as Tongasat. This private company acts as the leasing agency for the Tongan government and has registered seven orbital positions with Geneva's International Telecommunications Union since 1988, covering a large part of the Asia–Pacific region. Tongasat has the stated aims of not only boosting government revenue but also improving communications within Tonga. However the venture has experienced numerous setbacks and these aims are yet to be met.

Tonga's attempts to continue expanding its satellite empire may meet with some international resistance. Tonga initially attempted to register 31 orbital positions, causing an uproar in the international satellite communications community. Even the seven slots eventually registered by Tonga were seen as too many for such a small nation. In claiming these slots Tonga was metaphorically raising its flag in cyberspace, staking out territory, and the superpowers were not impressed that their ability to colonize space was being thwarted by this upstart nation.

Another Tongan venture into cyberspace, the Kalianet site, is promoted as claiming a different kind of territory. On the webpage the question, 'What is the meaning of Kalianet?' is answered: 'The kalia was ... used by early Tongans in their conquest of the South Pacific, now Tongans will conquer the world's imagination as this service and its local contributions grow.'

'Cyber-Polys'

Tonga's Kalianet site began operation in February 1997, and although Internet connection in Tonga is expensive relative to Tongan wages, within six months e-mail addresses for over 200 businesses, institutions and private individuals were listed on the site (still others chose not to be listed). The site, operated by Cable and Wireless, also provides a service in which e-mail messages to anyone in Tonga can be sent to a central address, which relays the message by telephone for a fee (cf. Dyrkton, 1996, on the introduction of e-mail into Jamaica). While these direct and indirect e-mail connections are, so far, the main function of Kalianet, Cable and Wireless claims that it intends to develop the site to provide resources for schools, businesses and institutions, to promote tourism, and generally to represent Tonga to the world.

Kalianet is but one example of the proliferation of Pacific-related sites on the Internet, and they continue to grow in number as well as sophistication, with existing sites being constantly reworked and improved. They include sites representing the Pacific Islands and their people to outsiders, those set up by academic and other associations concerned with the Pacific, news sites presenting up-to-the-minute reports from the region, and sites seeking to appeal primarily to islanders themselves. By August 1997 there were over 50 Pacific-related websites, in addition to many Pacific-oriented homepages created by individuals. There is a complex web of cross-links between all of these sites and some sites now exist just to list these numerous Pacific sites. Several Tongans have been prominent in the development of this Pacific network of Internet sites, including 'Alopi Lātūkefu, who established the Tongan History Association homepage and the South Pacific Information Network (SPIN) website, and Samiuela Taufa, creator of Tonga on the 'NET, a site he developed from Tonga. Another is Taholo Kami, who established Tonga Online while studying at Vanderbilt University in Tennessee. Tonga Online, the first site on Tonga by a Tongan, was extended in October 1995 to include the Kava Bowl discussion forum, which will be the focus of the remainder of this chapter.

The Kava Bowl Forum

The Kava Bowl opens with welcomes in Tongan, Fijian and Samoan and the claim that it is 'A Forum the Pacific Way!' An image of a *kumete*, or kava bowl, is a prominent reminder that the forum seeks to emulate the informal kava circles encountered in many Pacific societies, at which men gossip, argue and talk over issues while drinking the ground, dried root of the kava plant mixed with water. In this virtual version of the kava circle women are also welcome participants, as are males considered too young to drink kava, which has a mild narcotic effect.

Before the Kava Bowl (KB) discussion forum begins there is a section providing links to numerous other pages, including a Pacific Island directory with further links to sites on the Pacific. This preliminary section has a 'KB Feature' which changes at intervals, with humorous stories written by a Tongan living in America, to which people respond. In addition there is the 'KB Weekly Discussion Theme' with issues raised by the administrators of the site or suggested by participants, and responses by other participants. Some examples from August 1997 include 'Poly-owned businesses: are they given support sufficiently by other Polys?' ('Poly' is shorthand for 'Polynesian') and 'Sex education: when, how and by whom?' There are also links to pages facilitating the e-mailing of 'virtual island greeting cards' and listing advertisements and announcements.

The KB also has four 'chat rooms' which facilitate synchronistic communication, with participants online simultaneously, conversing about topics of their own choice. These conversations are frequently in Tongan, or a mixture of Tongan and English, and since they are not censored, participants sometimes use language that is sexually explicit, abusive, or otherwise considered inappropriate on the main forum. The chat rooms appear to have become very popular, and when technical problems temporarily closed them in late July 1997, a woman calling herself 'Tongan Sistah' posted a message on the KB begging to know what was wrong, saying she was 'desperate to talk to friends, family, and so on'.

The majority of participants in the KB forum are Tongan, with some other Pacific islanders, particularly Samoans. Participants other than islanders are usually interested in Tonga because they have previously worked or visited there, or are those who have (or previously had) a Tongan partner, as well as those planning to go to Tonga in the future. The last of these tend to enter the KB only briefly, to request relevant information. There are also occasional one-off posts by non-Tongans seeking specific information or calling for pen-pals. Very few participants who are obviously not Pacific islanders are involved simply because they happened upon the forum whilst 'surfing the net', and those who are do not stay for long.

Participants in the KB range from those who have been active from its beginning to those who participate only once. Some know each other in 'real life', while many do not personally know the other participants, although they often find they are connected to some of them in one way or another, for example through kinship or village of origin, or they establish a more personal relationship with one or more participants via e-mail or other forms of communication. Occasionally a message is posted announcing someone's intention to visit Tonga, inviting others who will be there at that time to meet up with them.

The ability to play with one's identity in computer-mediated communication (CMC) means that it is difficult to know just who is participating in the KB. However, it seems that the largest group of participants are tertiary students and young adults working at a professional or managerial level. Most participants are from the USA, particularly Utah (where many Tongan Mormons cluster) and California, but with many others scattered across North America. There are also posts from Tongans in Australia, New Zealand, Canada, Europe and other locations of the Tongan diaspora, and occasionally even from Tonga itself. Participants tend to be located in major cities, to be well-educated and of course to have access to computers and at least some computer literacy. As access to computers widens in schools, public libraries, workplaces and homes, the occupational range and number of participants are likely to increase, and the demographics of the forum may change over time. While it remains largely true that the Internet 'is still the territory of "Westerners" or those "Western"-educated elites in close contact with the "West" ' (Shields, 1996, p. 2), this is gradually changing as access increases.

Participants are not all highly literate in English; indeed numerous KB posts have indicated otherwise, with participants bemoaning or defending their poor English. The term FOB ('fresh off the boat') is sometimes used by posters to describe themselves or others whose English is poor. Some participants choose to post messages in Tongan only, or a mixture of Tongan and English ('Tonglish'). The language used on the KB reveals the variations in English used by Tongans overseas, with influences from different groups, such as Anglos, African-Americans, Hispanics, Rastafarians and others.

The flexibility of identities in CMC also makes it impossible to enumerate the participants on the KB. Many participants choose to use their own names, but many others use pseudonyms, often names that proclaim their Tongan identity. Of course, anyone can use any name, and individuals could use a different name each time they post a message. Although it is impossible to know how many individuals participate in the KB, the posts themselves can be enumerated and analysed. The KB has rapidly increased in popularity: in early 1996 the forum was receiving around 400 messages per month and by mid-March, 1997, a total of 583 messages were posted in just one week.[3] KB participants can post new messages and follow up previous posts, and while it involves asynchronous communication the posts often follow quite quickly so that a number of related messages are posted on one day or several consecutive days.

To indicate the kinds of communications involved in the KB forum, an analysis is given below of the 1670 initial posts for one year, from March 1996 to March 1997. The number of follow-ups to these initial posts varies considerably, from no posts to 143, and over the course of the year several thousand posts were made. In the week mentioned above, 13–20 March 1997, only 66 of the 583 messages (11 per cent) were initial posts.

Of 1670 initial posts analysed, 930 (56 per cent) identified the posters' gender. Of these 930 posts, 64 per cent were male and 36 per cent were female. Of course, given the free play with identity possible on the Internet, these figures may be somewhat inaccurate. They are also skewed by the fact that three of the most frequent posters are also male, as is the owner of the site, Taholo Kami. If the 166 posts made by these four males are removed from the equation, the male posts make up 56 per cent of the gender-identified posts. While these figures suggest there is some gender imbalance in participation in the KB, this does not indicate male domination of the forum, as the female participants speak out as freely and articulately as male participants.

The kinds of messages found on the KB vary enormously, and to simplify the discussion the 1670 initial posts have been grouped into categories:

- Greetings and personal messages sent by or exchanged between specific individuals or groups, messages seeking particular people, and general messages from individuals, such as Christmas greetings: 40.8 per cent
- Messages in which issues are discussed: 18.1 per cent
- Requests for specific information or assistance (not business-related) as when young people seek information on the availability of scholarships for colleges and universities: 10.8 per cent
- Messages from the KB administration about the workings of the forum, and comments from participants about the forum, including discussion about censorship: 8.4 per cent
- News of events, conferences, films, books, music, Internet sites, and so on; for example, the many clubs and associations that have been set up for Pacific Islanders in American colleges and universities announce events they have organized: 6.9 per cent
- Messages seeking contacts with groups, such as other Tongans in a particular country or state; or with non-specific individuals, such as requests for pen-pals: 6.8 per cent
- Sports news and issues: 3.5 per cent
- Jokes, funny stories, songs and poems: 2.6 per cent
- Business-related messages, such as requests for information related to establishing businesses in Tonga: 1.3 per cent
- News about Tonga and the Royal family, such as news of cyclone damage or a visit overseas by the king: 1.0 per cent

The Kava Bowl as a 'community'

As can be seen from this breakdown, the KB functions on several levels: establishing and maintaining a social network, facilitating the discussion of issues in a safe space, and providing information. This accords with Howard Rheingold's analysis of 'virtual communities' emerging from CMC: 'Virtual communities are places where people meet, and they are also tools; the place-like aspects and tool-like aspects only partially overlap' (1993, p. 56). Rheingold notes that as a tool such a community 'is like a living encyclopedia' (p. 57), and this is apparent in the use of the KB to request a vast range of information, not limited to Tonga or even to the Pacific. It is in this tool-like function that the KB is most often approached by non-Tongans, who are generally less likely to become involved in the social network established by its 'place-like' aspects.

Rheingold's use of the term 'virtual community' raises the question of how the KB can be described, beyond being simply a 'forum' or 'message board'. In the literature on CMC there is little agreement as yet regarding the appropriate terms to apply to the groups that form, and terms such as virtual community, or Steven Jones' term 'cybersociety' (1995), tend to be used rather cautiously. KB participants themselves describe the forum in various ways, including the obvious metaphor of a kava circle. One poster even commented that participants engage in 'cyber-kava drinking' (8/1/97). Metaphors of place are also used, such as 'a meeting place for Polynesians in cyber-space' (20/9/96) and 'our cyber neighborhood' (10/8/97).

At times the KB is referred to as a distinct social entity, as when it is called 'a cyber Poly culture' and 'social community' (6/11/96). This sense of being a community is enhanced by participants frequently addressing each other as 'Kbers' and 'Kavabowlers', and referring to 'our KB family'. Over time a core group of frequent and regular participants has emerged, which has become, as has been noted more generally for groups formed by CMC, 'a more stable and enduring aspect of community' (McLaughlin *et al.*, 1995, p. 92). Individuals who participate in the KB over time become more familiar with one another, having shared stories of their life experiences, argued over issues, exchanged news and greetings, and otherwise established social bonds.

Alongside these notions of the KB as a kava circle, meeting place, neighbourhood and community, there is a powerful sense of the forum as part of a much wider community of Pacific Islanders and in particular of Tongans. 'Alopi Lātūkefu has observed the KB since its inception and been an occasional participant. He commented,

> I would in some ways describe the Kava Bowl as a community, whether it is virtual or not is the key question. Virtual takes on a connotation of being unreal and immaterial, when I think in many ways the Kava Bowl is purely an extension of what Pacific communities have been doing for centuries, and that is communicating, integrating and sharing ideas over vast distances (personal communication, 10/7/97).

The Kava Bowl as Tongan

Wellman (1997) has observed that life 'online' is complexly connected to participants' 'real life', and this can be seen particularly clearly in the case of the Kava Bowl. The KB does not create an entirely new or isolated community; rather, it extends the pre-existing Tongan community – itself embedded in the wider Pacific community – and in so doing begins to transform it. The social network existing on the KB extends well beyond the individuals posting messages, and this becomes clear when Tongans posting messages for the first time are asked questions by other participants such as to whom they are related and where they are from. According to 'Alopi Lātūkefu, this 'reflects the fact that the internet is an extension of the normal processes of Pacific and Tongan communication' (personal communication, 10/7/97). These connections between the 'virtual' and 'real' Tongan communities often lead to exchanges of messages that reflect the gossip and rivalries characterizing Tongan social relations – as in the occasional heated exchanges between young migrants from rival villages in Tonga. Wellman has claimed that 'interactions on-line are a technologically supported continuation of a long-term shift to communities organized by shared interests (sometimes narrowly focused) rather than by shared place (neighborhood or village) or shared ancestry (kinship group)' (1997, p. 447). However, in the case of the KB, shared place, ancestry – and, more broadly, shared culture and history – to a great extent constitute those shared interests.

Wider connections with the 'real' community are also assumed in many messages that announce births, deaths, graduations, and other life events, involving people who may not even participate in the KB. One post referred to 'one of our sisters from the KB' who was about to undergo a serious operation, and called on 'all of the brothers and sisters out there *in and out of* the Kavabowl circle' to pray for her (6/8/97, my emphasis). This use of kinship terminology to signal a sense of community is common, as in 'my Tongan sister' or 'Polynesian brothas [sic]'. As the latter indicates, this wider community extends to other Pacific Islanders, and participants use the KB to strengthen this community by announcing events for islanders and encouraging unity amongst them, particularly Polynesians. After a violent incident involving rival gangs of Tongan and Samoan youth in Los Angeles there was a spate of messages supporting the sentiments in a message entitled, 'Keeping the peace with-in Polynesianz ... one love' (26/2/97).

Participants sometimes debate the boundaries of membership of the KB, and there have been attempts to both limit and expand these boundaries. Some wish to keep it as Tongan as possible, and occasionally make racist comments to non-Tongan participants or use the Tongan language to shut out non-Tongans. There have even been periodic calls to use the Tongan language exclusively on the site; however this idea is strongly resisted by most participants, who argue that many Tongans living overseas speak Tongan poorly, or not at all, and that in any case 'KBers' enjoy having non-Tongans involved in their discussions. One participant asserted that he had 'made friends from all Races' on the KB (20/9/96). Some participants who are not Tongan ask permission to be involved and give reasons for their interest – for example several women have been active participants who are or were in relationships with Tongan men and who seek contacts and information that will help with raising their children to appreciate Tongan culture. In any case, the nature of the KB necessarily limits full participation by anyone without some knowledge of Tongan culture and language, as without this knowledge the unspoken assumptions, assumed knowledge and use of Tongan terms and concepts would make many of the messages impossible to understand.

Little concern is expressed for 'netiquette' (network etiquette) on the KB and many participants do not appear to know the conventions commonly used on Internet message boards. Rather, the focus is primarily on the KB forum as part of the international Tongan community and therefore on whether to keep to 'the Tongan way' in the content of posts. Some participants express the belief that it is very important to uphold Tongan values on the forum in order to represent Tonga in a positive manner, in the knowledge that there may be many silent readers, or 'lurkers', who may be non-Tongans forming opinions of Tongans through the KB. This provides opportunities for explicitly discussing what it is to be 'Tongan'. In a post addressed to 'Dear Fellow Tongans!' and complaining about the use of bad language on the forum one participant wrote, 'I can only imagine what the world may think when visiting this fabilous [sic] site! ... We as Tongans must take responsibilities for our little Kingdom and ownership of our traditional values' (25/10/96). In another case the response to a verbally aggressive post in 'gang talk', expressing the rivalry between two villages in Tonga, asks: 'Are you Tongan? If so do you have any PRIDE or for that matter SHAME? Boy, the whole world is tapping into this "net" and what kind of impression are you setting up for your native brothers and sisters? ... Why is it that ya all have to kill each other like that on the net? You Tongans for God's sakes!!!!' (21/11/96, emphases in original). The KB administrators do occasionally delete posts that have inappropriate language or which slander or demean other individuals, and this censorship often initiates a flurry of messages about the issue of free speech versus the imposition of Tongan and Christian values. Some participants choose to uphold Tongan values out of respect for other Tongans participating; for example one young woman wrote 'this is a Tongan forum and I am a Tongan; therefore, I must always remember my place' (24/1/97).

Kava Bowl issues

Many participants use the anonymity afforded by CMC to express opinions and discuss topics that in other contexts would be discouraged or even forbidden (*tapu*). Individuals who effectively would be silenced in 'real' Tongan communities are able to speak up, freed of the constraints of the hierarchical Tongan social structure and value system. In 'the Tongan way' young people are expected to be unquestioningly respectful, obedient and deferent to their superiors, as are commoners in relation to the chiefly and royal 'classes' of Tonga (Morton, 1996). On the KB, individuals of any age, gender, rank or status can say what they wish, whether it be speaking frankly about sexuality or criticizing the Tongan monarchy. Other participants may argue with their views or attempt to shame them into silence, but, as the KB administration does not censor posts because of the ideas presented, they can continue to express their views. Women appear to have found this liberating; as 'Alopi Lātūkefu commented, 'there are clear developments in Pacific feminism of sorts over the internet, where women obviously feel safe to express their sexuality and other views without the hindrance of family disdain or pressure not to do so as it is not "the Pacific way" ' (personal communication, 10/7/97).

The anonymity of the Internet also enables individuals with personal problems to ask for advice, in a way that would be difficult and embarrassing in other contexts. As Rheingold has noted of CMC in general (1993, p. 27), some participants on the KB clearly find it easier to speak more intimately about themselves and their families than in other forms of communication. Whether or not pseudonyms are used, a continuing thread in the KB messages is the sharing of life experiences, as when part-Tongans describe growing up with parents of different ethnicity, or immigrant Tongans discuss their experiences of racism in their new countries.

Much of this sharing of experiences occurs within the category of messages in which issues are raised for discussion. Although these messages comprised only 18.1 per cent of total initial messages in the year analysed, they generated by far the most responses. In the week mentioned previously, 13–20 March 1997, messages responding to initial posts raising issues for discussion comprised 59.4 per cent of the total follow-up messages. The most popular issues discussed on the KB are those affecting Tongans living overseas. From March 1996 to March 1997, of the 302 posts raising issues, 83 were on topics such as access to education in the host country, views on mixed marriages and how best to raise the children of such marriages, problems of adapting to a Western lifestyle, experiences of racism, and language problems. Closely related to this category of topics were those discussing family relationships, childrearing, and the problems of youth (25 posts) as these topics were usually discussed in the context of adjustments to life overseas. Issues specific to particular places are seldom raised, and the discussions tend to deal more generally with life in 'the West'.

The second most popular category of topics (58 posts) were those discussing Tonga, particularly issues surrounding the political system, as well as views on tourism, cultural change, and social problems. Sexuality and relationships were also very popular topics (37) and led to many long discussions of homosexuality, appropriate behaviour in relationships, ideals of female beauty, and so forth. Other categories of topics included religion (36), the Pacific Islands (21 posts on a range of issues), American issues such as the taxation system and elections (14), as well as 28 miscellaneous posts on topics such as health, computers, music and cooking.

Some issues are particularly contentious and generate a large number of responses. During the year analysed, the most popular individual post was one on 'Poly attitudes to homosexuality' (27/10/96) which generated 143 responses. Another very popular topic was the question of Tongan men marrying *pālangi* (Western) women, which was raised several times and overall generated 256 responses. Other particularly popular topics included reactions to the publication of a Polynesian student newspaper in Utah (114 responses), the closure of Tongan wards of the Mormon church in America (83), the need for sex education in Tonga (72) and Tongan participation in gangs (62).

Discussions of issues concerning immigrant Tongans reveal the complexities of the experience of migration and settlement, not just for those participating on the KB but their friends, families and acquaintances. With a predominance of young people contributing to the discussions, a recurring theme is the problems they face growing up overseas. Posts often include reminders of why Tongan parents migrate from the islands, as in the following post made in relation to the need to encourage young people to further their education:

> Tongan youths need someone to say to them that they have the opportunity of a life-time, literally, before them and that they need to grab hold of it. They need to realize that the reason our parents are here is for the sole purpose of furthering the opportunities of the children, in order for them to have something that was not available in Tonga or Samoa (3/10/96).

Participants exchange stories of their experiences, such as their difficulties communicating with their parents. '[W]e had to obey them and shut up!' one young woman reflected (29/4/97). In this context the problems experienced by young women who are brought up in more 'traditional' Tongan homes are raised, as they describe their parents' strict discipline and insistence that they stay at home outside school hours and have a chaperone on any outings. As well as using the KB to share their problems, the forum is used as a means of social contact for young women feeling socially isolated. When one woman posted a message saying she was lonely and inviting people to contact her via the KB, another responded by pointing out that it can be particularly hard for such women to have contact with males, adding, 'And I think it's much more easier and comfortable conversing with someone via the net' (22/5/97).

Young people's problems are often discussed in terms of the clash of cultures experienced in the context of migration. Statements such as 'many teenage Tongans are stuck in two cultures that are contrary to each other' (28/4/97) are common, and participants explore the reasons for this clash and ways of coping with it. One participant saw blame on both sides, arguing 'it is not easy to keep your culture alive. But it is not easy due to the fact that most youths choose not to understand, let alone, analyze their culture'.

On the other hand, this young woman claimed, 'the parents [are] not giving the child the opportunity to know and then accept their culture' (9/4/97).

Problems stemming from parents' 'conservatism', lack of trust in their children, and the children's confusion about conflicting values are also discussed. The pros and cons of the strict physical punishment commonly used by Tongan parents are debated, as is the value of sending children back to Tonga for periods of time, to 'learn what it truly means to be Tongan' and 'gain discipline' (27/4/97). Questions are raised as to how younger Tongans could be encouraged to stay at school and discouraged from becoming involved in the gangs that are present in many cities. Parents sometimes offer their own views, and one participant, the mother of two teenage boys, wrote, 'We, as parents MUST prioritize our responsibilities in order for us to know our children and also TO HEAR them' (30/10/96, emphases in original).

Participants who discuss ways of coping with cultural conflicts often emphasize the need for balance: 'the new-age Tongan must balance what is good in the Tongan culture with what is good in the American culture' (27/4/97). This poster identified himself as a 21-year-old man, and after describing his adolescence he concluded, 'I had to be a chameleon and change my mind frame to fit the environment I was in.' He argued that parents need to help their children with this, by making sure they understand Tongan culture in order to take the best from it. Another participant said, 'it is not easy to "be a Tongan" in this society [USA]. It's like playing "roles". One minute you're a Tongan and the other, you are stepping out of that circle to join the rest of society' (26/10/96). Many of the discussions raise questions about which aspects of Tongan culture should be kept or discarded in the process of adapting to another culture.

It becomes clear in these discussions that participants are using the forum to carefully reassess 'Tongan culture'. This includes critiques of practices such as giving away money beyond a family's means to fulfil family and church obligations, and the demands placed on individuals and families by those considered superior to them in the hierarchy of kinship and rank. Issues such as sexism, homophobia, domestic violence, and sexual abuse as they occur in Tongan culture are discussed openly and often at length. On several occasions there have also been quite heated arguments about the male practice of kava-drinking. One angry woman berated men for wasting time drinking kava and added:

> There are countless numbers of young Polynesians that could use guidance, tutoring, and structure so that they can succeed within the culture here in America. The Polynesian Cultures here in the US are suffering because each of us are not taking responsibility to help each other and lift each other ... We must all become united in the effort to preserve our Polynesian Heritage while not living in the Islands (11/3/97).

Ideas about helping youth in the 'real' community in practical ways have been a recurring thread of the KB messages. One participant proposed a mentoring programme, using KB participants as a pool of volunteers who would provide role models and guidance to islander students in their home areas (30/10/96) and in later posts suggested other forms of assistance the KB could offer. However, there were few responses to his suggestions and there have been no moves to institute them. According to Taholo Kami, this may be because the KB participants are mainly young people, who in Tongan communities are not expected to express their opinions or to act on their own initiative (personal communication, 8/8/97). The more important role of the KB appears to lie in the social networks established through it, and the other aspects of 'community' it offers, such as sharing experiences, giving and receiving advice and support, and otherwise facilitating communication between participants.

Discussion

When Taholo Kami sought suggestions for making the KB more actively involved in helping Pacific Islanders both at home and overseas, he concluded, 'If this forum can act as a catalyst for change or a place to enhance vision then the benefits would extend beyond just "meetings in cyberspace"' (6/11/96). One participant quickly responded, 'What's wrong with meetings in cyberspace? I think KB like Polycafe [the website Polynesian Cafe] is good simply because it gives a forum for cyberpolys to meet and talk in the knowledge that the other people involved are "like me" ... my point is kavabowl is already doing good for our peoples by simply existing on the net' (6/11/96).

As has become clear, the KB is *not* simply a meeting place, and while its impact on people's lives cannot be accurately gauged, the messages on the forum demonstrate that its impact is significant. Apart from facilitating easier communication with people who are geographically dispersed, the forum can have a profound effect on participants, who can have their voices heard, their isolation broken, their values challenged, and their problems shared. Diasporic Tongans use the KB to explore their identities, question the status quo and establish themselves more firmly within the international Tongan community.

Given the global spread of contemporary Tongans, ventures such as Tongasat can be seen, according to Marcus, as attempts 'to preserve Tonga's position as *both* the economic and symbolic center of an internationalizing culture ... The development aims of the King are thus in part an effort to retain the Tongan state as the center of a society in centrifugal motion' (1993, p. 32, emphasis in original). These efforts have been significantly boosted by the existence of the Internet and fora such as the KB, which aid transnational communication and equally importantly enable their participants to maintain a focus on their homeland and their Tongan identity. With direct Internet links to Tonga now in place, there is further potential for a strengthening of ties between Tongans worldwide.

Of course, relatively few Tongans have access to computers or computer literacy as yet, so the impact of these developments has been uneven. This has been acknowledged by participants in the KB, as in a post on the problem of violence between Samoan and Tongan youth in America, in which a participant commented, 'I wish the people that have these differences had access to computers and the internet. It seems to me that the majority of us who are fortunate enough to be in school or can afford this kind of access are not the ones with the problem'(2/5/97). As computers become more accessible there will be increasing opportunities for Tongans to participate at least to some extent and the possibilities and limitations of this form of communication will become more apparent. In Tonga there are already moves to increase access to computers, with the Crown Prince promoting a computerization programme for schools in Tonga.[4] Although most Tongan schools have inadequate basic resources, such as textbooks, the fund-raising efforts by Tongans living overseas are increasingly being directed at providing computers and Internet connections for schools.

The long-term impact of Tongan participation on the Internet can only be imagined. David Elkins has argued of 'virtual ethnic communities' more generally that 'they have the potential to be just as fundamental to the identities of some people as the existing ethnic communities whose existence we have taken for granted for decades or even centuries' (1997, p. 141). The KB has already demonstrated this potential and shown that such groups can work to revitalize ethnic identity amongst dispersed populations. The discussions that take place on the KB constantly remind the participants of their homeland and 'heritage'. Perhaps the renewed enthusiasm for being Tongan generated by involvement with sites like the KB will encourage younger Tongans living overseas to renew and maintain their links with kin in Tonga, or inspire them to return to the islands to live. At the same time, by discussing settlement issues and sharing their experiences of migration, participants may be assisting one another with their adjustment to new cultures and the juggling of multiple identities.

On the other hand, the open cynicism about the Tongan political system and certain cultural values, and the freedom of expression possible on the KB, could have the opposite effect, discouraging young people from maintaining links with the islands and exacerbating the conflicts between those wishing to either challenge or retain the status quo. As the KB fosters closer contact between Tongans overseas, a stronger sense of the migrant community may emerge which lessens the need for continued links with Tonga. A division could also emerge between those with and without access to 'cyberculture', reflecting and exacerbating existing differences in socio-economic status. For Tongans still in the islands who can afford access, technology such as the Internet can strengthen ties with the diaspora and at the same time enable individuals' participation in all kinds of virtual communities. This can lead to 'a multiplication of identities and hence an expansion of options and choices' (Elkins, 1997, p. 149; and see Wellman, 1997), which in future may have significant implications for concepts of nationhood and cultural identity and for the relationship between the individual and the state.

Such implications have not, as yet, dampened the eagerness of the Tongan king and his government to embrace new forms of communications technology, nor that of the Tongans who are utilizing them. 'Taholo [Kami] has led our race into the future' proclaimed a Tongan participant in the KB (20/9/96). Just as the Tongan government has envisaged this new technology as a means of empowering the nation, the Tongan people using it clearly perceive it as empowering both to themselves and to their international community. As one participant asserted, 'Even by [the KB] just existing as an entity, Taholo has managed to gain some power for Tonga in it being there on the internet as a resource' (23/9/96). This message, entitled 'Knowledge is Power! Power is Knowledge!' (23/9/96) continued:

> We as Pacific Islanders are at the verge of a New era, and it is through sites like the Kavabowl that we not only get an insight of what others think, but what our own understandings of the region we live in are. By knowing clearly who we are in the world, we are empowered to stand up to any group with the power to say 'We are Pacific Islanders and we have a place in this world no matter how small'.

Acknowledgements

This chapter was written during an Australian Research Council Postdoctoral Research Fellowship, held at La Trobe University. My thanks to 'Alopi Lātākefu, Samiuela Loni Vea Taufa and Taholo Kami for helpful comments and discussions and to Russell King and John Connell for their useful editorial suggestions. Earlier versions of the paper were presented at the Anthropology seminar at the Australian National University in September 1997 and at the Australian Anthropological Society conference in October 1997; thanks to all those who gave valuable feedback on these occasions. My appreciation also to the many participants in the Kava Bowl discussion forum, who have greatly enriched my ongoing research on Tongan migration and cultural identity.

Notes

1. See the short article 'Tonga becomes a space-faring nation', *Matangi Tonga,* 9(3), 1994, pp. 14–15.

2. 'Tongasat stock valued at $25 million', *Matangi Tonga,* 9(4), 1994, p. 33.

3. Exponential growth continues: 510,000 hits were recorded in the month of January 1998.

4. See 'Kalianet links with Internet to service local subscribers', *Tonga Chronicle,* 23 January 1997, p. 3.

References

Ahlburg, D. (1991) *Remittances and Their Impact: A Study of Tonga and Western Samoa,* Pacific Policy Papers, No. 7. Canberra: Australian National University, National Centre for Development Studies.

Brown, R. and Connell, J. (1993) 'The global flea market: migration, remittances and the informal economy in Tonga', *Development and Change,* 24(4), pp. 611–47.

Connell, J. (1987) 'Paradise left? Pacific Island voyagers in the modern world', in J. Fawcett and B. Carino (eds), *Pacific Bridges: The New Immigration from Asia and the Pacific Islands.* New York: Center for Migration Studies, pp. 375–404.

Cowling, W. (1990) 'Motivations for contemporary Tongan migration', in P. Herda, J. Terrell and N. Gunson (eds), *Tongan Culture and History.* Canberra: Australian National University, Department of Pacific and Southeast Asian Studies, pp. 187–205.

Dyrkton, J. (1996) 'Cool runnings: the contradictions of cybereality in Jamaica', in R. Shields (ed.), *Cultures of Internet: Virtual Spaces, Real Histories, Living Bodies.* London: Sage, pp. 49–57.

Elkins, D. (1997) 'Globalization, telecommunication, and virtual ethnic communities', *International Political Science Review,* 18(2), pp. 139–52.

Faeamani, S. (1995) 'The impact of remittances on rural development in Tongan villages', *Asian and Pacific Migration Journal,* 4(1), pp. 139–55.

Gailey, C. (1992) 'A good man is hard to find: overseas migration and the decentered family in the Tongan Islands', *Critique of Anthropology,* 12(1), 47–74.

Hau'ofa, E. (1994) 'Our sea of islands', *The Contemporary Pacific,* 6(1), pp. 148–61.

James, K. (1991) 'Migration and remittances: a Tongan village perspective', *Pacific Viewpoint,* 32(1), pp. 1–23.

James, K. (1993) 'Cutting the ground from under them? Commercialization, cultivation, and conservation in Tonga', *The Contemporary Pacific,* 5(2), pp. 215–42.

Jones, S. (ed.) (1995) *CyberSociety: Computer-Mediated Communication and Community.* Thousand Oaks, CA: Sage.

McLaughlin, M., Osborne, K. and Smith, C. (1995) 'Standards of conduct on usenet', in S. Jones (ed.) *CyberSociety: Computer-Mediated Communication and Community.* Thousand Oaks, CA: Sage, pp. 90–111.

Marcus, G. (1993) 'Tonga's contemporary globalizing strategies: trading on sovereignty amidst international migration', in V. Lockwood, T. Harding and B. Wallace (eds), *Contemporary Pacific Societies: Studies in Development and Change.* Englewood Cliffs, NJ: Prentice Hall, pp. 21–33.

Morton, H. (1996) *Becoming Tongan: An Ethnography of Childhood.* Honolulu: University of Hawaii Press.

Rheingold, H. (1993) *The Virtual Community: Homesteading on the Electronic Frontier.* Reading, MA: Addison-Wesley.

Shields, R. (1996) 'Introduction', in R. Shields (ed.), *Cultures of the Internet: Virtual Spaces, Real Histories, Living Bodies.* London: Sage, pp. 1–10.

van der Grijp, P. (1993) *Islanders of the South: Production, Kinship and Ideology in the Polynesian Kingdom of Tonga.* Leiden: KITLV Press.

Vete, M.F. (1995) 'The determinants of remittances among Tongans in Auckland', *Asian and Pacific Migration Journal,* 4(1), pp. 55–68.

Wellman, B. (1997) 'The road to utopia and dystopia on the Information Highway', *Contemporary Sociology,* 26(4), pp. 445–9.

Online addresses

Kalianet: http://www.candw.to
Kava Bowl: http://www.pacificforum.com/kavabowl/
South Pacific Information Network: http://sunsite.anu.edu.au/spin/
Tonga Online: http://www.tongaonline.com/
Tonga on the 'NET: http://www.tongatapu.net.to
Tongan History Association:
      http://sunsite.anu.edu.au/region/spin/PACASSOC/TONGHIST/tonghist.htm

13

Spatial Population Mobility as Social Interaction: A Fijian Island's Multi-local Village Community

Carsten Felgentreff

Introduction

The Lau group in the eastern periphery of Fiji consists of some 60 atolls and islands with a total land mass of about 461 sq km, scattered over an area of 114,000 sq km (see Figure 13.1 which covers the northern part of Eastern Fiji). The eastern islands of Fiji exemplify some of the problems of fragmentation, smallness, remoteness and relative poverty of natural resources (including fresh water) that are typical of other parts of Oceania. Furthermore, these islands and their populations also exemplify 'all the common trends of South Pacific internal migration, that is from outer islands to inner islands, from small islands to large islands, from mountains to coast and from rural areas to urban areas' (Connell, 1987, p. 60). Since independence in 1970, many villages in Lau have witnessed a more or less steady decrease in the number of inhabitants. According to census figures, the Lau province contained 3.4 per cent of Fiji's total population in 1966, compared to 2.5 per cent in 1976 and 2.0 per cent in 1986 (Felgentreff, 1995, p. 51). At the time of Fiji's 1986 census, 26,363 persons stated that their place of birth was in Lau, but only 12,158 of them were actually enumerated in Lau. In other words, more than one out of two people born in Lau had left the province (Fiji Bureau of Statistics, 1989, pp. 90–1).

As Bryant (1990), Sofer (1993) and others have argued, this population redistribution is a reaction to uneven regional development. Given the absence of any perspectives for large-scale sustainable development beyond subsistence level, out-migration from such disadvantaged, outlying islands

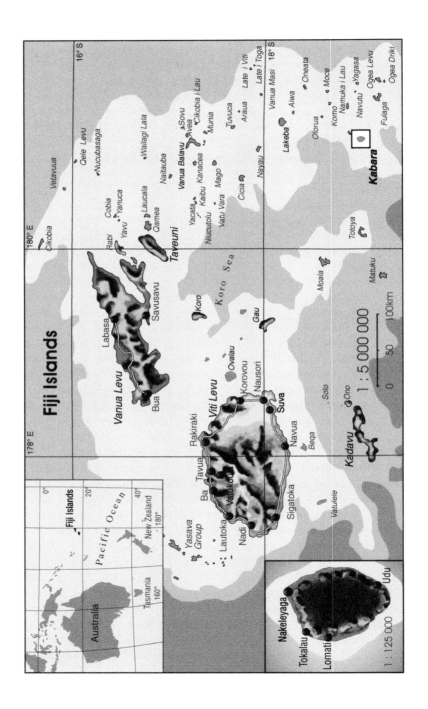

Figure 13.1 Map of Fiji showing location of Kabara

seems unavoidable. Corresponding to this pattern of out-migration from peripheral islands is a substantial in-migration into the Suva-Nausori urban area, especially into Suva's peri-urban area (Fiji Bureau of Statistics, 1989, p. 90). The fact that a large proportion of Fijians living in Suva once came from distant island provinces (Nair, 1985) clearly shows 'that, for Fijians, migration was a function of the relationship between remoteness and limited economic resources' (Connell, 1987, pp. 65–6).

According to the prevailing view, future prospects for the most marginal islands are grim: one day they will be depopulated. Villages on such peripheral islands will only survive where a sizeable monetary income can be generated locally or where people can commute to work. In fact, the overall material standard of living appears to be higher in villages with easier access to markets and wage-earning opportunities, for instance in the vicinity of Suva (see Overton, 1989 and 1993) in comparison to villages in Lau. Given these structural disparities, people will continue to leave their villages, and each decline in numbers of inhabitants is a substantial loss for the community left behind.

This pessimistic view is still prevailing, although some have offered different interpretations of the macro-level evidence (see Chapman, 1991; Hau'ofa, 1994). The aim of this chapter, which draws on my 1993 field research in Naikeleyaga village (Kabara island, Lau province) and Suva, is to challenge this common pessimism. In synthesis, the main arguments are as follows: firstly, insularity does not necessarily mean isolation; a remote island might be far better integrated than its position on the map would make us believe. This is because the geographical realm that the inhabitants of such tiny islands draw their resources from might be much wider than just their island. Secondly, a great deal of conceptual confusion arises from the term 'migration' when used in the context of contemporary internal population mobility in Fiji. The term was once suggested for 'changes of residence that involve a complete change and readjustment of the community affiliations of the individual' (Bogue, 1959, p. 489) – a constellation of processes which might be rather exceptional in contemporary Fijian internal migration. Much more appropriate are Bedford's words: 'Voluntary movement of any kind rarely leads to total severance of connections with people, events and places in a former residential environment' (1980, p. 32). In other words, a concept of migration that was created in last century's Euro-American context for, say, Germans who left their homeland and settled somewhere in the United States of America might – in analytical terms – be inappropriate for a villager from a peripheral island who sets sails for the next urban area. For the Pacific, micro-level analysis and longitudinal studies give ample evidence that local mobility patterns differ enormously from 'classic' rural-urban migration of the European kind (Bedford, 1980, p. 33; Felgentreff, 1995 and 1996).

Naikeleyaga and the Island of Kabara

Kabara is a raised limestone island with a size of some 50 sq km (Figure 13.1, inset). It is well characterized in the following description by Bayliss-Smith *et al.* (1988, p. 185):

> Except for tiny pockets, soils are everywhere thin and deficient in nutrients. Rainfall is extremely variable and drought a frequent hazard. Hurricanes are an ever-present seasonal possibility. Because of the porosity of the limestone fresh water is a scarce resource: there are no streams on Kabara. As a result of soil and water deficiencies the main subsistence crops in other parts of Fiji – yams, taro and different varieties of bananas – do not grow well. Marine resources in the reefs and waters surrounding Kabara are also poor, even in comparison with other islands in southern Lau.

To make it worse, the cost and unpredictability of getting there are both high: the one-way fare from Suva is enormously high by local standards, and it can take three or more weeks for the next ship to call.

Most households have to rely on subsistence production. Although Overton (1989, p. 11) has argued that a strict dichotomy of 'capitalist' versus 'non-capitalist mode of production' is of little analytical relevance for contemporary rural Fiji, economic life in a village like Naikeleyaga is still well described by the concept of the 'village mode of production' whose features were summarized by Sofer (1993, p. 308):

> These include the persistence of non-capitalist production forces and relations, a low level of technology mainly with use of traditional agricultural equipment, low cash returns from village agricultural production, a high proportion of subsistence production, a significant share of the end use of production directed to communal and household exchange, and a communal method of raising money for village projects.

Since prices for copra deteriorated in early 1993, the main sources of income for most village households in 1993 have been products made from tropical hardwood by local carvers, especially large bowls used for serving kava, a local mildly-intoxicating drink (see Felgentreff, 1996, Table 1).

The four villages on Kabara island (inset, Figure 13.1) are built on sandy flats between the steep cliff and the water line. Each of the villages has its own Methodist church building. Naikeleyaga has a government station and a hospital; a second nurse is based in Udu. Naikeleyaga, Tokalau and Udu each have a primary (classes one to eight) and a pre-school, a post office with radio telephone plus a local co-operative store, while villagers from Lomati use the services in Tokalau. There are no roads or motor vehicles on the island; the villages are connected by footpaths. As in the 1930s, when Laura Thompson conducted her studies in southern Lau, life is harder on

Kabara compared to living conditions on larger, more fertile Fijian islands. Still, in the 1990s Kabara is 'a hazardous world where the struggle for existence is keen, land is poor and food is scarce' (Thompson, 1971, p. vii). The structural disadvantages comprise ecological as well as economic factors.

Nevertheless, the villagers in Naikeleyaga, where I conducted my field work in 1993, appreciate their 'way of life' (*vakavanua*) – especially if compared to life in urban Fiji. Almost every adult in the village has been exposed to urban life during at least one stage in his or her life, and every adult in Naikeleyaga knows Suva from at least one visit. In 1993, 95 per cent of the male village population above 14 years had lived in urban Fiji for at least one month during their lifetime, either working for wages or receiving education or vocational training; the respective figure for the female village population aged 15 years or older was 36 per cent (Felgentreff, 1995, p. 126).

In general, living in Naikeleyaga is said to be more 'authentic'. By this is meant more satisfactory social relations; more leisure time; the absence of being bossed around by a foreworker; no rigid daily work routine; the certainty of being needed by others, the elderly and sick; the delight of doing work which is felt to be more useful and less alienated (i.e. working together in groups); and the opportunity to live according to tradition. Furthermore, nearly everybody stressed that one can, at least, survive without money in one's village; or that in the village one has fewer worries about money and where to get it. On the other hand, many villagers long for amenities which are only available outside the village. Moreover, money is needed for the purchase of imported food, tobacco, clothing, building materials, tools and chain saws, punts and outboard motors, fishing nets and hooks. Although all these and many more goods are eagerly welcomed in the village, they certainly do not constitute reasons to leave the island permanently. Those boarding a ship in Naikeleyaga usually seek an income-generating opportunity, higher education, medical treatment, or perhaps adventure (*gade*). They want to meet friends and relatives who are living away, and they have to look after their private or the village's affairs. Clearly these various specific motives differ in quality. It is questionable whether all of them are directly related to structural disparities or disadvantages as the underlying reason. One might expect, for example, that those leaving the island for some of these motives might return, the duration of their temporary absence being a function of the purpose initially intended – but of course unexpected events or unplanned opportunities might arise at the destination. Obviously, in not all these cases is the usage of the term 'migration' appropriate. Nevertheless, according to the 'Instructions to Enumerators', the 1986 National Fijian census was a strict *de facto* census (see Fiji Bureau of Statistics, 1988, pp. 17–18). That means that all Naikeleyagans absent on census night were enumerated as residents of other places in national statistics, no matter if they had just gone to see the next available dentist (who happens to practise in Suva), if they were just about to get a birth certificate (only issued in Suva), if they had just paid a visit to other family members residing elsewhere, or if they had left the island years before and were successfully established somewhere else.

Who Belongs to the Village's Population?

Conducting a *realistic* census of the present population of a village like Naikeleyaga turned out to be more difficult than anticipated. The *de facto* population of the village changes virtually each day; during my survey period there was probably not a single day which could be considered as representative. With each ship that called some villagers or visitors arrived, others left the island; almost every night Kabarans from other villages of the island stayed in Naikeleyaga while some Naikeleyagans were on a visit to other villages on Kabara or on neighbouring islands (see Felgentreff, 1995, pp. 100–1). The number of people sleeping in the village varied during the survey period, roughly, between 120 and 300 per night.

Although asked to give all names of household members currently physically present, nearly all household heads followed a – seemingly very wide – concept of their household's (*de jure*) population. In spite of obvious physical absence many names were given, mainly of unmarried children, younger siblings, spouses etc., usually of people who, probably also by European standards, could be regarded as (in the widest sense) dependent on the household head. The household heads insisted that these persons belonged to their household, regardless of the duration of their absence. In some cases the arguments given were convincing to the outsider: for instance the case of the unmarried son who was serving with the UN Peace Corps in the Middle East for one year only and expected to return soon. Some out of this group of persons who were so strongly felt to be present by people in the village actually returned during the survey period, others not. For example, the wife of one household head had left Naikeleyaga for Suva some eighteen months before, intending to assist her daughter with the preparations for her wedding. Nobody in the village was exactly sure when the daughter's wedding was expected to take place, but everybody was convinced that, thereafter, the wife would soon return. The fact that she had to be regarded as a present inhabitant was beyond question for everybody in the village. From an emic point of view, the current physical whereabouts of a person are of little or no relevance. This example and many others clearly show that the actual or planned duration of absence is not necessarily relevant. What matters is the degree to which one is affiliated with a household (cf. Overton, 1993, p. 57).

On the other hand, the concept of *de jure* population which is so important to household heads in Naikeleyaga meant that some individuals (plus a complete household) were *not* mentioned when I asked for all persons currently physically present in Naikeleyaga. No matter how actively these persons participated in communal work, social life and general village affairs, no matter how much time they had already spent in a household or in the village, some were simply not regarded as 'present village population'. Let me give just one example. A man in his 50s had spent many years outside Naikeleyaga, got divorced, become unemployed and then returned to the village some months prior to the survey. He had no clear intentions concerning his future; he stayed in his brother's house, did some carving and helped his brother with the garden work. Although he was respected as leader of the village choir, he was not regarded as a member of his brother's household and, consequently, was not a regular inhabitant of the village.

In the course of a micro-level study, such apparent differences in emic and etic concepts of 'present' and 'absent' population should not be neglected. In the end I conceptualized my own census in the following way. The 'present' population comprises all those mentioned as present household members and actually present for at least one day during the survey period, plus everybody residing in Naikeleyaga for at least three months. The sum total of this population was 191 individuals, 100 females and 91 males, belonging to 32 households. All were ethnic (Melanesian) Fijians. In addition, those who were mentioned as regular village inhabitants or household members but who failed to be present on at least one day during the survey period were recorded as being 'absent'. This group comprised one household of three persons plus 19 individuals (Felgentreff, 1995, pp. 101–5). It is obvious that, according to this census conceptualization, the number of the present population should increase with the duration of the survey period. Indeed, due to their return, more of the Naikeleyagans now counted as absent would have been counted as present if I had had the opportunity to extend the survey period. There is a functional relationship between the duration of the survey period and the number of Naikeleyagans one can meet in the village, both in reality and in the chosen census conceptualization; the key variables are the (widely varying) number of inhabitants absent at the same time and the (similarly widely varying) duration of their absence.

However, the results of the census thus conceptualized are closer to the villagers' perceived number of Naikeleyaga's inhabitants, although each estimate included a number of different persons. The calculations given by village elders ranged from 'about 200' to '224'. Both the estimates by villagers and my own census results indicate that, at each point in time (and even in each time span of some weeks), at least 10 per cent (and possibly 20 per cent) of the village population temporarily stay outside the village. Other micro-level studies have shown similar results (for Naikeleyaga in 1975 see Bedford, 1976), but such evidence – namely that a village's population fluctuates and is therefore hard to conceptualize and measure – is rarely taken into account in macro-level analysis.[1]

The Village's Social Structure

Of much more importance than the rather unstable categories of 'absent' and 'present' population is, from the villagers' point of view, the dichotomy of *taukei* and *vulagi*. While everyone is free to change position in space by his or her own choice, not everybody is entitled to regard themselves and to act as members of the village's community, its *yavusa*. The Fijian term *yavusa* refers to 'a social unit of agnatically-related members larger than the *mataqali* and the members of which claim descent from a common founding male ancestor' (Ravuvu, 1983, p. 123). Each *yavusa* consists of differing numbers of sub-groups (*mataqali*, literally one blood, one kind) who, again, are usually divided into sub-sub-groups (*tokatoka*, translated often as sub-lineage). Each Melanesian Fijian is from birth a member of a kin group, usually that of the father.

Besides a more or less uniform mythology, a *yavusa* is defined by a place, the place of descent, the place one belongs to (Chung, 1991, p. 88; Hooper, 1982, pp.16–23; Ravuvu, 1983). It is the privilege of the members of a *yavusa* to rightfully settle at the place the kin group is connected with, to use part of the kin group's communal garden land for personal use and to actively take part in village affairs. Full rights – and this is what the term *taukei* implies – are accorded only at the place one belongs to by affiliation, while at other places one is more or less *vulagi* (a visitor, guest, or stranger). Even nowadays, when probably more ethnic Fijians are born in towns and in places other than their kin group's village, the affiliation between the ethnic Fijian individual and 'his place' is very, very strong. A migrant's son or grandson, no matter where he was born, is a *kai Kabara* (man from Kabara) if his father or father's father belongs to a kin group originating in Kabara. The person will act according to his descent, as *kai Kabara*, even though he or she might never have set foot on the island. As a member of his group of descent, the individual (and all his descendants) can 'return' to his place of descent, to his village. Many urban Fijians I interviewed referred to this customary practice as a kind of social security for themselves and their offspring (see also Overton, 1993, p. 46), although in practice it might be difficult to retain full rights as *taukei* for a person who had lost the contact to 'his place' and its community.

As Table 13.1 indicates, in Naikeleyaga 148 out of 191 present inhabitants (1993) had full rights as *taukei* (77.5 per cent of population present). Some *taukei* can be regarded as lifetime migrants whose place of birth was somewhere else, but the majority of them were born in the village. The remaining 22.5 per cent of the village population were non-*taukei*. Nearly all of them were born in other places, but the reason why they were not *taukei* is that they were not members of one of the village's kin groups. They are – more or less – *vulagi*. This group comprises wives who married into the village; they stay members of their own kin groups although they are expected to

Table 13.1 Naikeleyaga's present population: *yavusa* membership, 1993

	Taukei	Non-taukei	Total
Males	76	15	91
Females	72	28	100
Total	148	43	191

Source: Felgentreff (1995, pp. 101–5), based on field survey, 1993.

act as members of their husband's group (see Ravuvu, 1983). Many non-*taukei*, especially the state employees, were referred to as *vulagi dokai*, literally 'respected (or welcomed) strangers' (Hooper, 1982). Usually these *vulagi* have a specific reason why they stay in the village, mostly temporarily: the nurse and her family, the postmaster and his family, one of the teachers plus the church catechist and his household. Further *vulagi* were a family belonging to a neighbouring village but temporarily residing in Naikeleyaga, plus some other individuals (see Felgentreff, 1995, pp. 101–5 and Table 30). None of them has full rights: it would be inconceivable that one of them could work his own garden plot, contribute to a discussion of important village affairs without being asked or perhaps act as the village spokesman – these and many other privileges are reserved for *taukei*.

Apart from the wives from 'outside', all these non-*taukei* are of no real concern to village affairs: one might make friends with them, but it is taken for granted that they will leave the village one day. They are not related; they have their own affiliations and obligations towards their own kin and towards their own place of descent. Very many people are much closer to the heart although they are physically absent: relatives who belong to the *yavusa* but reside – temporarily or permanently – outside the village. Kinship ties are extensive, and everybody in Naikeleyaga has either siblings, parents or children who do not reside there, besides all the others of 'one blood', of one common descent. It is these relatives who are of major concern for the *taukei* residing in the village, not only in terms of emotions and memories. Although circumstances prevent all of them from residing at their common place of descent (even if they wanted to), they nevertheless belong together and are perceived as one entity.

The Multi-local Village Community

Asked about relatives and friends residing outside the village that he or she is in regular contact with, each Naikeleyagan can give dozens of names. The list usually starts with an individual's siblings, children and/or parents, followed by further members of the informant's kin group. Thereafter members of his wife's kin group are mentioned, followed by relatives of his mother, his brother's wife, and many more. Although the agnatic relations are, in general, of primary concern in Fijian society (see Ravuvu, 1983), the affinal affiliations (via marriage, e.g. the individual's wife, his mother, his father's mother, his mother's mother etc.) are very important for the creation of new bonds. As Rutz (1987, p. 546) has put it, in Fiji 'to marry a woman is to form an alliance with her brothers; to marry a person from another lineage is to marry into the village of that lineage; to marry a person from another village is to marry someone from that village's island.' In Suva sayings like 'When we see a man marry a Lau woman, we know that he has married an island' (Rutz, 1987, p. 547) can be heard, and they are definitely not jokes!

Interviews which I subsequently conducted with some of the relatives in the capital, Suva, proved that the information given by informants in Naikeleyaga was exact – for instance where the relatives stayed, what kind of job they currently had, and many details about the household they belong to. Although such ego-centred networks were partly congruent among villagers belonging to the same kin group, the sum total of consanguinal and affinal relatives residing outside the village must be, in the case of Naikeleyaga's whole population, not just hundreds, but thousands. It soon turned out to be impractical to capture data on all relatives residing outside the village with whom villagers said they were in close contact.

However, for Fiji, another set of data is available at the Native Lands Commission (NLC) in Suva. As early as the last century the colonial administration began to register the Melanesian Fijian population according to the individual's membership in kin groups (Chung, 1991, p. 91). Additionally, the connection between every kin group in the whole country and each area of land claimed by these kin groups was codified once and for all. Today more than 80 per cent of Fiji is in the communal possession of such autochthonous kin groups (Overton, 1993, p. 45). These files (*Vola ni Kawa Bula*; literally 'book of descent' and henceforth abbreviated VKB) should contain for each kin group the following information for each registered member: identification number, first and second name, sex, date of birth, name and number of the father (or, alternatively, the mother), and the date of death. It must be kept in mind that these lists only contain members of the kin group in question and give no reference to wives from 'outside' (and all others not agnatically related). No reference is made to places of birth or to the present places of residence, nor to whether a person ever lived in the village or even visited it. For the 'book of descent' all these types of information are irrelevant; the only fact that needs to be documented is the membership of each registered Fijian in the kin group in question. As Chung (1991) asserts, the population registered in the files represents (or is very close to) the 'genealogical population' of a place. It is generally accepted that the registers are not complete (see Bedford, 1980, p. 39; Overton, 1993, p. 45), but they are nonetheless regarded as being able to give a good idea of the *de jure* population of a place.[2]

For the *yavusa* of Naikeleyaga and its five *tokatoka*, in 1993 these files[3] included between three and five generations. The generations of the living were relatively complete, though some children born during the last few years were missing. As a result of many discussions with Naikeleyagans in the village and in Suva about these files, a 'missing rate' of roughly 10 per cent seems to be likely (Felgentreff, 1995, p. 130). On the other hand, the registers contain names none of my informants had ever heard of, and many of those enrolled had passed away without notice of the NLC. In total, the files of the kin groups affiliated with Naikeleyaga village included 1114 persons, of whom 21 were unknown to my informants or not valid (e.g. registration in the wrong kin group, different entries of the same individual), and 138 were dead. For the remaining 955 living registered individuals (471 females and 484 males), my informants were able to name the place where they currently stayed or resided, besides further information.

As Table 13.2 shows, of the 955 registered *taukei* in October–November 1993, only 136 belonged to Naikeleyaga's 'present' population (of the 148 present *taukei* mentioned above only 136 were registered in the files). That means that in 1993 less than 15 per cent actually resided at the place they 'belong' to. This can be compared with figures for October 1975, when the VKB for Naikeleyaga comprised 552 individuals but Bedford encountered only 152 (27.5 per cent) of them in the village (Bedford, 1976, p. 64). Hence the *yavusa's* total population (as registered in the VKB) has roughly doubled from 1975 (552) to 1993 (1114) while, over the same period of time, the proportion of registered Naikeleyagans actually staying in the village has halved. Or, speaking more generally, the number of the village's 'present' inhabitants was decreasing from 1975 to 1993, while during the same time the total 'genealogical population' of the village doubled.

Of the other 819 registered Naikeleyagans (85.8 per cent), the majority probably intend to reside outside the village permanently. Many of them were born outside the village, especially over the last two or three decades,

Table 13.2 Spatial distribution of the registered members of the *yavusa* of Naikeleyaga village, October–November 1993

	No.	%
Living registered Naikeleyagans whose current places of residence were known to informants	955	100
of whom staying in Naikeleyaga village	136	14.2
in other rural Fiji	123	12.9
in Suva	575	60.2
in other urban Fiji	85	8.9
abroad	36	3.8

Source: Felgentreff (1995, p. 134) and National Land Commission files, Suva.

and have never actually been to 'their village'. The younger the age group under examination, the smaller the proportion of those registered in the VKB who ever spent some time in the village. Consequently, not all of them can be regarded as movers or as migrants. Yet the fact that informants in the village were well informed about these 819 absent relatives, and that the information given in the village could be verified by informants in Suva, indicates that they are not totally 'gone' and, perhaps, not completely lost to those currently residing at the place of all their common ancestors.

From a demographic point of view, the sheer number of 575 Naikeleyagans (60.2 per cent of all living registered *yavusa* members) residing in Suva, as compared to the number of village inhabitants, suggests that the village community centre is located in the capital, away from its common territorial base. Individual life histories of older informants reveal that the settlement of Naikeleyagans in Suva began at least as early as 1910: since then villagers from Kabara have had relatives residing there, caring for travellers from the island.

A further 85 (8.9 per cent) of the *yavusa* members lived in other towns in Fiji, mainly in Labasa (43) and Lautoka (12). The 'settlement' of Naikeleyagans in Labasa (on Vanau Levu – see Figure 13.1) started in the early 1950s when one villager established himself in this town, working as a boat-builder. Attracted by these new perspectives and by his economic success, others of his kin group followed so that, in 1993, this 'outlier' of one of Naikeleyaga's *mataqali* comprised 43 individuals registered in the VKB (Felgentreff, 1995, p. 134). The elders in Labasa are still remembered in the village, but only some of the Labasa-born offspring were known personally; they had never come to the village, but some had met them while on a visit in Suva. In total, some 70 per cent of the registered members of the *yavusa* of Naikeleyaga resided in urban Fiji in 1993 – quite a high proportion, especially if compared with a situation where only 41 per cent of the national population live in urban areas. In 1993, 123 registered members of the *yavusa* (12.9 per cent) stayed in other rural places than Naikeleyaga. A further 3.8 per cent of the *yavusa*'s members resided abroad; some probably permanently, others – such as students and soldiers – perhaps temporarily. The overall spatial distribution of the *yavusa*'s members within Fiji is illustrated in Figure 13.2. While only 136 resided in Naikeleyaga in October–November 1993, 575 stayed in Suva, 83 elsewhere in Viti Levu, 68 in Vanua Levu, 33 in other places in Lau than Naikeleyaga, and 24 in the area ranging from Kadavu to Taveuni, including the province of Lomaiviti.

The intensity of villagers' contacts with relatives residing outside the villages differs, probably according to accessibility. While very few people have information about those residing abroad or living in remote villages, each villager has many close relations in Suva, and in other places within Fiji. Many indications (and further evidence, e.g. individual life histories of villagers) clearly show that the contact between Naikeleyaga and Suva is

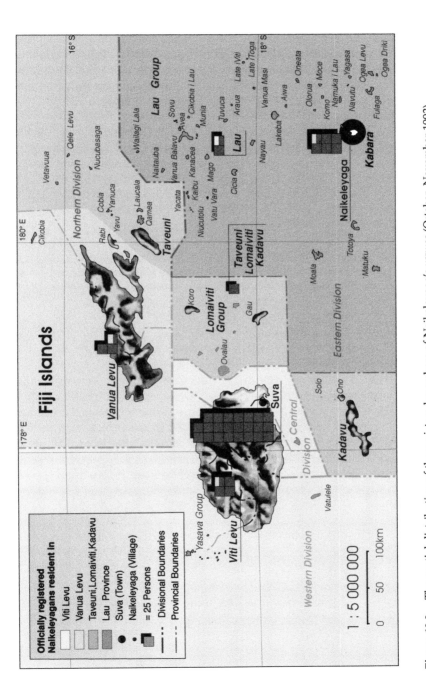

Figure 13.2 The spatial distribution of the registered members of Naikeleyaga's *yavusa* (October–November 1993)

Source: Felgentreff (1995, p. 136).

much closer than with other places. The number of agnatically-related kin beyond the village and their spatial distribution give an idea of the *yavusa*'s success in considerably extending its territorial base over time. Even if affinite relations are neglected in the analysis, and even if a fraction of agnatic kin residing outside the village may have 'forgotten' the guardians of their roots, it should become clear that the villagers do not have to rely solely on their narrow local resource base. By no means are they just at the mercy of their tiny island's resources, as many *vulagi* believe.

Internal Relations of the Multi-level Village Community

The Fijian conception of kinship is much more extensive compared to the ones most foreign analysts were exposed to during their own socialisation. To give just one example, formulated by a Melanesian Fijian author:

> One's father's brother's children ... are treated as brothers and sisters and their mothers are *for some purposes* like one's mother. Likewise, one's mother's sisters, real or assumed, are treated as 'Mothers', and their children ... as sisters and brothers. Much the same behaviour as displayed to one's real father or mother is also expected by classificatory relatives. By the same token, one expects from them similar behaviour to that normally displayed by one's own parents. Through blood and marriage, the Fijian kinship system is thus extended outwards and it is therefore not unusual to find a Fijian referring to, or addressing, quite distant relatives with the same kinship terms used within the immediate family circle (Ravuvu, 1983, p. 4; italics in the original).

Taking the Fijian system of extended kinship into account, distance in space – as between the relatives in question – might be seen in another light. In contemporary Fiji it is not uncommon that the above-mentioned children of a villager's father's brother are born and raised in Suva. Nevertheless, they are not felt to be distant relatives. The fact of being related to each other also implies another degree of being *obliged* to each other. Irrespective of spatial distance, these far-ranging obligations have to be met. While discussing the formal principles of a couple's residence after marriage, the same author explains:

> Today, ... there is more tolerance and understanding if for one reason or another the married couple decides to live elsewhere, provided they continue to satisfy the social and economic demands made upon them by the husband's group (Ravuvu, 1983, p. 3).

Another Fijian author puts the principle determining the relationship between relatives residing at different locations into the following words:

> There is, however, no complete freedom for any member of the Fijian community irrespective of where one lives. The burden of custom is sometimes much lighter

away from the village but obligations to kin, especially financial ones, must still be met (Tubuna, 1985, p. 219).

Statements like this perhaps make clear that villages like Naikeleyaga, with such a vast and wide-ranging kinship network, are far from being isolated. Indeed, as Overton suggests, '...villages must be seen not as isolated social and economic entities but as part of larger national and even global systems' (1993, p. 47).

In common with all other Fijian villages, not only those on outlying tiny islands like Kabara, Naikeleyaga is interrelated and interwoven with the wider world by lines of blood and kinship, which are the main base of social networks in Fiji. However, such bonds are useless if they get forgotten; they have constantly to be renewed. Therefore, a 'dialogue' (Chapman, 1991) within the network is necessary. Phone calls are one means of keeping in touch, as can be observed each day the local post office is open. Nevertheless, the alliance between the villagers and the relatives beyond is mostly maintained by personal contacts. The main vehicle of social interaction between the *yavusa*'s members residing at different localities is spatial mobility. For various reasons it is usually the villager's role to pay visits to kin residing elsewhere; visits vice versa are less common. Many villagers take any opportunity to see their relatives outside the village as often as possible. In fact, 30 per cent of the 191 present inhabitants of Naikeleyaga were in Suva at least once during the twelve months prior to my survey: 35 out of 91 males (38 per cent) and 22 out of 100 females (22 per cent), including all age groups (Felgentreff, 1995, pp. 126–8). All of them had motives that led them to Suva, but whatever the reason for their presence in the capital, everybody used the opportunity to see as many relatives and friends as possible. In this way, each visit of a villager certainly has the (side-) effect that personal ties get renewed and re-established. Temporary spatial mobility of members of the *yavusa* plays a central part in the social interaction between the island and locations beyond.

The same can be said about the permanent relocation of villagers. The likelihood that someone who has lived in the village for a while and settles elsewhere will keep in close contact with those left behind is fairly high – higher than for second- or third-generation migrants. He or she will be more reliable, and more understanding towards the concerns of the village's inhabitants than other relatives who were never exposed to village life. And, if they successfully settle in other places, they open up new resources, parts of which can be transferred to those left behind. With reference to the whole multi-local *yavusa* and its internal relations, it seems that out-migration from the village is partly a necessity, and partly the foundation of the future prosperity of the villagers. While temporary circulation and permanent population relocation are totally different categories for most analysts, they appear as complementary strategies in maintaining the dialogue between the village and the wider world. Both kinds of mobility were needed for maintaining contact.

This perhaps explains why it is not felt as a tragedy by the villagers when one of them sets sail for Suva. As long as someone's duties towards his or her household can be met by others in the village, he or she can gain experience in the outside world. At the destination, relatives will look after him or her. If someone is economically successful, they are not really a loss for the community in the village; on the contrary, they have enlarged the world of those he has left behind. What Hau'ofa describes with general reference to Pacific islanders' international migration is also valid for Naikeleyagans leaving their village:

> Everywhere they go, to Australia, New Zealand, Hawai'i, mainland USA, Canada and even Europe, they strike roots in new resource areas, securing employment and overseas family property, expanding kinship networks through which they circulate themselves, their relatives, their material goods, and their stories all across their ocean.... The resources of Samoans, Cook Islanders, Niueans, Tokelauans, Tuvaluans, I-Kiribatis, Fijians, Indo-Fijians and Tongans are no longer confined to their national boundaries; they are located wherever these people are living permanently or otherwise (Hau'ofa, 1994, p. 153).

In the opposite case, if someone was unable to establish themselves elsewhere or if the person's labour was needed in the village, he or she would sooner or later return. That person would be welcome; it is not seen as a failure to come back, even empty-handed. There is nothing wrong with being mobile, as long as obligations and duties towards others are kept in mind. People from Naikeleyaga were probably always mobile, as other Pacific Islanders were: 'From one island to another they sailed to trade, to marry, thereby expanding social networks for greater flow of wealth (Hau'ofa, 1994, p. 152). This is what the elders did with their outrigger canoes, and it is now being done by the younger generation travelling to Suva, going to school or looking for a job.

Villagers view all this mobility with calmness. There are no doubts that the parents, elders and chiefs have the power to make a person return to the village – individual life histories are rich in examples of this (Felgentreff, 1995, pp. 176–202). Compulsion is needed only in exceptional cases; it is usually sufficient for a father to say that he intends to stay in the village forever. When he gets older, one of his sons will feel obliged to stay with him. According to Fijian custom, words or even arguments about this point are futile. From the villagers' point of view, continuous out-migration therefore seems less worrying and frightening than it does for many analysts, regional and development planners, politicians and others. Villagers see clearly that out-migration and all other kinds of spatial mobility to and from the village are essential and necessary ingredients of the way of life that they – in general – appreciate. With respect to numbers, the village is the giving end, but in exchange it receives help and support, in cash and in kind, in goods and in services.

Indeed, many examples can be given showing that people in the village can rely on their kin residing beyond. There is a substantial gap between the overall monetary income and expenditure within the village which can only be attributed to the assistance of others. Nobody in the village really denies that the contemporary material standard of living in Naikeleyaga relies on money orders and remittances (Felgentreff, 1996). In one way or another (with the likely exception of the *vulagi dokai*, the state employees' households not belonging to the *yavusa*), every household in the village probably relies on some sponsoring by relatives residing elsewhere. One household head in Naikeleyaga estimated that in some months up to 90 per cent of his expenditure (consumption and contributions to the manifold village funds) were met with money sent by relatives, while he himself contributed some 10 per cent earned by carving. Due to communal work (up to three days per week), two daily church services and the time to be spent on agriculture, many men argued there was no time left for income-generating activities.

Beyond personal relations between individuals residing in the village and kin at other locations, the dialogue between these places can be shown on another level. If concerted action is needed, the multi-local community acts as one entity. With respect to finances, this can be seen when larger investments in village infrastructure have to be made: the purchase of the village generator and the last roofing of the village church were only made possible by contributions collected among Naikeleyagans residing in Suva. Perhaps even more important is the unity of the multi-local *yavusa* in village politics. The delegation from Naikeleyaga at the Lau provincial meetings (for reasons of convenience and accessibility held in Suva) comprises political authorities, softball and rugby players, both from the village and from Suva, acting together as one team. Important village affairs, especially if they concern relations with other groups, are always negotiated not only by village elders, but also by elders in Suva. The latter meet monthly in the house of the Suva-based chief, sometimes involving up to 60 heads of households belonging to the *yavusa*. Indeed, it might be an adaptation to the spatial distribution of its members that the *yavusa* of Naikeleyaga now has two chiefs, but if circumstances demand, the *yavusa* is one entity. This can be said both of external and of internal relations.

Reciprocity and Asymmetric Relations

Most statements in this chapter indicate that it is mainly the villagers who are the beneficiaries of the multi-locality of the *yavusa*. In material terms, the relationships within the multi-local *yavusa* hardly fit the basic principles of reciprocity so central to Fijian culture. In general, the material inequality between urban- and village-based kin is strongly perceived. It is certainly not the will to reciprocate that is missing but the material base that would allow it. The kin in the city can definitely rely on the villagers, provided they have kept in contact and supported the village when necessary and possible. More than one house in Suva erected by Naikeleyagans is made from wood cut on the island. Each Suva-based household belonging to the *yavusa* can have as many *tanoa*-bowls as its members want – villagers would happily provide twice as many instantly. And certainly, visitors coming from the island carry as many locally made handicrafts as possible, plus coconuts and fresh fish. These are all the goods available to them on the island which could be of any material significance to the urban kin. The asymmetry of the relationship is neither the fault of the urban nor of the rural protagonists: it is a consequence of uneven development and the structural disadvantage of the peripheral island. People realize these differences.

Furthermore, reciprocity was probably never bound to strict accounting and exact compensation (Ravuvu, 1983, pp. 11–12). Those who give are compensated for their material loss by social gain. And, besides the prestige of the giver, there is another central, non-material aspect involved: those staying on in the village

> maintain ancestral roots and lands for everyone, homes with warmed hearths for travellers to return to at the end of the day, or to re-strengthen their bonds, their souls and their identities before they move on again. This is not dependence but interdependence, which is purportedly the essence of the global system (Hau'ofa, 1994, p. 156).

Hau'ofa discusses the reciprocity in relations between international migrants and their Pacific homelands, but the point he makes is also true for the relationship between Naikeleyaga and the village's offspring residing elsewhere. It also happens that some Naikeleyagans over time stop seeing the point as to why they should feel responsible and obliged to their ancestral roots, their land, their village and its inhabitants. These people can, perhaps, be regarded as migrants in the term's pure sense, having severed their connections, or, as some Fijians would say, 'forgotten that they are Fijians'. The social costs in the form of total ignorance by the rest of the *yavusa*'s members might be welcome for some, but for the majority this is definitely not the case.

This is not to suggest that Naikeleyagans residing outside the village only keep in contact because they worry about sanctions. Instead, for many it is more a social need. The land of the ancestors is the base of their personal identity; keeping in touch with its guardians is self-evident, supporting them being the most natural thing in the world. Assisting kin might be seen as a concession to one's personal history by some, while others regard it as an investment in one's own and one's children's future. Like the villagers themselves, many urban Naikeleyagans not only think in terms of reciprocity in cash and kind, they also remember the precept to share. As town-dwellers they can draw from a different resource base than their relatives back 'home', and they are happy that they can. Like people from other outer Fijian islands, a substantial number of Naikeleyagans residing outside the village 'continue to maintain strong links with the island and like to play some part in what is happening on the island' (Bryant, 1990, p. 150).

Conclusion

As leading migration theorists such as Massey (1990) have shown, complex migration processes can only be understood in terms of complex bundles of underlying reasons. In cases of tiny, remote islands like Kabara, the deficiency of local resources is especially evident. Estimates given by elders in Naikeleyaga suggest that the communal garden land would be sufficient for the subsistence needs of a maximum of only ten additional households. The islanders are aware that their natural environment cannot feed and carry all of them. Asked what would happen if all *de jure* members of Naikeleyaga returned one day, the answer was laughter about this totally absurd idea – 'Kabara would probably sink into the ocean.' Obviously, the conclusion must be that out-migration from Kabara is a necessity, due to the island's limited natural resources.

Similarly, out-migration from Kabara can be ascribed to economic factors. The few wage-earning opportunities offered in the village are mostly held by immigrants not belonging to the village community, having been temporarily transferred by their employer, the government. Income locally derived from copra and carving is low and insecure, compared to a regular salary or wage. Finding a wage-earning opportunity usually means leaving the island. Scholars stressing the importance of economic disparities between structurally disadvantaged islands like Kabara and more 'central' places are perfectly right: economic factors certainly play a most vital part in spatial mobility, be it temporary or permanent population relocation. This conclusion is probably uncontestable, no matter if one argues in terms of micro- or macro-level analysis. But it is not the whole story.

Another strong factor of out-migration from Kabara is education. There is no high school on the island, consequently pupils interested in secondary education have to leave. Most of them actually do, and it has become a fixed element in many rural-born Fijians' life-cycles to spend secondary schooling in Suva. Some establish themselves outside the village thereafter, while others might return, sooner or later. Again, spatial structure and disparities in local infrastructure can be blamed for out-migration.

However, such conclusions are, at least partially, somewhat self-evident. Location theory holds that there are structural disparities between virtually all given locations – no matter if people actually move between them or not.

Pointing at these structural disparities and relating them to migration behaviour gives only a weak picture of the manifold processes involved. With respect to the analysis of spatial population mobility, the questions of how people cope with the given disparities and what actually happens when they move or stay, seem to be more interesting. Naikeleyagans tend not to behave solely as victims of spatial structures. Instead of mechanical stimulus-response reactions, they tend to act creatively, by constantly shaping and fashioning their position within given structural disparities according to their needs, and within the context of their traditional social networks.

The villagers on the outer islands are well aware of the advantages of urban life and its amenities. Although Naikeleyagans in Suva give ample evidence that shifting to an urban area does not necessarily mean a 'full' urbanization (in the sense that personal life-styles change significantly), living in the village is perceived by many as having more advantages. The decision to stay (or to return) always leaves enough room to leave – and vice versa: those who leave are always welcome to return. Migrants who break all links with the island, its inhabitants and all other agnatically-related kin, are clearly an exception. There do exist in Fiji migrants who, for one reason or another, have lost their roots, who voluntarily or by force have detached themselves from their origins. They exist without obligations towards their hundreds of relatives; but the majority of Melanesian Fijians probably feel sorry for them, for they are not part of a larger, wide-spanning, rich social unit which provides almost everything one could wish for, in material and non-material terms.

From the villagers' point of view, the structural differences between their island and urban Fiji are perceived more as a challenge. The core of their problem is how to combine the advantages of their life-style (*vakavanua*) with the amenities which are available only for money and only in the urban centres. Reducing differences in infrastructure would mean that large-scale industry would have to be established on Kabara for the provision of local job opportunities, plus a high school. None of the village elders would allow such developments, and nobody wishes them to take place. The absence of such an infrastructure is the very essence of their rural, island way of life. Any measures towards that kind of development must be small-scale and gradual, and the elders are aware of it. Those who want to live in places like Suva are free to go, but nobody must try to turn Naikeleyaga into a place like Suva. In this respect, the location problems resulting from the remoteness, smallness and relative poverty of natural resources of islands like Kabara are perhaps a real stroke of luck, at least for those who prefer life to be a bit different to that in the global village.

The solution to the question of how to transfer 'modern' amenities into the 'traditional' village is keeping in contact with the wider world. It is mainly the villager's responsibility to keep that contact. To do this, all means of communication are useful but of central importance is social interaction via spatial mobility. Temporary and permanent spatial relocations are thereby not necessarily opposed to each other. Instead, if handled flexibly, both are powerful tools which can be seen as complementary in network formation. In one case the network gets extended, in the other it gets maintained and strengthened. This is how migrant networks work, and they will work as long as a sufficient number of participants perceive them as necessary, advantageous and/or profitable. Migrant networks are often overlooked: one cannot identify them by statistical analysis of census data, but they nevertheless exist. Given the rapid improvement in transport and communication, their global importance will probably increase.

Notes

1. For Naikeleyaga, the number of inhabitants increased from 121 in 1921 to 288 in 1966 (National census results quoted by Bedford, 1976, p. 58). Since then, a more or less steadily decreasing number of inhabitants were counted by myself and by other enumerators (Felgentreff, 1996, p. 106).

2. For the whole island of Kabara, Bedford found that in October 1975 the island's *de facto* population was less than a third of its *de jure* population as registered in the VKB. In the case of Batiki Island (Lomaiviti Province) it was only 23 per cent in 1974 (Bedford, 1980, p. 39). In comparison, for her study area on the much larger, more fertile and more accessible island of Kadavu, Chung (1991) estimates for the late 1980s that about half of the *de jure* population (as registered in the VKB) resided permanently elsewhere (see also Overton, 1993, p. 55).

3. NLC No 2246, 2247, 2248, 2249, 2250 of 18 November 1993.

References

Bayliss-Smith, T., Bedford, R., Brookfield, H. and Latham, M. (1988) *Islands, Islanders and the World. The Colonial and Post-Colonial Experience of Eastern Fiji*. Cambridge: Cambridge University Press.

Bedford, R. (1976) *Kabara in the 1970s: Dimensions of Dependence in a Contemporary Lauan Society*, Project Working Paper No. 3. Canberra: Australian National University.

Bedford, R. (1980) 'Demographic processes in small islands: the case of internal migration', in H.C. Brookfield (ed.), *Population–Environmental Relations in Tropical Islands: The Case of Eastern Fiji*, MAB Technical Note No. 13, pp. 29–59. Paris: Unesco Press.

Bogue, D. (1959) *The Study of Population*. Chicago: University of Chicago Press.

Bryant, J.J. (1990) 'Rotuman migration and Fiji', in J. Connell (ed.), *Migration and Development in the South Pacific*, Pacific Research Monograph 24, pp. 136–50. Canberra: National Centre for Development Studies.

Chapman, M. (1991) 'Pacific Island movement and socioeconomic change: metaphors of misunderstanding', *Population and Development Review*, 17(2), pp. 263–92.

Chung, M. (1991) 'Politics, Tradition and Structural Change: Fijian Fertility in the Twentieth Century', unpublished PhD thesis. Canberra: Australian National University.

Connell, J. (1987) *Migration, Employment and Development in the South Pacific*, Country Report No. 4: Fiji. Noumea: South Pacific Commission.

Felgentreff, C. (1995) *Räumliche Bevölkerungsmobilität in Fiji: Eine Exemplarische Untersuchung der Dorfgemeinschaft von Naikeleyaga (Kabara Island, Lau-Province)*, Potsdamer Geographische Forschungen 11. Potsdam: Selbstverlag des Institutes für Geographie und Geoökologie der Universität Potsdam.

Felgentreff, C. (1996) 'Migration and remittances in the eastern periphery of Fiji: the case of Naikeleyaga village', *Asian Migrant*, 9(1), pp. 21–7.

Fiji Bureau of Statistics (1988) *Report on Fiji Population Census 1986. Vol. 1: General Tables*, Parliamentary Paper No. 4. Suva: Government Printer.

Fiji Bureau of Statistics (1989) *Report on Fiji Population Census 1986: Analytical Report on the Demographic, Social and Economic Characteristics of the Population*. Suva: Bureau of Statistics.

Hau'ofa, E. (1994) 'Our sea of islands', *The Contemporary Pacific*, 6(1), pp. 148–61.

Hooper, S.J.P. (1982) 'A Study of Valuables in the Chiefdom of Lau, Fiji', unpublished PhD thesis. Cambridge: University of Cambridge.

Massey, D. (1990) 'The social and economic origins of immigration', *Annals of the American Academy of Political and Social Sciences*, 510, pp. 60–72.

Nair, S. (1985) 'Fijians and Indo-Fijians in Suva: rural–urban movements and linkages', in M. Chapman and R.M. Prothero (eds), *Circulation in Population Movement*. London: Routledge & Kegan Paul, pp. 306–30.

Overton, J. (1989) *Land and Differentiation in Rural Fiji*, Pacific Research Monograph 19. Canberra: National Centre for Development Studies.

Overton, J. (1993) 'Farms, suburbs, or retirement homes? The transformation of village Fiji', *The Contemporary Pacific*, 5(1), pp. 45–74.

Ravuvu, A. (1983) *Vaka i Taukei: The Fijian Way of Life*. Suva: Institute of Pacific Studies.

Rutz, H.J. (1987) 'Capitalizing on culture: moral ironies in urban Fiji', *Comparative Studies in Society and History*, 29(3), pp. 533–57.

Sofer, M. (1993) 'Uneven regional development and internal labor migration in Fiji', *World Development*, 21(2), pp. 301–10.

Thompson, L. (1971) *Southern Lau: An Ethnography*. New York: Klaus Reprint (originally published in 1940 as Bernice P. Bishop Museum Bulletin 162).

Tubuna, S. (1985) 'Patterns of return migration in the Wainibuka River valley, Viti Levu, Fiji', in M. Chapman and R.M. Prothero (eds), *Circulation in Population Movement*. London: Routledge & Kegan Paul, pp. 213–24.

14

The Changing Contours of Migrant Samoan Kinship

Cluny and La'avasa Macpherson

Introduction

In the past fifty years a new wave of migration has occurred within the Pacific region. This has seen large numbers of people leave small, relatively poorer, tropical islands such as Samoa, Fiji, Tonga, the Cook Islands, Niue and the Tokelau Islands to seek new opportunities in larger more industrialized ones such as New Zealand, Australia and Hawaii. This movement has generated the development of significant migrant communities in the cities. While these migrant populations may be physically separated from their homelands, they are generally seen as socially continuous by Pacific Islanders who, as Tongan sociologist Epeli Hau'ofa (1994) notes, perceive the ocean as something which connects rather than separates islands. The migrant enclaves in New Zealand, Australia, and the western United States, then, are seen as extensions of families and of villages. They are places where members of families and villages go periodically and for various lengths of time and from whence they serve their families in various ways.

These movements have had profound consequences for the economic and social organization of the source islands. They have, along with other factors, distorted Pacific Island microstates' economies (Bertram and Watters, 1985) and created increasing degrees of economic dependency in the Pacific islands (Connell, 1991; Hayes, 1991; Shankman, 1976). These new 'villages in the city' also play a significant role in shaping contemporary social organization in the islands of origin. In the case of Niue, where there are now only 2000 Niueans on the island compared to 11,500 people of Niuean descent in New Zealand and more in Australia, the movement has led to a shift in the centre of Niue's 'social gravity'. Even in other, less extreme, cases the flow of people, ideas and resources between home and diaspora settlements have profound effects on the social organization of each.

The new urban settlements are, firstly, venues where 'traditional forms' of organization can be, and are, modified to meet the changing needs and circumstances of the migrants. Thus, new forms reflect the fact that many involved are now engaged in wage employment and cannot spend as much time in traditional activities as they might have spent in the islands. New social practices reflect the fact that many islanders have married out and that new people who may not understand island languages and customs are now increasingly involved in, and must be incorporated into, 'traditional forms and activities'.

These 'islands in the city', then, are venues where rapid change in forms of 'traditional' social organization can occur. Thus, one may see in these migrant communities a glimpse of the future of the smaller islands as they too become increasingly incorporated into the world capitalist system. 'Islands in the city' are also settings where new 'traditions' and 'protocols' are created to meet new social realities and where these are progressively refined until they appear to have been a permanent part of social organization!

The emigrant societies are, secondly, the source of social innovation in the origin societies. There is lot of movement between source and migrant communities and each draws on the other for personnel, resources and ideas. Samoans in Samoa borrow video-cameras and tape recorders to record ceremonies which surround the exhumation and reburial of ancestors' bones and send the videotapes to their relations in Auckland, New Zealand, to show that the event was conducted properly and that money which they contributed was well used. Relatives in Auckland, in turn, borrow video-cameras to record the ceremonies which surround the graduation of a young Samoan with a PhD in organic chemistry and send the videotapes to Samoa to show that they are also working to maintain and enhance the family's, and the village's, reputation.

Somewhere between these two sets of communities, and as a consequence of their respective activities, a meta-culture, which embodies elements of each of their contributions, emerges. Samoans training migrant cultural-performance groups for competitions in New Zealand scan videos of recent performances in Samoa to ensure that their performances are 'authentic' in every respect. Samoans in Samoa in turn scan videos of cultural competitions in New Zealand for innovations which might be incorporated in their performances. Eventually the meta-culture, which develops in the space between the sets of communities, becomes the source of social 'reality' and boundaries between 'authentic' and 'inauthentic' dissolve. Performances, and indeed life, in each place increasingly begin to resemble each other.

This chapter is about a part of this process of transformation. It examines the portability of forms of social organization which originally evolved in small islands and how these function when the social and economic conditions in which they develop are altered. Using data from a thirty-year longitudinal study, we outline the significance of kinship in Samoan society and consider the changing role and form of kin-based social organization among Samoan migrants in New Zealand since migration commenced in the 1950s. The chapter traces the transition in kinship functions from their limited roles in the 1950s and 1960s to their expansion and elaboration in the 1970s and early 1980s and then their contraction from the mid-1980s to the present. It seeks to identify social and economic factors which have shaped these forms at various times and concludes that this is not an outcome but part of a process of continuous change.

Kinship in Pre-Colonial Samoa

Some 3500 years ago, Polynesian descendants of voyagers whose journey began many centuries earlier in South China settled in the Samoan archipelago in the south-east Pacific Ocean. Samoan society evolved largely uninterrupted in the small, isolated but well-endowed group of islands.

The central feature of pre-European Samoan social organization was kinship or *aiga* (Kramer, 1994). Society consisted of clusters of localized, co-resident kin corporations or *aiga*. Rights over land and sea in particular areas were vested in the various, usually related, *aiga* which lived in villages in an area. A chief, or *matai*, selected by senior members of the *aiga*, held each corporation's chiefly title and 'managed' its resources on behalf of its members. Effective *matai* could use the human and material resources of their *aiga* to enhance its material and socio-political wealth; ineffective *matai* could, equally, squander its resources. Well-led *aiga* tended to attract members and poorly led ones tended to lose them which meant that at any given time the relative size and power of *aiga* varied with quality of leadership. Over time certain *aiga* consolidated gains and came to control more land and to enjoy more permanent social prestige and political influence. This led eventually to more or less fixed relations between *aiga* of a village or district and to the emergence of relatively stable local polities.

Every Samoan had links with the *aiga* of their four grandparents. Of those, their links, and identity, were most vitally connected with the *aiga* with which they opted to reside at a given time and which became their 'strong side' or *itu malosi*. For individuals, *aiga* membership established social location and significance, conferred rights to use agricultural land and have a house site, and to enjoy protection by the *aiga*. These rights were offset by a set of parallel obligations which included the requirement to serve, *tautua*, one's *matai*, to protect the land of the *aiga* and to defend its honour. The rewards of service were a degree of psychological and material security, and, for males at least, the prospect of eventual leadership and power.[1]

Kinship determined who one could and could not marry. Those who married within their villages could expect to spend most of their lives living, working, fishing, fighting and relaxing with kin. The isolation in which this form of social organization evolved ended with the arrival of Europeans.

The Colonial Presence

European beachcombers and deserters began to drift to Samoa toward the end of the eighteenth century. Samoans exploited the boat-building and trading skills of some of this group but their impact on Samoan social organization was generally limited. They were followed early in the nineteenth century first by missionaries and later, as was so often the case, by traders and settlers. In other Pacific settings similar combinations of events had profound effects on existing forms of social organization. But these were to have a somewhat less profound effect on Samoan society.

Christianity was introduced in 1830 by John Williams and Polynesian teachers of the London Mission Society in fortuitous circumstances (Gilson, 1970; Meleisea and Schoeffel-Meleisea, 1987). Williams had been advised by a chief, Fauea, that Samoa was ready for Christianity but that the mission would fail if it engaged in widespread condemnation or proscription of Samoan social organization (Moyle, 1984). This advice tempered the mission's approach, as did Williams' early appreciation of its dependence on powerful chiefs. Malietoa Vainuupo, who accepted, protected and promoted the mission, did so, at least in part, for his own political reasons. Malietoa believed that the mission was the key to fulfilment of a prophecy of the goddess Nafanua that he would one day control the Samoas. The mission's early association with chiefs, and their practice of directing their *aiga* to convert *en masse*, led to rapid conversion of most of the population (Gunson, 1978).

Samoans were not passive receptors of the new faith. They assumed early and significant influence in the Christian church by virtue of their dominant role in the interpretation and dissemination of the new faith. They subtly samoanized its doctrines and structures 'using indigenous beliefs to interpret and contextualize new ones, and embodying new forms of religious conduct in Samoan social practice' so that, 'over the past one hundred and fifty years of Christian Samoan history, Christian and *papalagi* [European] customs and institutions have been made distinctively Samoan' (Meleisea and Schoeffel-Meleisea, 1987, pp. 67–8). Thus a social movement, which could have caused significant social disjunction in Samoan society, was instead incorporated into it and has become a central element of Samoan culture or *fa'asamoa*.

The same is true of the consequences of Samoa's incorporation into the capitalist world economy which followed the establishment of the mission. Samoans, with assistance from missionaries, established and farmed cash crops on customary land under the authority of the traditional chieftaincy to meet the routine cash demands associated with religion. Settlers who arrived in Samoa hoping to find cheap land and labour found that both were effectively controlled by a relatively small number of powerful chiefs who apparently saw little to be gained from alienating either. Only occasionally, to obtain cash necessary to meet the extraordinary demands of religious activity, periodic civil wars and other political activity, did they increase cash crop production or sell land and labour to settlers from Germany, Britain and the United States. This they could do at times of their choosing (usually when normal trade was interrupted by war) and on their terms because of their ability to produce commodities and to trade within traditional social relations of production.[2]

Colonial powers' attempts to gain control of Samoa from the mid-nineteenth century met with limited success. Extension of their authority depended on undermining or co-opting the Samoan polity but their efforts were largely fruitless. This was the consequence, firstly, of an imperfect understanding of the complexity of Samoan politics, which led, in turn, to fundamental disagreement between British, German and US consular representatives on an appropriate colonial strategy. Secondly, there was a lack of resources with which to carry out those initiatives on which they could agree (Gilson, 1970; Meleisea and Schoeffel-Meleisea, 1987). Finally, the absence of valuable natural resources was reflected in a lack of determination on the part of any one colonial power to pursue annexation. This disunity allowed Samoans to live without significant interruption to their social and economic organization until 1899 when the colonial powers attempted to rationalize their interests in order to end the stalemate that had dogged their efforts to dominate and exploit the Samoas.

The Treaty of Berlin of 1899 vested control of the eastern islands of Samoa in the United States and those in the west to Germany. Samoans now found themselves facing determined colonial authorities in the US Naval Administration and the German Imperial Government. From here on the story focuses on the larger group in the west which found itself facing the formidable Dr Solf who, while respecting the Samoans, was determined to limit the power of their traditional polity. Given more time, he might have brought about fundamental change in Samoan society, but World War I broke out and brought his experiment to a premature end.

New Zealand, acting on a British request, occupied Western Samoa until its control was formalized in a mandate conferred by the League of Nations in 1919. New Zealand's administrators, like the Germans before them, believed that the key to their plans was the reduction of the influence of the traditional polity. They too sought to change fundamental elements of Samoan social organization. The New Zealand administration enacted an ordinance designed to force Samoan authorities to comply with regulations which the administration imposed upon the districts and villages.

However, New Zealand failed to recognize the resilience of Samoan social organization and the strength of their determination to preserve their autonomy. Thus, its village development programme not only failed but, worse, generated widespread resistance to its power. As Meleisea and Schoeffel-Meleisea (1987, p. 133) noted,

> It was not that Samoans did not want peace, good health and prosperity. They did, but they felt strongly that they should have a voice in planning and policy-making. Village authorities throughout Samoa deeply resented the imposition of rules and regulations to which they had been unable to make any contribution... the paternalism of the New Zealand authorities, lack of consultation and power sharing, ... and, most of all, the interference with traditional authority and rights over titles.

This resistance movement, the Mau, grew and eventually drew additional support from local European and part-Samoan populations in most districts of Samoa. Its policy of peaceful non-co-operation with the New Zealand administration took a range of forms.

> Committees and councils established by the Administration stopped meeting, villages ignored visiting New Zealand officials, courts of law were avoided by disputing parties, children were withdrawn from government schools, and officially promoted copra and banana projects were abandoned. In many districts all New Zealand imposed village regulations...were disregarded, and, instead of paying taxes money was raised and collected for the Mau. (Meleisea and Schoeffel-Meleisea, 1987, p. 135)

New Zealand's tragic mishandling of a Mau demonstration in 1929 resulted in the deaths of 11 Samoans and led the movement to declare its goals to be independence and self-government. Resistance continued until 1936 when a Labour government which seemed open to the idea of self-government came to power in New Zealand. But the Second World War intervened and in its aftermath the United Nations assigned trusteeship of Western Samoa to New Zealand.

New Zealand was bound by its trusteeship to move as quickly as was practicable to prepare Western Samoa for self-government and throughout the 1950s steady progress was made in the areas of educational, economic and political development. A constitution was drafted which incorporated and validated significant elements of the Samoan world view and social organization. The new constitution, and the decision to become independent, were confirmed by plebiscite in May 1961 and Samoa became independent on 1 January 1962.

Kinship in Post-Colonial Samoa

At independence, after 130 years of exposure to European society, the fundamental elements of Samoan social organization remained largely intact despite the appearance of change. A polytheistic religion had been replaced by a monotheistic one but it had been co-opted into Samoan society and used to bolster its traditions. A national polity, based on a modified Westminster system, had been created but existed alongside the traditional polities which continued to provide local government and to enjoy considerable autonomy. The leaders of the new state – the joint heads of state and the council of deputies – were chosen because of their traditional social status and the new parliament comprised *matai* elected by *matai* suffrage. New social movements such as the women's committees came to exist alongside 'traditional' women's organizations within villages[3] to promote new initiatives, but did so with the consent and support of village polities and had overlapping memberships.

Most significantly for our purposes, a non-monetized subsistence economy had been replaced by a monetized, mixed subsistence and cash-cropping one producing commodities for export to purchase consumer goods which had by that time become part of Samoan life. But this had occurred without significant transformation of either the land tenure system or the lineage mode of production. Most Samoans continued to live on and farm *aiga* land under the authority of family heads and to contribute labour and part of their production, as a form of rent, to the *matai* who used it on behalf of the *aiga* to enhance its socio-political status within the village. As a consequence, kinship remained a fundamental organizing principle of social and economic life in independent Samoa.

Kinship in Action

The extended kin group constituted a matrix within which goods and services were constantly exchanged in a series of more-or-less public events. The least public involved goods (such as money and food), services (such as labour and advice), and equipment (such as canoes and machetes), which were loaned and exchanged daily and informally among related individuals and households. The most public involved the mobilization of larger numbers of kin and quantities of resources in more formalized exchanges which were usually associated with rites of passage and, less frequently, with physical capital creation.

As long as people derived a livelihood from the use of the group's resources they were bound to participate in both of these forms of activity. Participation was, in each case, underpinned by two related beliefs: that kinship confers on one the obligation to give to kin who make legitimate requests and the right to expect that at some time in the future the goods and services given to others will be returned. At any given time there was then a level of residual indebtedness within an *aiga* which bound members to one another.

Both forms of exchange reproduce and reaffirm kinship and kin-groups. In one respect both have similar outcomes. The exchanges are sites where some discharge pre-existing 'debts' and others incur 'debts' to their kin. The net result of the exchange process is a residual level of indebtedness which binds kin to one another over time. In other respects the two forms of exchange have quite different outcomes for the reproduction and reaffirmation of kinship. The daily, informal exchanges remind individuals of their dependence on kin for a range of essential equipment, goods and services. The larger, more formal mobilizations of kin groups' resources reaffirm the importance of co-operation of larger groups of kin and remind members of the socio-political benefits of active membership of a larger corporate entity. In fact, the latter are also essential to the maintenance of collective identity because on these occasions groups demonstrate publicly their ability to co-operate and mobilize resources, and their right to public attention and respect both to those inside and outside of the *aiga*.

Winds of Change

After independence the conditions for rapid social change started emerging both inside and outside Western Samoa. The new government was divided between groups with somewhat different visions of the country's future.

> The first envisaged little change, and saw Samoa's future in terms of the continuity of the past, of Samoan traditional institutions and a 'plural economy' in which a few Samoans and part-Europeans would operate plantations and stores while the majority of Samoans would live 'as they had always lived'. The other view envisaged growth, a gradually expanding economy which would permit increasing numbers of Samoans to become involved in commercial agriculture and other kinds of enterprise. (Meleisea and Schoeffel-Meleisea, 1987, p. 158)

The latter group, persuaded by the analyses of modernization theorists, believed that 'traditional' forms of social organization were obstacles to economic development and national progress. This group began to consider significant reforms of 'traditional' land tenure systems and other structures of authority.

A more powerful force for change was, however, developing beyond Western Samoa's shores. Throughout the 1950s and 1960s New Zealand governments promoted industrial growth as part of a programme to diversify the country's economy. This soon ran into labour shortages as a consequence of low pre-war population growth rates and losses in the Second World War. To resolve these, successive New Zealand governments promoted immigration from Europe and the Pacific Islands (Ongley, 1991). People were drawn in relatively large numbers from rural, semi-subsistence lifestyles in kin-based villages in Western Samoa, the Cook Islands, Niue and the Tokelau Islands. They took up residence in the socially and ethnically heterogeneous urban milieu of wage work in industrial centres. The Samoan migrant population grew rapidly and became the largest of the expatriate Pacific Island communities in New Zealand.

Foundations of Migrant Samoan Kinship

It is difficult to imagine a set of circumstances more likely to deal a fatal blow to a system of extended kinship based on common ownership of resources and which had evolved in a small, rural village-based society. Yet early studies of the migrants showed that kinship turned up as a central feature of migrant community organization (Macpherson, 1973; Pitt and Macpherson, 1974). Several factors shaped the composition of the Samoan migrant stream and may explain, at least in part, why kinship remained significant in the Samoan emigrant enclaves.

Firstly, Samoans, like other Pacific Island migrants, were generally recruited into semi-skilled and unskilled positions in the secondary and tertiary sectors of the economy and became concentrated both in occupational terms and in low-income residential areas in and around the major cities where the economic diversification process was occurring (Ongley, 1991, 1996). Enclave formation was hastened by the process of 'chain migration' in which early migrants opened the way for other migrants from their family or locality, resulting in clustering. While the resultant spatial and occupational concentrations provided critical masses of Samoans in which elements of a Samoan world-view and life-style were most likely to survive, their formation is not a sufficient explanation.

New Zealand's immigration policy favoured less 'expensive', better 'educated', younger, single migrants. Families in Samoa also encouraged younger, single people who had demonstrated commitment to service, *tautua*, to their kin group to migrate in the belief that they would continue to acknowledge their membership of the *aiga*, their obligations to it, and would remit money and goods (Pitt and Macpherson, 1974). Migrants who met the costs of sponsoring the migration of kin did so to bring others who would share the costs of supporting the non-migrant *aiga* and, therefore, also had good reason for identifying people committed to Samoan values and practices.

To ensure that new migrants remained committed to these values, most were sent to live, at least initially, in established households with older relatives who had demonstrated continuing commitment to them. Many of these people were integrated into Samoan migrant communities and encouraged new migrants to associate with other Samoans. Pressure from household heads, and from their new peers, combined to ensure that new arrivals remained committed to 'service' family in the island. The consequence was the rapid growth of socially 'traditional' Samoan populations in New Zealand cities in the 1960s (Pitt and Macpherson, 1974).

Migrant Kinship: The Early Period

It was hardly surprising then that, in the early phases of Samoan settlement, kinship remained an important feature of migrant social organization. Migrants were chosen by kin to migrate to fill job opportunities located by kin in the belief that they would accept their responsibilities to 'serve' non-migrant kin. On arrival most went to live with relatives with whom they worked, worshipped, and spent much of their leisure time.

Within this migrant enclave related individuals and households exchanged goods and services informally in much the same ways as they had in Samoa. This pattern was partly the consequence of choice and familiarity and partly of a lack of familiarity with the organization and institutions of the dominant society. For whatever reason, kinship remained a matrix within which goods, services and information were routinely exchanged. The daily, informal exchanges among migrant kin had much the same consequences as they had in Samoa: reproducing dependence on kin and preferences for dealing with relatives in a range of matters.

In these early stages of settlement, however, the second role of kin groups, mobilization of larger amounts of human and physical resources to commemorate life crises, or *fa'alavelave*, was more difficult to reproduce, for several reasons. Firstly, there were generally too few members of *aiga* present to be able to fund the celebration of life crises such as baptisms, weddings, funerals, the conferring of chiefly titles, in ways considered appropriate. Secondly, young migrants were acutely aware of their lack of familiarity with the cultural knowledge required to manage such events and of the absence in the emigrant community of older people with the necessary skills. Thirdly, a lack of familiarity with international air travel discouraged many older people from travelling to New Zealand. Since their presence was crucial to the success of the event, their reluctance to travel to ceremonies in New Zealand meant that , in many cases, the ceremonies were taken instead to them. Fourthly, the fine mats, or *'ie toga*, which are a central part of these exchanges, were not readily available in New Zealand. Finally, the absence of appropriate Samoan-owned or controlled venues for the celebration of such events, and a reluctance of dominant groups to hire their facilities to unknown migrants, made it difficult to find places in which to commemorate these life crises.

As a consequence, throughout the 1950s and early 1960s, when life crises occurred the migrants often transferred the celebrations to Samoa where the critical mass of most *aiga* remained and where the necessary personnel, venues and fine mats were readily available. There they were performed by chiefs and elders of the *aiga* according to 'traditional' protocols. As air services improved, large parties of people moved back and forward between Samoa and New Zealand to ensure that achievements of members were appropriately commemorated and that the kin group demonstrated its ability to mobilize and claim recognition.

This arrangement contained the seeds of its own demise. Migrant *aiga* continued to exchange goods, services and information informally but were unable to find the resources to carry out the other activities which are so important to the reproduction of commitment to kin groups: the periodic, public demonstration of the group's power and claims to recognition. This too was about to change.

The Elaboration of Migrant Kinship

Between the mid-1960s and the mid-1980s, the New Zealand economy, and the country's Samoan population, grew rapidly. Kinship was changing. Whereas earlier, kinship had been primarily a matrix within which goods, information and services were exchanged informally and routinely, it was now becoming the basis both for the organization of larger, more complex traditional events and for the celebration of new types of events. Several factors lay behind this transformation (Macpherson, 1984, 1991).

Firstly, migrants who attended events in Samoa began to acquire the necessary cultural knowledge, skills and confidence to organize and manage them. Those returning from ceremonies in Samoa also brought with them the fine mats which they were given in Samoa as recognition of their service to their families and which created a supply of a crucial element of these ceremonies.[4]

Secondly, as a consequence of continuing migration, expatriate *aiga* were starting to grow and the critical masses of relatives necessary to fund and manage major celebrations were developing in New Zealand. These later migrants included chiefs with the necessary experience to manage life crises and migrants with both the knowledge and desire to take control of events.[5]

Thirdly, growing familiarity with international air travel led increasing numbers of older people to travel to events in New Zealand. Indeed, many older Samoans visited expatriate children and grandchildren during the summer and returned to Samoa at the onset of winter. Their presence meant more events could go ahead in New Zealand.

Finally, a surge of Samoan church-building, which began in the late 1960s and has continued to the present, resulted in the creation of Samoan-owned and controlled buildings in centres of Samoan population concentration. These offered venues for a range of activities. This combination of factors transformed the pattern of migrant kinship over the next two decades.

The same principles as in Samoa were used to encourage members to participate in these events. These principles appealed to their sense of pride in, and responsibility to, their *aiga* as well as creating possible indirect benefits from increasing the group's socio-political status in both New Zealand and Samoa. The same strategies were used to plan and manage. Leaders of the various sub-lineages discussed the resource requirements of the gathering. The *aiga* was, typically, divided into traditional sub-sections which were assigned responsibility for labour and other resources needed. Leaders of each sub-section then became responsible for communicating with members and mobilizing its resources for the occasion.

Extended kinship provided an ideal vehicle for organizing and funding events which were beyond the capacity of individual households. Participation, while now effectively voluntary,[6] continued to be underpinned by the same principles as it had been in Samoa: namely that kinship confers on one the obligation to give to kin who make legitimate requests and the right to expect that at some time in the future the goods and services given to others will be returned. It also maintained a level of residual indebtedness within an *aiga* which bound members to one another over time.

The use of kin-based organization to replicate these major events had some interesting latent social consequences. It periodically reaffirmed the existence and potential benefits of membership of extended kin groups. Participation in the large events renewed bonds between members of the *aiga* who did not normally meet, and extended individuals' social networks. The celebration of these large events, which were typically reserved for members with a record of unstinting service to the *aiga*, reminded members of the benefits of active participation in its affairs. Effectively-managed events won extended kin groups considerable socio-political prestige within the Samoan migrant community and indeed in Samoa, and all benefited by association. Indeed, the events' latent functions were as important as their manifest ones for the reproduction of extended kinship and kin-based social organization.

Early in this period these events were controlled by *matai* who were either resident in New Zealand or were brought from the island for an occasion and who faithfully followed the protocols with which they were familiar. In terms of both scale and form, such events were replications of similar events which took place in Samoa. But several factors were about to change this forever.

The Emergence of 'Ceremonial Inflation'

As a consequence of continuous chain migration, *aiga* had become bigger and wealthier. Early migrants who had become established sought and claimed opportunities for leadership. They now had more disposable income and also had access, in many cases, to the income of their New

Zealand-born children with which to back their claims. New migrants, who were generally firmly committed to a Samoan world-view and lifestyle, were also willing contributors to, and participants in, ceremonial events.

Probably the single most significant factor, however, was the re-emergence of the competitive tendency in migrant Samoan society. This was, in large part, a consequence of the growing number of *matai* whose role was to use resources to enhance a kin-group's socio-political status, and a growing pool of resources with which to work. As kin-groups sought to establish their claim to status and recognition in the migrant enclave by demonstrations of collective strength and ability to mobilize resources, a form of 'ceremonial inflation' occurred.

This process was intensified by the fact that migration had transformed traditional limits on families' access to resources. Families which in Samoa controlled lesser titles and had access to limited numbers of people and resources were no longer constrained by these facts. The only limits to the potential of migrant families were the numbers of wage-earners and the ability of leaders to mobilize their support. This provided both opportunity, and incentive, for some families to seek upward mobility by challenging larger, better-established ones in ways which would not have been possible in Samoa. This inflation manifested itself in two ways: a growth in the scale of these formal events, and the creation of new ones.

The elaboration of 'traditional' ceremonies

Weddings, arguably the most frequent of these celebrations, serve as a useful illustration of the way in which these new dynamics transformed such events. Over twenty years typical weddings became larger and more expensive. The guest lists grew from around 50 people to 400–500. The wedding photos of the period, displayed in many homes, reflect the steady growth of bridal parties from four to around 16.[7] They moved in stages from small family-catered celebrations held in church halls to commercially-catered events held in reception lounges, then to medium-priced hotels and finally, in the early 1980s, to the banquet halls in the most expensive hotels in the city.

When there were no more exclusive venues to 'conquer', competition focused on more and more expensive menus, mounting larger bridal parties, booking the most popular and expensive Samoan bands, and further elaboration of the Samoan exchanges which went alongside the wedding. The role of kin involved moved from sewing, cooking and serving to the raising of money to purchase specialist services such as bridal gown-making, limousine hire and video-recording of the event.

New factors combined to increase the size and social complexity of these events. In Samoa, guests and chiefs were typically from either the bride's or groom's families and well-known to all involved; the necessary protocols could be planned with some confidence in advance. But in New Zealand guests included workmates, fellow church-goers or unrelated friends and it became increasingly difficult to plan for these. The uncertainty raised the possibility of denying guests due deference and of appearing ignorant of Samoan etiquette.

As guest lists grew so did the number of *matai* who attended the weddings in an 'unofficial' capacity. At a certain point in proceedings, each of these chiefs could claim the right to speak and could reasonably expect a gift for having contributed his or her prestige to the event. Since no one knew in advance how many might attend it was difficult to plan for, and yet the consequences of under-estimation were socially very serious. Failure to provide appropriately generous gifts for all who claimed or were entitled to them left a family with a reputation for meanness or, worse still, for attempting something for which they lacked the knowledge and resources.

There were not only more *matai* but more complex relations between those who were present as social networks extended to include increasing numbers of non-Samoans who were invited to weddings. The increasing number of European, or *papalagi*, guests raised questions of how they were to be integrated into ceremonial sequences and their presence acknowledged. Those controlling the events needed ever more skill to ensure that all who were entitled to either social deference or gifts were identified and acknowledged publicly.

Formal exchanges of gifts, which occurred between the bride's and groom's families at the conclusion of the ceremonies, also became more complex and more expensive. As the bride's and groom's families competed to acknowledge the importance of the other by conspicuous demonstrations of their own wealth, pressure for ever-more formal and more generous gifts grew. The numbers of fine mats given by women's families at large weddings grew from around 20 to over 2000. The value of gifts of goods, or *oloa*, usually food and money, given by grooms' families to brides' families also grew dramatically as did the formality which surrounded the exchanges.

Finally, the gifts received by the representatives of the bride's and groom's *aiga* on these occasions had then to be redistributed among those members of their *aiga* who had contributed in some way to the wedding as acknowledgement of their part in the proceedings. As the numbers of sub-lineages contributing grew so too did the political complexity of redistribution. An undiplomatic redistribution meant that the family at the centre of the occasion would have considerable difficulty in finding support on any later occasion from those whom they offended.

But despite the risks inherent in poor 'performances' on these occasions, *aiga* continued to mount larger and more complex weddings because the socio-political 'rewards' of good 'performances' were high. The same could be said of the growth of another category of new kin-sponsored occasions which emerged in New Zealand over the same period.

The creation of new ceremonies

Graduation ceremonies, which began to occur in the 1960s and 1970s as New Zealand-born Samoans started to graduate from New Zealand universities, serve to illustrate this process. Since such events had not occurred in Samoa there was no established protocol and yet the occasion was of great significance to families who sought to commemorate their members' successes publicly.

The first graduation ceremonies were small, informal and relatively unstructured events, held usually in homes or church halls, in which families and friends gathered and shared a meal, held a brief service of thanksgiving and made some speeches in which graduates were commended for their efforts, thanked for bringing honour to the family, and younger members of the family were encouraged to emulate them. These were usually reserved for the celebration of university graduations.

But as *matai* and *aiga* sought new social sites in which to demonstrate their growing human and material resources, graduations became an opportunity to extend and formalize the ceremony.[8] Families started to invite guests whom they wanted, ostensibly, to thank for contributions to graduates' success and the 'top table' came to include ministers of religion, Sunday School teachers, music teachers, university faculty, sports coaches, colleagues, members of parliament and so on.[9] They too were increasingly invited to speak on graduates' achievements and potential and to congratulate the graduates and their families. In many cases, these speakers were given gifts, often fine mats, as recognition of their contribution.

Increasing numbers of Samoan graduates from an ever-widening range of tertiary institutions and courses provided an increasing number of opportunities for evolution and refinement of a 'protocol' of sorts. Over time an extended format started to emerge in which thanksgiving services were followed by formal meals, speeches by graduates and representatives from both sides of the graduate's family, dances, formal photographs and gift-giving. As with weddings, a competitive element led to pressure for larger and more elaborate versions of graduation ceremonies.

Limits of Growth

By the mid-1980s the limits to expansion and elaboration were becoming apparent. Some of these new constraints were the consequence of expansion itself, others products of demographic and cultural factors, and still others were the result of a more general economic contraction in the New Zealand economy. While all contributed to the current situation, they are treated separately here for analytical purposes.

As kin-based events became larger and the numbers of contributing relatives increased, so too did the number of 'debts' incurred by those who sought to mobilize *aiga*. A further category of 'debts' was created when friends and workmates from outside the *aiga* chose to contribute to the events. The expectation of reciprocity generated large numbers of debts which honour demanded were satisfied. This created problems as people set out to repay existing debts by contributing to more frequent and expensive ceremonies put on by those to whom they were indebted. There was simply not enough time or resources to settle them. Individuals who borrowed from loan companies and others were able to meet short-term obligations but created more serious longer-term problems by replacing existing debts with more expensive ones.

This situation was exacerbated by the contraction of the New Zealand economy as a consequence of the restructuring which began in 1984 and accelerated throughout the early 1990s (O'Brien and Wilkes, 1993). Heavy job losses occurred in industries in which Samoans had become concentrated as tariff protection was removed and companies were forced to close or to move offshore (Ongley, 1991, 1996). This produced rises in Samoan unemployment rates from historic levels of less than 1 per cent in the 1960s and early 1970s to over 20 per cent by 1991. Significantly, long-term unemployment increased from 28 per cent in 1988 to 54 per cent in 1995,[10] and was accompanied by decreases in full-time and increases in part-time employment (Krishnan *et al.*, 1994).

This situation was in turn made worse in the early 1990s by three factors. Firstly, the deregulation of the labour market and restructuring of labour relations in the form of the Employment Contracts Act (1991) led to loss of union protection for many workers and to falls in real wages.[11] Secondly, the general reduction of welfare benefit levels, on which Samoans were becoming increasingly dependent, placed further downward pressure on incomes (Macpherson, 1992). Thirdly, institution of a range of 'user pays' charges, which were only partly offset by reductions in direct taxation, placed further pressure on discretionary income.

These factors combined to produce a situation in which the numbers of debts being incurred by mobilization of the extended kin group were increasing rapidly when resources available to settle these were contracting very quickly. Failure to meet due debts led to embarrassment, *ma*, and humiliation, or *masiasi*, and made people reluctant to incur debts which they would be unable to repay, for, as Samoans argue somewhat starkly, *ua sili le oti i lo le masiasi*, death is preferable to humiliation. Paradoxically, the same circumstances led to increasing dependence on kin for the informal, routine exchanges of goods and services necessary for daily existence.

Even if this crisis had not occurred, other factors might have constrained the elaboration of kin-based activity in the migrant community. More migrants were reviewing the costs and benefits of various forms of *aiga* participation as their circumstances changed and they were forced to choose between the respective needs of their immediate and extended families.

More people became cautious about mobilizing their extended *aiga* and incurring new debts. High levels of intermarriage, particularly with more individualistic Europeans, brought into play new sets of considerations and constraints on participation.

New Zealand-born Samoans, raised in very different circumstances, almost inevitably view kinship differently from their parents. The difference stems in part from socio-economic values stressed in formal education, higher levels of formal education and greater economic and job security, changing composition of social networks and reference groups, high levels of intermarriage, changing patterns of religious affiliation, the emergence of new forms of ethnic identity, and critical reflection on the costs and benefits of participation in the full range of extended kin-based activities.

While many New Zealand-born Samoans continue to participate in extended kin group activities they do so at the request of, and out of a sense of responsibility to, their parents. Their participation is, in many cases, indirect either because of a lack of language and of familiarity with ceremonial forms, or because the demands of work prevent or discourage their direct involvement. Whether their participation will continue after their parents' death is a matter for conjecture. These people, however, continue to value and to exchange information and services and to spend leisure time with family.

Conclusion

These social and economic factors have reduced the numbers of occasions on which the extended family is mobilized and the scale and complexity of those events for which it is. This in turn has attenuated the competitive pressures which drove the earlier social inflation. As economic times get harder *aiga*'s primary role, as a social matrix within which goods and services are exchanged informally and as a 'shelter' which provides some protection from harsh realities of life in a vulnerable position on the edge of a volatile economy, has again become increasingly important. In this respect the role of *aiga* among migrants can be said to have almost turned full circle in the last 35 years. This should not be regarded as the final phase of a process but rather as another stage. If, for instance, the economic circumstances which have produced the current situation were to change, the role of kinship and the form of kin-based organization could easily begin to expand once more. Indeed, with the benefit of experience of unchecked growth and the increasing influence of New Zealand-born Samoans, it could be that those who control kin groups may move back to more sustainable forms of kinship mobilization. This process has influenced the ways in which kinship is organized in Samoa itself for, as noted at the outset, these migrants do not constitute discrete, distant communities but rather a venue in which modification and experimentation take place, legitimated by the necessity of finding Samoan solutions to new social, economic and political realities. Those modifications, which seem to meet new needs, find their way back into Samoan 'traditional' forms. In the social space between these two settlements the meta-culture of contemporary Samoan culture is emerging, neither purely 'traditional' nor purely 'migrant'.

Notes

1. While nothing prevented women from holding power, and in fact women were immensely important, in practice men held most *matai* titles.

2. In fact, settlers were forced to recruit and use more expensive indentured labour from China and Melanesia because of Samoan reluctance to become involved in plantation wage labour.

3. The traditional organizations were the wives of chiefs and orators, *faletua ma tausi*, and the association of village women, *aualuma*, which comprised women born in the village but not in-marrying women.

4. Only a small pool was necessary because they were constantly in circulation and because during this time only relatively small numbers were being exchanged at any ceremony.

5. More chiefs were created after 1962 as a result of changes in electoral legislation, and new chiefs were not necessarily concerned with day-to-day *aiga* management and were able to work overseas.

6. Because migrants' livelihoods were no longer derived directly from the use of land or sea vested in the *aiga* they were no longer legally bound by the authority of the *matai*.

7. This was only partly the consequence of conspicuous display. More members from various sub-lineages had to be included in the wedding parties as recognition of the contributions of their sections of the family.

8. Graduates who discouraged families from spending large amounts of money and who resisted the increasing formality and ceremony were reminded that this was an occasion *for the family*!

9. A number of graduates noted that these occasions gave their parents an opportunity to show friends, family and guests 'how many important people they knew'.

10. See the data in the *New Zealand Official Yearbook 1997*. Wellington: Statistics New Zealand, p. 346; also *Samoan People in New Zealand: A Statistical Profile*. Wellington: Statistics New Zealand, 1995.

11. Between 1991 and 1995 union membership fell from 514,325 to 362,200 and coverage from 35.4 per cent to 21.7 per cent; see *New Zealand Official Yearbook 1997*, pp. 356–7.

References

Bertram, G. and Watters, R. (1985) 'The MIRAB economy in South Pacific microstates', *Pacific Viewpoint*, 26(2), pp. 497–520.

Connell, J. (1991) 'Island microstates: the mirage of development', *The Contemporary Pacific*, 3(2), pp. 251–87.

Gilson, R.P. (1970) *Samoa 1830–900: The Politics of a Multi-Cultural Community*. Melbourne: Oxford University Press.

Gunson, N. (1978) *Messengers of Grace: Evangelical Missionaries in the South Seas 1797–1860*. Melbourne: Oxford University Press.

Hau'ofa, E. (1994) 'Our sea of islands', *The Contemporary Pacific*, 6(1), pp. 148–61.

Hayes, G. (1991) 'Migration, metascience and development policy in island Polynesia', *The Contemporary Pacific*, 3(1), pp. 1–58.

Kramer, A. (1994) *The Samoan Islands*, Volumes 1 and 2. Auckland: Polynesian Press; originally published in 1902 as *Die Inseln Samoa*.

Krishnan, V., Schoeffel, P. and Warren, J. (1994) *The Challenge of Change: Pacific Island Communities in Change 1986–1993*. Wellington: NZISRD.

Macpherson, C. (1973) 'Extended Kinship Among Urban Samoan Migrants: Toward an Explanation of its Persistence', unpublished DPhil thesis, University of Waikato, Department of Sociology.

Macpherson, C. (1984) 'On the future of Samoan ethnicity', in P. Spoonley, C. Macpherson, C. Sedgwick and D. Pearson (eds), *Tauiwi: Racism and Ethnicity in New Zealand*. Palmerston North: Dunmore Press, pp. 107–27.

Macpherson, C. (1991) 'The changing contours of Samoan ethnicity', in P. Spoonley, D. Pearson and C. Macpherson (eds), *Nga Take: Ethnic Relations and Racism in Aotearoa/New Zealand*. Palmerston North: Dunmore Press, pp. 67–84.

Macpherson, C. (1992) 'Economic and political restructuring and the sustainability of migrant remittances', *The Contemporary Pacific*, 4(1), pp. 109–35.

Meleisea, M. and Schoeffel-Meleisea, P. (1987) *Lagaga: A Short History of Western Samoa*. Suva: University of the South Pacific, Institute of Pacific Studies.

Moyle, R. (ed.) (1984) *The Samoan Journals of John Williams*. Canberra: Australian National University Press.

O'Brien, M. and Wilkes, C. (1993) *The Tragedy of the Market: A Social Experiment in New Zealand*. Palmerston North: Dunmore Press.

Ongley, P. (1991) '"Pacific Islands" migration and the New Zealand labour market', in P. Spoonley, D. Pearson and C. Macpherson (eds), *Nga Take: Ethnic Relations and Racism in Aotearoa/New Zealand*. Palmerston North: Dunmore Press, pp. 17–36.

Ongley, P. (1996) 'Immigration, employment and ethnic relations', in P. Spoonley, D. Pearson and C. Macpherson (eds), *Nga Patai: Racism and Ethnic Relations in Aotearoa/New Zealand*. Palmerston North: Dunmore Press, 13–34.

Pitt, D. and Macpherson, C. (1974) *Emerging Pluralism: Samoan Migrants in Urban New Zealand*. Auckland: Longman-Paul.

Shankman, P. (1976) *Migration and Underdevelopment: The Case of Western Samoa*. Boulder, CO: Westview.

15

Islands as Havens for Retirement Migration: Finding a Place in Sunny Corfu

Gabriella Lazaridis, Joanna Poyago-Theotoky and Russell King

Introduction

Amongst the several migration processes affecting various islands around the world, the phenomenon of retirement migration has received little attention, except where the 'retirees' are islanders returning home after living and working abroad. What we are interested in exploring in this chapter is the use of islands as 'retirement havens' on the part of outsiders, usually nationals of wealthy industrialized countries, who seek a relaxing, scenically attractive and climatically benign environment in which to live out the last stage of their lives.

Although little studied, the inward retirement migration of 'outsiders' and foreign nationals to islands is a well-known phenomenon in some parts of the world. For the British there is a well-established retirement migration to places such as the Isle of Man, Isle of Wight and the Channel Islands. More recently, since the 1970s, the British, together with other North European nationals such as the Germans, Dutch and Scandinavians, have settled in the warm and sunny holiday islands of the Mediterranean where many of the retirees had accumulated prior touristic experience. Majorca, Malta and Cyprus are the major island destinations involved in this movement, together with the Canary Islands further south. But many other Mediterranean islands are also affected to a lesser extent, such as Corsica, Elba, Capri and various Greek islands, including the setting for this chapter, Corfu. Before we examine the Corfu case in more detail, we also briefly note the Caribbean case and consider some general research questions concerning the nature of international retirement migration as it affects islands. These issues are then applied to our field research on Corfu.

The Caribbean case has been reviewed by McElroy and de Albuquerque (1988) who have also published what is probably the only detailed field analysis of the economic and social impact of international retirement migration on a small island community – Montserrat (McElroy and de Albuquerque, 1992). In many respects the Caribbean stands in a similar relation to North America as the Mediterranean does to Europe; both of these island-rich basins have functioned over the past forty or so years as 'pleasure peripheries' where sunshine and a warm sea have led to the development of large-scale tourism. And in both cases mass (and elite) tourism has led in a natural progression to the establishment of a more permanent residence on the islands of foreigners who wish to escape the cold north and perpetuate their holiday experiences throughout their later years.

McElroy and de Albuquerque (1988) describe as 'migration transition societies' those Caribbean islands which have seen a 'population substitution' resulting from a decline in decades-old labour-exporting migration and an increase in tourism and in foreign immigration for work and, increasingly, for retirement purposes. This transition has been accompanied by relatively rapid demographic shifts in the faster-growing, high-impact tourist islands (Antigua, Bahamas, Bermuda, Cayman Islands, US Virgin Islands); but the trend can also be observed in low-impact, newer tourist islands such as the British Virgin Islands, Montserrat, Nevis and St Vincent/Grenadines. In the case of the 'Emerald Isle' of Montserrat, the green landscape and the peaceful, safe lifestyle have led, according to McElroy and de Albuquerque (1992), to an harmonious insertion of the retirees into the main residential areas. The authors concluded that the economic impact of low-density long-stay retirement migration was significant and sustainable, resulting in minimal cultural and environmental disturbance: sustainable that is, in the case of Montserrat, until the next major environmental disaster![1]

Several questions arise from McElroy and de Albuquerque's work, and from the other more partial analyses of the impact of foreign retirement migration on small islands (on Malta for instance see Boissevain and Serracino Inglott, 1979; Warnes and Patterson, 1998). Six key issues are set out below and will be used to help frame our empirical data from Corfu.

First, who are the retirees? In addition to the obvious variables of nationality and ethnicity, what are their social backgrounds, and how relevant is a prior experience of visiting the island on holiday in leading to their more permanent settlement there? If such a link is strongly indicated, are all islands which trade on tourism destined to play host to settled communities of retired and semi-retired foreigners?

Second, what appear to be the precise influences and mechanisms by which retired foreigners decide to settle on an island such as Corfu? In terms of the potential attractions of Corfu, what is the relative importance of factors like climate, sea and mountain views, the tranquil pace of life, the friendliness of the local population, and the quality of services such as shops and medical facilities?

Third, what are the retirees' lifestyles on the island? Do they interact in a meaningful way with the islanders, or do they tend to remain within their own group, perhaps developing an 'enclave mentality' based on a kind of neo-colonial sense of superiority towards 'the natives'? Alternatively, are elderly British on Corfu creating a new identity built around the search for 'authentic living' in an idyllic island setting? These notions of the romantic 'island gaze' derive from the pioneering work of Urry (1990, 1992) but must also be built around a deconstruction of retired migrants' narratives which, as we shall see, are often patronising and reveal a neo-colonial mentality (Aldridge, 1995).

Fourth, what are the retirees' precise residential and mobility strategies? Cohen (1974, p. 537) conceptualizes retirees as 'permanent tourists' because they are culturally non-indigenous and their income comes from abroad, but this seems an over-simplistic view which not only denies them the possibility of moving beyond the standard tourist–host interaction (one of temporariness, superficiality and commodification) but also ignores a range of mixed residential and migration/travel strategies such as seasonal residence and regular contact with the home country.

Fifth, how do foreign retired migrants see their future? Is their island home 'forever'? Or do they anticipate a return to their country of origin as various possible later-life events (failing health, loss of a partner) overtake them?

And finally, what are the local environmental, economic and cultural impacts of foreign retirement settlement, and how are these viewed locally, for instance in terms of institutional policies for welcoming, encouraging (or otherwise) this type of in-movement?

Corfu and Retirement Migration

Corfu (in Greek, Kerkyra) is the northernmost of the seven major Ionian islands located off the west coast of Greece. It lies opposite the Albanian frontier and the Greek coast of Epirus. With a total area of 593 sq km, it is 58 km long and its greatest breadth is 27 km. The island is mountainous in the north, lower in the south. Corfu is fertile, well-watered and reputed to have the most attractive countryside of the Greek islands. Some lyrical descriptions of the island's landscapes are contained in the writings of one of its most famous early foreign residents, Lawrence Durrell (1945; 1978, pp. 14–35; 1988, pp. 26–56). The island is quite densely populated with a total population (1991) of 107,592 spread between Corfu town (36,293) and around 250 villages. Figure 15.1 shows some of these features as well as the distribution of retired British residents who are the main subject of this chapter.

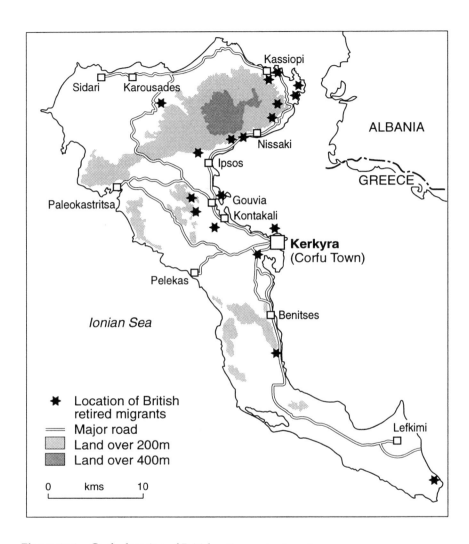

Figure 15.1   Corfu: location of British retirement migrants

Like most of the larger Mediterranean islands, Corfu has had a complex history. In ancient times it was colonized by the Corinthians in 734 BC and then in 435 BC it petitioned for the help of Athens against Corinth, a request which became one of the primary causes of the Peloponnesian War. It then changed hands several times, passing from the Romans to the Goths, Lombards, Saracens, Normans, Genoese and Venetians. From 1401 to 1797 it was an important part of the Venetian Republic (a fact still very much evident in its culture and architecture). Upon the demise of the Venetian Republic in 1797 it was assigned to the French and was part of the Napoleonic Empire until 1815. Subsequently it became a British Protectorate and remained so until 1864 when it was given as a wedding present to King Constantine I of Greece and thus became part of the Greek State. During World War II it was occupied by the Italians and Germans and suffered considerably from their bombings which destroyed many of its fine buildings and landmarks.

The economy of Corfu is mostly concentrated on the agricultural (olive oil, fruits, wine) and service sectors, but for the past 30 years the main contribution to the economy has come from tourism. Corfu island has been a tourist destination since the 1950s and, together with the island of Rhodes and the region of Attica, has become established as one of the main tourist centres in the whole of Greece. One can distinguish two phases in the history of tourist development in Corfu: up to 1975 the island attracted holiday-makers of high and medium incomes; while from 1975 onwards it has been more intensively visited by medium- and low-income tourists, mainly in the form of package holidays, and 75–80 per cent of these tourists come from the United Kingdom (Ionian Islands Region, 1988, p. 7). Corfu has become a popular tourist destination through the involvement of the large tour operators and also because of the active promotion of tourism as a means of regional development followed by the Greek government. Since the 1980s, however, the effects of mass tourism on Corfu have led to a deterioration of the quality of the environment, both built and physical, and of some of the services provided.

Official statistics give only a partial insight into the structure of the island's economy and employment. According to 1980's data in a report authored by Tsartas *et al.* (1990), the contribution of the tertiary sector to the island's GDP is by far the largest (58 per cent, cf. industry 20 per cent and agriculture 22 per cent) despite the fact that most people (43 per cent) are employed in the primary sector. This apparent mismatch can be explained both in terms of the dynamic development of mass tourism and in terms of the phenomenon of multi-occupation whereby people declare that their main occupation is in the primary sector whereas in reality a large proportion of their income derives from undeclared activities in the tertiary sector. For example, in 1988 there were 95,000 beds for tourists in Corfu; 40,000 of these were illegally operated which generated undeclared earnings (Tsartas *et al.*, 1990, p. 47).

In recent years, Corfu has attracted two different types of long-term visitors. The first type consists of labour migrants who come to Corfu in search of work, mainly males from Albania (working in the primary sector, the construction industry and taking up low-skill employment in the tourist industry) and females from the Philippines (domestic workers). The second type are retired migrants from Western Europe who come to Corfu in search of 'a place in the sun' (Williams *et al.*, 1997). It is this latter group of migrants which this study focuses upon.

Because the topic of retired migrants in Greece has not been systematically researched, there are serious data inadequacies which make it difficult to establish a clear picture of the scale of the phenomenon. There were major disparities between the little statistical data made available to us from the British Consulate in Corfu, the data provided by the editor of a local British magazine, and the actual numbers found 'on the ground' during our fieldwork.[2] The British Consul estimated (in summer 1997) that out of a total British community of about 3000 there are approximately 800 British retired migrants in Corfu; the editor of the British magazine estimated the retirees to be less than 100 (this figure was acknowledged to possibly include people who have left the country or died); and the actual number estimated during fieldwork dropped down to 45–50! One reason for these wide discrepancies is the fact that the higher figures included many older people who came to Corfu via marriage or at a young age in search of better work prospects, whereas our own estimate is restricted to those Britons who migrated to Corfu at or just after retirement. Another reason is that, under the provisions of the Treaty of Rome, the Single European Act of 1986 and the Maastricht Treaty, citizens of the EU have the right to enter and reside in any member state. There are no checks at the frontiers. It is therefore difficult for the British Embassy to keep track of the numbers of Britons coming in and out of Greece as consulate registration is no longer obligatory. Desire to avoid administrative procedures which are often time-consuming, unclear and complicated might also induce settlers to remain 'clandestine', although during the winter, when all the tourists have gone, such 'hidden' residents would be easier to spot. And resident status is essential for having access to health services in Corfu, for having a car with Greek plates or for being able to purchase a house. The above factors underline the reasons why there are no reliable official estimates of retired Britons living in Corfu and of changes in their numbers over recent years. Conversations with locals, however, indicate that the popularity of Corfu as a destination for retired migrants has probably decreased in recent years; some of the reasons for this will become apparent later.

There are also definitional difficulties which are fundamental here. Who are these migrants? Is someone who retires at the age of 40 and migrates to Corfu and sets up his/her own business a retired migrant? Is a retired person who bought a house in Corfu and resides there for a few months each year a retired migrant, or is he/she a tourist, or maybe something between the two? These – and other – types of circumstance make this a rather difficult population group to define and identify.

The next sections of the chapter examine the experiences of the retired migrants themselves. We start by considering migration motives and then move on to issues of health care, citizenship, integration and language. In the conclusion we return to the questions posed in the introduction and essay some further general comments about the 'island gaze' of foreign retired settlers.

Motivations for Migration to Corfu

Older Britons in Corfu are mainly married couples. They are socially diverse in that some are rather well-off with well-funded pensions and relatively high levels of savings, whereas others have limited financial resources. The majority are early-retired, still active, people. The lack of attraction of Corfu to the 'old elderly' can be attributed to the lack of health and care facilities to cater for long-term physical and/or mental incapacities, a point that will be elaborated upon.

The decision of the young-elderly to migrate to Corfu, as this is reconstructed by the retired migrants themselves, is the result of multiple and interconnecting reasons related to both push and pull factors. Amongst such factors are shifts in earnings (including unemployment or redundancy), changes in life-cycle (for example, no more mortgages to pay or family to support), connections with people who had visited the island, a warm climate, and the tranquil pace of life. One interviewee summed up some of the positive attractions in the following words:

> It is a wonderful place to be. It is still in my eyes one of the few remaining civilized parts of Europe. You have a lot of personal freedom, a great deal of safety, you don't feel threatened and I am afraid I can't say that for many other places in Europe now. And the weather is good. I think it is wonderful!

Primarily, however, there seems to be a natural progression from being a tourist paying holiday visits to the island, to buying a 'second home' in Corfu in order to visit the island more frequently and escape the cold winters in Britain, to then deciding to move permanently to the warmer climate, picturesque landscape and slower pace of life. In other words, pre-retirement visits to the island pave the way to retiring in Corfu. One of our interviewees described this staged evolution from an initial holiday on Corfu in the mid-1970s to an eventual permanent settlement in the following words:

> My wife went to Crete and she loved it – she went with my sister because we had a shop and it was very difficult to have a holiday together. She came back from Crete very happy with it ... fell in love with Greece generally ... Next year, she booked a holiday for the whole family in Crete but when it came to the crunch the flight wasn't full so they had to cancel it, the whole holiday, and they said 'You can go to Corfu if you like'. So we thought 'Better than nothing' and they sent us to Ipsos ... There were only three tavernas in those days ... we had a very

enjoyable time with lots of interesting experiences with locals because they weren't quite so commercially minded then; they were still very Greek ... On the first night I said 'I'd love to live here' ... I was very reluctant to go home at the end of the two weeks which is most unusual for me because I was a workaholic in England and so we started working towards buying a house here ... We are not like many people here with huge bank balances to help them out ... we put away as much money as we could towards this aim ... When we got home Sally said to me 'I think you should go back and not return until you've bought a house' ... So I went out with Yannis one day and drove around the island and the first place we stopped was K. It was really dilapidated. We went on around the island but everything else I looked at wasn't very interesting except K. which came a bit as a surprise. So I bought it without my wife ever seeing it ... bless her ... we used to come out for two to three weeks every year ... So that was how we bought our first house and eventually when I retired we came to live here.

Similar experiences were narrated to us by several other retired migrants:

We first came in 1965 ... the following year we went to Cyprus but we didn't like Cyprus because in the summer it is very brown and very bare ... so we came back to Corfu and after that wherever else we went we got in at least one week in Corfu. And gradually we said that we always come back here and so maybe we will buy a bit of Corfu. And in 1972 we bought the land ... we looked around for a year or two, we went to Akis who was then in business acquiring land for the English people and we got this through him ... and the house was finally built in 1974 ... In between we let it to friends. When I retired at the end of 1976, we came here for good.

Yet another interviewee said:

The reason why we decided to come is because we wanted a summer house in the Mediterranean area ... we preferred that to the Caribbean and other places. We liked Majorca very much because we had friends there and it is a pretty island and then by chance we came to Corfu and we found it simply lovely ... in those days it was unspoilt ... Then a friend said 'Why don't you try and buy something?'... we bought this for holidays and then when my husband retired we came and spent more time ... I don't live in England any more.

And finally:

We came here for three weeks in 1976 and we have never been anywhere else ... In 1979 we had the opportunity to buy the house up in the mountain. I was an engineer ... I moved here because of the miner's strike, which made a big difference to my business. I decided to sell it and come and work in my house.

These typical accounts show that purchase of a house, often initially as a second home, is one of the side-effects of the growth of mass tourism in Greece since the 1960s, and this constitutes the common step towards more permanent migration. From short-stay tourists moving to Corfu seeking pleasure and escaping everyday life, they become 'residential tourists' (Betty and Cahill, 1999) committed to living for part of the year in Corfu; subsequently they become 'retired migrants' living almost permanently in Corfu. Whilst this sequence represents the progressive realization of an 'island dream', it also has a more pragmatic basis: initial holiday visits allow acquisition of experience of the local facilities available to residents and tourists. Also important are the role of family and friends who make recommendations to buy accommodation, and the help of local people in searching for a property in an attractive location.

Whilst most of the retired people had spent their working lives exclusively in Britain, a minority had lived abroad for many years before retirement.[3] Only a small number of Britons living in Corfu continue to maintain a residence in Britain, thus sustaining a mobile lifestyle by dividing their time between the two. The small number can be largely explained in terms of the high cost of air fares from Corfu to the UK and the complications with travelling. Direct scheduled flights from Corfu to the UK were stopped a few years ago, thus compelling people to go to Athens and change plane, which often means several hours of waiting. Direct, and cheaper, charter flights are only available during the summer, but these are difficult to book at the Corfu end of the trip. Many of the interviewees drew an unfavourable comparison with Spain: 'Spain has always been probably the most popular country to migrate to from England because it is easier to get to; you get there for about £110 return, so it is a quicker journey and all around the year. One of the biggest problems with Corfu is that it has got no winter flights....'

Corfu has never been marketed as a retirement destination; indeed until relatively recently there was a discouragement of foreigners buying property in the area because it is considered to be *paramethoria periohi* – a strategically sensitive frontier area (in Corfu's case, close to Albania). In order to buy a house or land, a company had to be formed with a Greek partner and then the company could make the purchase. This process often took some effort to set up and undoubtedly acted as a barrier to the mass settlement of foreigners in Corfu. Another complication was the position of the Greek partner if the house was sold on. These obstacles meant that, prior to the Single European Act of 1986, foreign settlement on the island was mainly restricted to semi-derelict houses in the inland areas. This investment in old rural dwellings in need of restoration and refurbishment finds parallels in the British purchase of cottages and farmhouses in Tuscany (King and Patterson, 1998) and in parts of rural France (Hoggart and Buller, 1995), but is in contrast to the situation in the Balearic Islands or the Costa del Sol where many retired Britons live in purpose-built residential estates (Williams *et al.*, 1997). The economic and ethical dimensions of foreigners 'buying up' part of the island patrimony of Corfu will be briefly touched upon in the conclusion.

Health and Care Issues

Retirement migration to an island like Corfu remains potentially an attainable dream as long as the migrants enjoy reasonable health. When illness and physical fragility become a problem, the perceived limitations of medical and other forms of care on the island induce a radical reassessment of the situation. We found that when retired migrants become frail they usually return to their former country of residence, mainly for family support and in order to make use of more familiar and better health services. 'I am quite fit; I can do anything that I want to. If I was not fit then we would probably spend less time here,' one retiree said. A woman whose husband died a couple of years ago said: 'This is my home; there is no other place I can possibly live, unless I were incapacitated'. When asked where she would go if she became infirm she indicated that she would probably move within reach of a daughter who lived in Australia. Another woman said: 'I think that if you are very ill, you need to be able to feel comfortable with the nurses and the doctors, understand what they are saying to you. And it worries me that maybe I am going to have to go to hospital and I am not going to be able to explain ... this will push us to go'.

This pattern of planned return or onward migration alleviates part of the burden on health and care services available for Corfu's elderly population. Like other parts of Mediterranean Europe, Greece has adopted a 'residual' or 'rudimentary' approach to welfare provision characterized by 'underdeveloped social services, entitlements related to the employment and contribution record and an emphasis on the role of the family as the core unit of social care' (Katrougalos, 1996, p. 41). Within the context of the present economic recession in Greece and the static or declining tourist flows to Corfu, there is little chance that extra resources will become available. Therefore elderly residents will continue to have to rely on family support. Clearly this is not an option available to the British, most of whom live far from family, friends and other support networks. These are people who have not only moved away from potential sources of informal care but have also moved away from a system with more generous social provision for the elderly, with community and day-care services, into a country with minimal public provision. Thus family and friends are important in the retired migrants' decision to stay or to move back to the UK, which indicates a chain return migration influence. On the other hand, their links with the UK in terms of familial and cultural ties may have diminished over the years, making return migration more problematic.

As the ability of Greek families to look after their more elderly and frail members decreases due to urbanization, decline of extended family networks and increase in women's employment, other types of social services such as charities may have to fill the gap which exists between the lack of state welfare provision and the inability of family to compensate. At the moment, however, such charities exist in Corfu at a superficial level and cater only for the very poor and (mostly orphaned) children. There are two residential homes, one privately run by the Catholic (Maltese) nuns and one state-owned, but there is no home help or nursing care offered apart from domestic workers available to those who can afford to buy their services. The state-owned home's reputation is not a good one in so far as the quality of services offered is concerned. One may ask, then, what about those who cannot afford to utilize private services? The outlook for these pensioners is rather bleak.

With regard to the primary and specialist health care provision available locally, the British Consul in Corfu commented as follows: 'The general hospital here is very good; the doctors are fantastic, the nurses are not bad at all. If you have private insurance you can go to the clinic here but they lack equipment so they have to refer you to the general hospital. If you are really ill, they will transfer you to Athens.' Retirees' experiences, however, suggest that there are some problems with the health care offered locally at the state (general) hospital. One man said: 'My wife was sick ... We got to the hospital here and I was a bit disgusted ... she was lying in a corridor, fortunately she was unconscious ... the visitors were all around this area where she was, no screens ... They were all sitting around smoking, coughing'. The service offered by the private clinic seems better. A retiree whose wife broke her leg said: 'The service at the polyclinic was extremely good; I don't know about the hospital.' Others said: 'We went to the polyclinic for the first time and we were satisfied; they were very good for emergency treatment ... we heard tremendous horror stories about the state hospital ...' Another retiree characterized his experience with the state hospital as 'colourful' and he continued:

> To cut a long story short, I think the next time I am ill here I'll just lie down on the pavement and die quietly ... The private clinic is not much better either. It is a nicer building, you have TV and air-conditioning, but the nurses leave a lot to be desired ... It is the one thing that frightens everybody here irrespective of nationality: serious illness ... what on earth do you do?

Others had had better experiences: in the words of one woman: 'I visited people in the hospital and I was shocked by the situation. By the same token, last November I had a serious operation and everybody recommended me to go to England. I decided to have it here in Corfu and I was very, very pleased'. In fact, British retirees have access to local family doctors, the local health centre, the polyclinic and the state hospital. Their health requirements are covered by form E111, which enables citizens of the EU to receive emergency treatment whilst on holiday in an EU country (see Betty and Cahill, 1999 for details), or by private insurance. All retirees we interviewed have private insurance in the UK and make use of either the private clinic or a local doctor. Most seemed impressed by the fact that they got immediate treatment.

Access to medical care is by and large seen as quicker than in the UK. The majority of interviewees mentioned that healthcare provision has changed markedly in recent years in the UK and has departed from the assumption that the state is the main provider of services. So the emphasis in Greece upon payment for services in the private and state sectors does not come as a major surprise to them. Indeed some of them had opted for private health insurance because of the long waiting lists for seeing a medical specialist and for operations in the UK. Because, however, the local health system lacks up-to-date technology and expertise, in serious cases patients have to fly to Athens for medical treatment or cross over to the mainland city of Ioannina. As mentioned above, although the medical attention in the private clinic is generally described as being good, the care in the state hospital is generally regarded as rather poor. Most respondents were concerned about the inadequate availability of nursing care and the lack of home-based care services for the elderly. Social service provision too is inadequate, with very few and rather unqualified social workers in the island. This reflects the emphasis on the family in caring for elderly relatives in Greece.

In sum, the majority of the retired migrants we interviewed do not make heavy demands on the public funds of the host state; they are covered by sickness insurance and have financial autonomy, since they are recipients of a retirement pension or old-age benefits, and the amount they receive is sufficient to avoid them becoming a burden on the social security system of Corfu.

Citizenship and Identity

Although the attractions of the island setting of Corfu – its landscape, climate and way of life – are seen as major reasons for settling there by the majority of foreign retirees, the people we interviewed were strikingly indifferent to, and ignorant of, EU regulations which would allow them to exercise their rights to vote and stand for office in local elections – in other words, to participate more fully in the society they seem in many respects to admire. Some were unaware of their new rights to vote under the Treaty of Maastricht, whilst others maintained that they did not know enough about Greek politics and local affairs – and moreover seemed uninterested in acquiring more knowledge. 'I can't vote, I'm not Greek,' was a frequent answer. On the other hand, there were signs of a rising awareness. One retired Englishwoman said: 'I know about the European elections, but I didn't know about local elections ... it's far more interesting voting for the local mayor than voting for a European Member of Parliament ... I think I'll make enquiries.' Only one couple – ex-civil servants – did exercise their rights to vote: 'We made it our business to find out how to do it ... we registered at the local office here a couple of months before the elections ... it was quite an experience: we felt proud of that!' For the time being the majority of the British retirees in Corfu remain a passive force since the majority voluntarily abstain even from discussions about local politics. One reason they gave was that 'Most of our English friends are not interested in politics; our Greek friends, well, if we were not sure what their political persuasion was, we wouldn't say anything'. Another reason that some gave was related to complications anticipated with obtaining the correct papers required for enrolling themselves on the electoral register.

The above discussion and quotations from interviewees raises interesting questions about identity, integration and 'home'. What are the self-images of these migrants? Most said that they are British but feel at home in Corfu. Characteristically, one retired migrant said: 'For the last eight years it has been home ... when we have been in England on holiday, we've said "We are going back home tomorrow".' Interestingly, none of the interviewees said that they are *Europeans* living in a member state of the EU; they do not seem to embrace a concept of European citizenship. 'Do you feel integrated?', we asked. A typical response was: 'Ah! I don't think that one is ever integrated. You will always be a *xenos* [a foreigner], won't you?' Another person said: 'If you live in a place you do not necessarily feel that you become part of it. You are the guest of the country you choose to be in ... but I don't think that you could be considered of that country'. The same point was developed further, but in a rather patronising tone, by an English woman who came to live in Corfu a couple of years ago:

> I don't feel Greek. I am very British at heart really but I am a British person living in a foreign country and I accept that I am a foreigner and that I've got to understand the Greek way of life, which is what I've done all the way through, you know ... I never said 'Oh, that wouldn't have happened in England', or 'I wouldn't do that' ... yet I am not threatened by being a foreigner ... I am sort of enfolded in a way with the Greek people ... I speak to them, they speak to me, they come and knock at the door, they give us eggs and things like that, they are lovely.

Integration: A Multi-faceted Process

Unlike Britons who concentrate in retirement havens in other parts of Southern Europe, often residing in enclaves in leisure-oriented coastal locations surrounded by other English-speaking residents, marked spatial concentrations of Britons in Corfu are not evident. Most of the retirees whom we interviewed live scattered in rural areas of the island, but nevertheless within easy reach of the sea. As the map (Figure 15.1) shows, they mainly live on the sheltered east side of the island and tend to avoid the more remote western and interior districts; though it appears that most of the retirees live along the coast, in fact nearly all live at least a couple of kilometres inland.

Coastal areas of Corfu have not attracted a large number of retirees, first because property prices along the Corfu coastline are exorbitant. Second, these are very noisy areas during the tourist season. Third, in the inland villages one can find a rural idyllic setting and live in a relaxed lifestyle, whereas living in a larger town might be more socially isolating. A possible fourth reason was given to us by a retired migrant woman in the following words:

> I have always felt that the English people that do come here to live are a different type of person than those that go to Spain or Portugal to live. They are very often loners. They do not want to be involved with other English and make ghettos of English ... they want to do their own thing. They like Greece, they like Greek things and they want to be here ... they do not want to create a little England. Whereas the people who go to stay in Portugal, maybe, this is only maybe, still want to take that bit of England with them ... they still want to be English, so they gravitate together ... English people do not do that in Corfu ... it is a certain type of person that comes to Corfu.

Another retired migrant attributed this difference to a different reason: 'It is entirely due to how nice the Greeks are; they don't make you feel like an outcast so that you have to seek out your own kind. I would put that entirely down to their courtesy, *filotimo*.' Unlike retirees in Majorca, Tenerife and other places, Britons in rural Corfu are not surrounded by English restaurants and pubs and recreational clubs. British residents in Corfu have not tried to organize social clubs to cater for their needs, help them settle in the new environment and offer help in times of need or emotional upset. Only the Anglican Church in Corfu can be said to provide a framework where friendship links can be made with other expatriates; but the church in Corfu is not active in providing community services and care assistance for retired migrants.

Many retirees told us that they value the Greek *filoxenia* (hospitality) and *filema* (the gift of hospitality). Some retirees have built close friendships with fellow villagers. This point was illustrated by a woman who had been widowed while in Corfu: 'I couldn't leave here after he died, this is my home. I love it. The people are so nice, always friendly. When I lost my husband ... they were so kind. Every day outside my door there was a pie, something was brought as a little gesture of friendship ... all my friends are Greek. Well, mostly!' Amongst the Corfu islanders, once a friendship is built, then one has the right to ask for a *hari* (a casual favour). So a network of reciprocities is gradually established and little by little foreign residents get more and more incorporated into the Corfiot society.

However, despite the 'geographical integration' with the Corfiots described above, difficulties arise from linguistic differences in terms of the amount and the quality of communication between the Britons and the islanders. 'The language is a barrier for me; I tried to learn it when I was younger but it was difficult,' a retiree told us. 'I've never really mastered the language. I can shop, I can speak a little bit of Greek, but I can't hold a conversation. I never say too much because if you say too much they think you can speak.' In fact, most of the people we interviewed mentioned that they felt that it was necessary to learn Greek in order to integrate more fully into the local society, but they found the task daunting. Another retired migrant said: 'It's not that we don't want to communicate. To sit down and study a language is too late for us; if I have to communicate with anybody here, English first, if that doesn't work then German, then French, then I go to my little Italian and then to my even smaller amount of Greek and then I draw pictures and somehow I get what I want.'

This lack of fluency in the Greek language seems to be of great concern to British residents on the island and influences their capacity to adapt to the new environment. It may mean that access to local services and communication with local bureaucrats is difficult, though this is not always the case as many local services in Corfu are oriented towards the tourist industry and many locals like to practise their English language skills. Indeed some retired migrants try to talk to local people in Greek and they often receive a reply in English. As one elderly English woman said: 'The first time that Vassilis replied to me in Greek, I thought he is paying me a great compliment by doing that; now he always talks to me in Greek.'

Retirees and the Local Bureaucracy

Foreign residents in Corfu have contact with the local authorities when having to deal with issues related to purchase of property, residence permits, health services, obtaining Greek registration plates for their car, and so on. Our research has elicited mixed feelings and experiences; encounters with some officials are positive, with others extremely negative. 'When it came to things like dealing with the customs authority, they were not exactly bending over to help us,' one recently retired couple told us. And the wife went on to relate a long and complicated story about how their car had been impounded pending payment of a hefty import fee. In fact it was remarkable that problems regarding the importation and registration of cars by British retirees were stressed as a major difficulty of coming to live in Corfu by virtually all the people we interviewed. Other interviewees had similar stories to tell about the difficulties of bringing in furniture and other possessions. Another woman described her problems with the local bureaucrats as follows: 'Bureaucracy is slow here. When you go to the tax office you seem to have to go from here to there to there. You haven't got a central person who can deal with all your problems ... that's it, Corfu is Corfu.'

Whilst most departments within the local government offices now have people who speak English, retired migrants often do not know what is going on as prior information received from the Greek Embassy in London is often misleading or insufficient, and this also applies to the much-lamented problems of importing vehicles. Confusion resulting from problems with local bureaucrats and police often leads to feelings of discrimination; some British said that they could not get away with things a Greek would be allowed to. The retirees felt generally unable to conduct their own affairs, to gain access to the offices of senior officials and to make suggestions of offering secret gifts and money bribes without embarrassment in exchange for a favour (*efkolia*). Yet of course it is well-known in Corfu and in other parts of Greece that such 'presents' are commonplace, part of the normal currency of everyday life. In a purely practical sense, they are necessary to avoid interminable delays and to get public servants to relax certain regulations whose administration and interpretation lie in their own hands. Yet outsiders always seem to be at a disadvantage in this system. Direct bribery is dangerous as one does not know who is corruptible. Moreover, a government official sees no reason why he or she should assist a foreigner, someone without friends of some standing in the island who in turn might be able to help them. So the problem is not necessarily lack of command of the Greek language as such, but rather lack of command of the patronage system. The best way to get through the bureaucracy is to have someone Greek on your side who has contacts in the right offices and can find out what needs to be done. Most frequently this person is a lawyer who amongst other things acts as an intermediary between the retiree and the Greek authorities. Some respondents said that they developed a close friendship with their lawyer.

Concluding Remarks

In this conclusion we return to the six key questions posed in the introduction to this chapter and attempt to synthesize some answers from the evidence we have accumulated from the field research on Corfu. In addition we draw out some more general conclusions about the nature of insider–outsider relations in the context of foreign residential settlement on islands, and about the economic and ethical dimensions of the phenomenon.

First, we have a clear picture of the social background of the retired British migrants on Corfu. They are generally well off but not necessarily rich, and come from an employment background in business, the professions and other middle-class occupations. They were attracted to settle on Corfu by a prior experience of holidays there, often via the intermediate step of buying a holiday home which then became a permanent residence after retirement. Many retirees have had a long involvement with Corfu, with holiday visits starting as early as the mid-1960s.

Second, the motivations for retirement migration to Corfu are likewise clear from our interviews. Eschewing the densely settled, noisy and expensive coastal strip, retirees have opted for quiet lifestyles in the peaceful rural interior, living in or close to villages where they can also enjoy the friendliness and hospitality of the local Corfiots. Climate, of course, retains an overall importance in their migration decision-making, but this factor is combined with more human and aesthetic quality-of-life variables – not unlike the retired British settlers in Tuscany (King and Patterson, 1998) or in rural France (Hoggart and Buller, 1995). Whilst retirees' reactions to local services and the bureaucracy often constitute a negative aspect of their overall experience of living on Corfu, nearly all are on balance satisfied with their decision to come and settle on the island.[4]

The third question listed in the introduction concerns retirees' lifestyles: here there are several dimensions to draw out. The spatial dispersion of retirees amongst the local population encourages a degree of social interaction with the Corfiots, despite the social and cultural differences between the retirees and the host community. Although individual experiences naturally vary, some barriers to full integration are posed by language and by the retirees' inability or unwillingness to get involved in community affairs and local politics. This links back to questions related to the original motivation for migration, and to issues of identity and self-image (cf. Aldridge, 1995).

Most of the elderly British who have settled in Corfu are people who already had a 'feeling' for Greece and the Greek way of life, having holidayed in Corfu and perhaps also in other parts of Greece. Whilst they tend to consider Corfu now as their 'home', they retain a lot of their British cultural background, partaking of the local culture only rather selectively. For them Corfu is a rural island idyll which they are certainly able to be a part of, but only to the extent of enjoying the climate, sea, landscape and selected aspects of local society such as the islanders' genuine friendliness. Quite apart from the language barrier, they can never be fully immersed into the local family-centred life or into local politics. Partial and superficial integration combined with a peaceful coexistence with the islanders is, perhaps, the best that can be achieved. In the words of one of the interviewees, an elderly lady:

> I wanted to integrate with the local community but I've been unable to do so because of my inability to learn Greek. I exclude myself because I am shy and because I can't speak the language ... The local people are helpful and generous ... We know most of them, we've been in their houses ... but we don't socialise with them to a large degree ... But that's not their fault ... we feel that the language is the biggest barrier.

And, of course, language difficulties constitute a major problem in negotiations with the local authorities, against whom retirees feel powerless and disadvantaged. A further layer of exclusion is produced by migrants' inability to manipulate local patronage systems and the culture of 'gifts'. These problems persist despite the ability of many retirees to speak rudimentary Greek and the growing fluency in English of many Corfiots, especially those who are involved in the tourist industry and those who have a good level of education, such as professionals and younger people.

From what has been said above, it is obvious that the elderly British in Corfu do not have an enclave mentality or behaviour. They do not deliberately seek out the company of other British expatriates and they do not live in the kind of residential concentrations that are known to exist in the islands of Cyprus and Malta, or in southern Spain and the Canaries (Williams *et al.*, 1997). The British in Corfu are not reconstructing 'a little England in the sun' in the way the British in southern Spain have done (Betty and Cahill, 1999).

In a similar way, retired Britons in Corfu are not networked through British clubs and associations which, elsewhere (in Malta, Cyprus, Tenerife etc.), provide a forum for interaction and solidarity. In the Costa del Sol it is possible to observe that 'club life' enables British residents to consolidate their Britishness whilst simultaneously enjoying the advantages of living in a foreign country – climate, relaxed way of life, cheaper cost of living etc. (Betty and Cahill, 1999; O'Reilly, 1995; Rodríguez *et al.*, 1998). In Corfu, whilst it is true that elderly British settlers do bring with them their values, concepts and other aspects of their accumulated experience of being British, they retain these at an individual level; such values are not reinforced by the production of a 'club identity' as elsewhere. On the other hand, it is also true that Corfu is in many ways (and not just geographically) closer to Britain than other parts of Greece and has had already, even before the post-war boom in mass tourism, an infusion of British culture – including a cricket pitch!

Our fourth question was about mobility strategies. As O'Reilly (1995) has pointed out with reference to the British living in Fuengirola in southern Spain, a whole spectrum of mobility types can be theoretically identified, ranging from permanent residence overseas to various divisions of time between the country of origin and the destination, including regular seasonal migration. The elderly British living in Corfu are mainly permanent or virtually permanent residents. Only a minority retain a home in Britain to which they might move for some parts of the year, although in the past, before retirement, some had originally obtained their Corfu property as a seasonal or holiday residence.

This links to the fifth question, concerning the future. Problems with the lack of specialist health care on Corfu seemed to lead most respondents to the view that one day in the future they will probably be forced by illness or extreme old age to return to the UK or to move to a close relative elsewhere.[5] This expectation seems to conflict with the fact that most have 'sold up' in Britain and hence have no base of their own to return to. For most, however, the intention is to 'hang on' in Corfu as long as they can and then to return to sheltered housing or a nursing home in Britain, drawing on savings and private health insurance. Hence the shift from active to frail old age is likely to be accompanied by a return to Britain because of poor nursing and rehabilitative care. Decisions to return on the part of the British in Corfu do not emanate from a retention of a 'myth of returning home' nor 'an idealized view of home'. For them, home is Corfu. On the other hand, the peace and security of living on Corfu has recently been threatened somewhat by the violent situation in nearby Albania.[6]

Finally, there are the wider economic, environmental and cultural impacts of the retired British in Corfu. The economic issues are fairly simple to set out. The purchase and restoration of semi-derelict rural houses brings money into the hands of the owners – mostly local farming families – and creates a certain amount of local employment in the construction trades. Such properties are saved from complete ruin and, if the restoration is carried out sympathetically, there is an overall contribution to conserving the rural architectural patrimony that is tending to disappear with the preference of local Corfiots for building new houses. Next, there is an inflow into the island of British pensions. The scale of these transfers is likely to be rather modest given the small number of retirees living in Corfu, but the exact amount of this income is unknown. Retirees' impact on the island's health and welfare provision is also limited because of the small number of foreign retirees and because most migrants rely on self-provision by means of private health insurance schemes; those who are very ill or require long-term care usually return to the UK. There is probably scope for the formation of a British migrant association which would help to cater for the needs of the older British residents. Such an organization could co-operate with other voluntary organizations such as the church or the Red Cross to provide care and assistance with transport, interpreting and dealing with the local bureaucracy, and hence to compensate for the inadequacies of the local welfare system. This would also help to redress a fundamental dilemma in the behaviour of foreign retirees which our interviewees seemed only vaguely aware of. On the one hand they have chosen to migrate to the tranquillity and friendliness of a beautiful, sunny island; on the other hand, they have not fully grasped the fact that the same 'dream island' setting implies small-scale and often rudimentary services, especially health care, which their increasingly advanced years are likely to need.

Corfu is one of many Mediterranean islands which have experienced mass tourism leading to a significant residential settlement of foreigners. Some of them come to retire, others to open businesses trading off the tourists and foreign residents, yet others as a result of settling down with local partners following holiday romances. Sizeable numbers of British residents are known to exist on several other Greek islands such as Zakinthos and Cephalonia (Ionian islands to the south of Corfu), Crete, Mykonos and

Rhodes, not to mention other Mediterranean and Atlantic islands like Cyprus, Malta, Majorca and the Canary Islands. In an increasingly affluent and leisure-oriented society (at least in Europe and North America), it seems that such retirement colonization of islands is destined to grow (Williams *et al.*, 1997) – despite some evidence of a decline in interest in Corfu.

How to interpret this colonization from a sociological and an ethical point of view is more problematic. Certainly there are complex moral and ideological questions surrounding the phenomenon of settling on an island in order to 'possess' part of what might be termed the 'island gaze'. Ownership of a complete small island – off the west cost of Ireland, or on the coast of Maine, or in the Caribbean – is one of the foibles of the super-rich who may come to regard 'their' islands as mere playthings. On the other hand the behaviour of retired persons who choose to settle on islands is rather different and in a sense more deeply meaningful. Whilst it is true that they are 'buying in' to the 'island experience', their cultural objectives have more to do with the characteristics of the 'romantic' and 'anthropological' gazes described by Urry (1992). Urry has pointed out the growing significance of the romantic tourist gaze as a search for authenticity and distinctiveness, a distancing from the mass experience of packaged tourism. Interestingly Corfu (like some other Mediterranean islands) has both these elements of the tourist experience: mass packaging along some coastal strips where the island 'has indulged in grotesque acts of self-mutilation ... a gauntlet to be run only by the aesthetically blind' (Mayes, 1996, p. 53); and scattered settlement in the interior hinterlands where terraced olive groves, peasant agriculture and 'authentic' village life are the main (perceived) constituents of the rural scene. And similarly – again following Urry (1990, 1992) – we can distinguish in Corfu the 'collective, spectatorial' gaze of the package tourist, whose main experience is of the beach, bars, discos and tavernas of the coastal strip, from the 'solitary, romantic' gaze of the retired resident in the hills whose vision embraces not only the detailed scanning of the beautiful countryside (often from a 'superior' position on an elevated slope) but also, in the dealings with the 'friendly villagers' and the regard for the 'quaint' vernacular architecture, a quasi-anthropological participation in the richness of human culture (cf. Aldridge, 1995, p. 431). A close reading of the interview quotes reveals a sense of self-righteousness on the part of people who are certain (or who like to give the impression that they are certain) that they have done the 'right thing', as well as a rather patronising attitude towards the locals who are seen either in the mould of 'helpers' or 'friendly peasants' or as troublesome people who get in the way of the British achieving their private ambitions (importing a car) or getting the best health care.

But despite this distinction between the two 'styles' of tourism – visitor package and permanent residential – both ultimately involve the commodification of the place that is the island of Corfu. The political economy of tourism, including retirement settlement, likens international tourism to a form of colonialism which is ultimately destructive of the places which are 'consumed' and where 'local voices' are largely silent: a rather different interpretation of the phenomenon of foreign retirement settlement on islands than the earlier-noted conclusion reached by McElroy and de Albuquerque (1992) for Montserrat. For the time being the Corfu authorities welcome British retirees who are seen as bringing benefits to the island and who cause no problems, unlike illegal immigrants from Albania and other 'third countries' (Lazaridis and Romaniszyn, 1998). Only when the Albanian situation stabilizes will Corfu be seen as an unproblematic destination for foreign retirement migration. Even then, it will surely take some time before the foreign settlers are perceived as a real economic and cultural threat – as recent events have shown to be the case with the German residential and business colonization of Majorca.[7]

In fact the Majorcan case shows more clearly than that of Corfu how the relationships between islands and migration, mediated through ties of dependency and peripherality, have come full circle. At an early phase islands, as overpopulated peripheral spaces whose economic systems have been distorted by dependency on colonial or metropolitan regimes, have few options but to fall back on 'trading' their remaining resource, people, by emigration. Later, islands discover that they have another tradeable resource, their insularity. Hence they appeal to tourists (including Cohen's residential 'permanent tourists') seeking an idyllic away-from-it-all environment in which to relax, especially if the island has an attractive landscape, warm sea and pleasant climate. Like the affluent English portrayed by Peter Mayle (1990, 1992) in Provence, for the well-off (and not-so-wealthy) Britons sequestered in their villas and restored cottages in the Corfu hills, the island is a 'positional good' whose value is lessened by sharing it with others, such as tourists (cf. Aldridge, 1995, p. 416). This leads to the incorporation of islands into the pleasure periphery of metropolitan economies and cultures and to a form of neocolonialism which involves an uneasy compromise between preservation of island authenticity as part of the 'island gaze', and the modernization, consumption and possession of islands as a result of tourism and the residential settlement of immigrants.

Notes

1. As McElroy and de Albuquerque note in a postscript to their 1992 article, Hurricane Hugo destroyed or damaged 95 per cent of the island's buildings in September 1989. The majority of the foreign retirees departed for North America or other Caribbean islands. Some had returned to resettle in Montserrat when the recent cycle of cataclysmic volcanic eruptions rendered much of the island uninhabitable. For further details on the recent eruptions see Stuart Philpott's chapter (Chapter 7) in this book.

2. Research on Corfu took place in 1997 and comprised a range of field techniques, the principal one being in-depth biographical interviews with a sample of eighteen elderly British residents. The interviews were taped and transcribed, allowing verbatim quotation of British residents' experiences in Corfu. We view these interviews as research which relies on the validity of the analysis rather than on the representativeness of the sample and of the events. Additional interviews were conducted with key informants such as the priest of the Anglican Church in Corfu, the British Consul, the editor of a British magazine, a Greek lawyer who looks after the British community's affairs, a British tour operator and an English teacher. The field work was carried out by Joanna Poyago-Theotoky and Gabriella Lazaridis, and incorporated some of the research questions derived from another project developed by Russell King (joint with Allan Williams and Anthony Warnes) on international retirement migration to Spain, Portugal, Italy and Malta, the first results of which are now being published (see King *et al.*, 1998; King and Patterson, 1998; Warnes and Patterson, 1998; Williams *et al.*, 1997; Williams and Patterson, 1998). Hence the research in Corfu is a complement to this larger comparative project. In the account of Corfu which follows, the identities of all the people interviewed have been concealed in order to protect their anonymity.

3. Although the small size of our Corfu sample prevents a confident judgement, we have the impression that the retired British population on Corfu contains a smaller proportion of 'lifetime expatriates' than other research has found in Tuscany (King and Patterson, 1998), Malta (Warnes and Patterson, 1998) and the Algarve (Williams and Patterson, 1998).

4. Of course, we acknowledge the fact that those who became dissatisfied are likely to have returned to the UK, and hence our impression might be biased. However, few such cases were mentioned to us by our informants.

5. In this context it is tragically ironic that a young English woman who recently collapsed on Corfu and who was given an emergency operation to save her life was found by the Corfiot surgeon to have died because of surgical clips and a large piece of gauze left in her stomach by a previous operation in England. See the many reports in the English press on this in May 1998, e.g. N. Bunyan: 'Surgical clips were inside Corfu death girl for 7 years', *Daily Telegraph*, 19 May 1998.

6. Some of this violence seems to have impacted directly on elderly Britons holidaying or living on the Ionian Islands. In September 1996 a Briton was murdered by Albanians stealing his boat moored at Gouvia marina near Corfu town; in May 1997 two Britons were robbed at gunpoint by Albanian pirates as they sailed off the coast of Corfu; and in March 1998 a retired British couple were murdered in their house on Cephalonia, south of Corfu. Press reports taken from the *Guardian*, 30 September 1996, 21 May 1997 and 20 June 1998; *The Times*, 14 March 1998. These British press reports, however, failed to mention certain salient details, namely that the British owner of the moored boat was also armed, and that the plundered yacht was sailing in Albanian waters and had ignored warnings.

7. Majorca has been a major 'sun, sand and sea' destination for German tourists since the 1960s, but in recent years a major development of German residential settlement has occurred, as well as business ownership of tourist enterprises, and, now, of land in all parts of the island, including the interior. According to one

source, there are 50,000 German-owned properties on the island. Most recently, the German residents have established their own political party to contest the 1999 local elections. See M. De la Pau Janer: 'Mallorca, objectivo alemán', *Periódico*, 29 December 1997, p. 5; also I. Traynor: 'German Balearic bliss fades', *Guardian*, 22 November 1997, p. 17.

References

Aldridge, A. (1995) 'The English as they see others: England revealed in Provence', *Sociological Review*, 43(4), pp. 415–34.

Betty, C. and Cahill, M. (1999) 'British expatriates' experiences of health and social services on the Costa del Sol', in F. Anthias and G. Lazaridis (eds), *Into the Margins: Migration and Social Exclusion in Southern Europe*. Aldershot: Avebury.

Boissevain, J. and Serracino Inglott, P. (1979) 'Tourism in Malta', in E. Kadt (ed.), *Tourism: Passport to Development?* New York: Oxford University Press, pp. 265–84.

Cohen, E. (1974) 'Who is a tourist? A conceptual clarification', *Sociological Review*, 22(4), pp. 527–55.

Durrell, L. (1945) *Prospero's Cell*. London: Faber & Faber.

Durrell, L. (1978) *The Greek Islands*. London: Faber & Faber.

Durrell, L. (1988) *Spirit of Place: Mediterranean Writings*. London: Faber & Faber.

Hoggart, K. and Buller, H. (1995) 'Retired British home owners in rural France', *Ageing and Society*, 15(3), pp. 325–53.

Ionian Islands Region (1988) *Regional Development Plan*. Corfu: Ionian Islands Region.

Katrougalos, G.S. (1996) 'The southern welfare model: the Greek welfare system', *Journal of European Social Policy*, 6(1), pp. 39–61.

King, R. and Patterson, G. (1998) 'Diverse paths: the elderly British in Tuscany', *International Journal of Population Geography*, 4(2), pp. 157–82.

King, R., Warnes, A.M. and Williams, A.M. (1998) 'International retirement migration in Europe', *International Journal of Population Geography*, 4(2), 91–111.

Lazaridis, G. and Romaniszyn, C. (1998) 'The experience of Albanian and Polish undocumented workers in Greece', *Journal of European Social Policy*, 8(1), pp. 1–22.

Mayes, I. (1996) 'Saints alive', *Guardian Weekend Magazine*, 4 May 1996, pp. 52–5.

Mayle, P. (1990) *A Year in Provence*. London: Pan.

Mayle, P. (1992) *Toujours Provence*. London: Pan.

McElroy, J. and de Albuquerque, K. (1988) 'Migration transition in small northern and eastern Caribbean states', *International Migration Review*, 22(1), 30–58.

McElroy, J. and de Albuquerque, K. (1992) 'The economic impact of retirement tourism in Montserrat: some provisional evidence', *Social and Economic Studies*, 41(2), pp. 127–52.

O'Reilly, K. (1995) 'A new trend in European migration: contemporary British migration to Fuengirola, Costa del Sol', *Geographical Viewpoint*, 23, pp. 25–36.

Rodríguez, V., Fernández-Mayoralas, G. and Rojo, F. (1998) 'European retirees on the Costa del Sol: a cross-national comparison', *International Journal of Population Geography*, 4(2), pp. 183–200.

Tsartas, P. *et al.* (1990) *Social and Economic Effects of Mass Tourism in Corfu and Lasithi*. Athens: EKKE-EOT. [in Greek]

Urry, J. (1990) *The Tourist Gaze: Leisure and Travel in Contemporary Societies*. Sage: London.

Urry, J. (1992) 'The tourist gaze and the environment', *Theory, Culture and Society*, 9(3), pp. 1–26.

Warnes, A.M. and Patterson, G. (1998) 'British retirees in Malta: components of the crossnational relationship', *International Journal of Population Geography*, 4(2), pp. 113–33.

Williams, A. M., King, R. and Warnes, A.M. (1997) 'A place in the sun: international retirement migration from northern to southern Europe', *European Urban and Regional Studies*, 4(2), pp. 115–34.

Williams, A.M. and Patterson, G. (1998) 'An empire lost but a province gained: a cohort analysis of British international retirement migration in the Algarve', *International Journal of Population Geography*, 4(2), pp. 135–55.

Index

acculturation 67
Achill Island 12–13
Africa 77, 79, 83, 154, 168
agriculture 6, 9–10, 12, 15–16, 29, 35, 48, 59–60, 63, 77–8, 118–20, 131, 138–41, 149–50, 198, 228–9, 236, 258
aid 14, 44, 83, 94 n.1, 151–2, 156, 180
Albania 301, 305, 315, 317, 318 n.6
Alcatraz 21
alcohol 207–8
alienation 173
ambiguity 20–1, 173, 182–3, 188–91, 274–5, 316
American Samoa 237
Anguilla 6, 22, 125
Antigua 19, 152, 154, 165, 177–94, 298
Anu, Christine 21, 195–212
Aran Islands 29, 35, 39, 46–7
art 182, 185
Aruba 5, 141–2, 145, 147, 166
Asia 154
Atlantic 6, 16
Australia 1, 4, 23 n.1, 28, 195–212, 216, 219–21, 231, 236, 237, 241, 306
authenticity 204, 221, 259, 278, 317
autobiography 19–21, 38–43, 121–2, 163–74, 177–93
Azores 6, 16, 55–76

Bahamas 298
Bahrain 4
Banaba (Ocean Island) 4
Barbados 17, 19, 121, 125, 130, 161–2, 166–7, 171–2
Bermuda 6, 22, 118, 121, 133, 298
Bhabha, Homi 191
Blasket Islands 9, 11–12, 19, 36, 38–43, 45, 50
Bonaire 6
Bonin 221–2
Bougainville 5
Braudel, Fernand 3
Brazil 59
Britain see United Kingdom
Buffong, Jean 165–6
bureaucracy 7, 14, 147–8, 197–9, 311–12

California 61, 64–7
Canada 5, 19, 56–7, 62, 66–8, 72, 95–113, 130, 133, 141, 150–1, 155–6, 170–3, 181, 241
Canal Zone 121, 167
Canary Islands 6, 8, 16, 28, 314, 316
Cape Verde 8, 11, 14, 16–17, 76–94
Capri 20, 297
Caribbean 5–8, 10–11, 16, 19–20, 22, 115–94, 298, 304
see also names of individual islands
Caribs 180

cartography   216
Cayman Islands   8, 18, 22, 298
Channel Islands   6, 8, 297
Christmas Island (Indian Ocean)   12, 202
citizenship   18, 117, 122, 125, 152, 158, 210,
     308–9
Clarke, Austin   170, 172–3
class   102–5, 139, 143, 147–9, 317
Clipperton (Pacific)   4
Cocos (Keeling) Islands   12, 202, 220
colonialism   3–7, 15–16, 19, 21, 57, 59, 77–8,
     117–21, 138–58, 168–74, 178–93, 198–9,
     202, 208–10, 216–29, 280–2, 300–1,
     316–17
community   50, 97–8, 102–11, 208–9, 242–4,
     262–71
Cook Islands   8, 14, 17, 21, 285
cooperatives   43–4
Corfu   6, 8, 297–319
Corsica   3, 297
Costa Rica   121, 164
Cuba   8, 121, 134 n.5, 141, 154
culture   8, 96, 132, 146, 153–8, 209, 244–6
Cyprus   297, 304, 314, 316

demography   6–8, 10–13, 79–93, 115
Denmark   7, 18
diaspora   13, 20, 22, 27, 117, 154–7, 236–8,
     249–51, 277–9
disasters *see* natural disasters
Dominica   165
Dominican Republic   118, 120–1, 130, 133,
     134 n.5, 141

ecology   227–9
economic change   6–7, 11, 14–15, 45–6, 73,
     92, 259, 277, 301
education   18–19, 94 n.1, 143, 147–8, 169–70,
     172–4, 179, 219–20, 224–5, 241, 250, 273,
     291
Elba   297
employment   6–7, 9–10, 12, 15, 18, 46, 66–7,
     72–3, 104, 117, 127, 130–1, 149–50,
     196–201, 259, 292, 301
England *see* United Kingdom
environment   2–3, 7, 9, 16, 77, 103, 178,
     227–8, 230, 258–9, 301, 313
ethnicity   66–7
Europe   8, 64, 79, 216, 241
European Union   44, 73, 302, 308–9
exile   162, 190, 192, 193 n.5, 210, 216

family   143–4, 186–7, 193, 246–9, 260, 285–6
famine   27, 78, 80, 82
Faroe Islands   5, 7
Fiji   4, 13, 255–76
fishing   5, 16, 35, 42, 46, 48, 78, 100–8, 111,
     197, 201–2
food   77, 128, 131–2, 147, 156–7, 198
France   4, 7, 82, 138, 305, 313, 317
friends   109–10, 309–11, 313

gender   20, 109–10, 124, 129, 178–94, 241–2,
     245–7
     see also women
Germany   280–1, 317, 318 n.7
Glissant, Edouard   180
Gozo (Malta)   17
Grand Manan (New Brunswick)   16–19,
     95–113
Greece   3, 6, 8, 297–319
Greenland   1, 7, 18
Grenada   165–6
Guadeloupe   7, 19, 121, 162, 165
Guam   22
Guernsey *see* Channel Islands
Guiana   118, 139

Haiti   8, 20
Hau'ofa, Epeli   23, 235, 270, 272, 277
Hawaii   4, 61, 65–7
health   11, 306–8, 314–15
home   20, 204, 210, 221, 309
houses   14, 68, 71, 117, 124, 127, 129, 149,
     151, 304–5, 315

Iceland   5
identity
     cultural   17, 20, 66–7, 206–9, 219, 243–5,
          271–3
     ethnic   17, 155, 181–93, 207, 216–17, 224,
          250, 293
     individual   17–19, 97–110, 166, 178,
          181–93, 204–5, 245–51
     national   154–5, 157, 163, 178, 203,
          209–10, 220–1, 237–8, 243–5, 249,
          309–11
     spatial   48, 96–8, 118, 154–5, 158, 168,
          203–4, 209, 262–3
ideology   17, 56, 59, 68–9, 129, 134, 148,
     178–80, 210, 214, 216, 229
illness   124, 140, 153, 198, 306–8
imagination   162, 182–3
income   84, 120, 129, 198, 200–1, 271, 273

Indian Ocean   3–4, 6–7, 125
industry   45, 78, 131
insularity   1–2, 17–18, 21–2
Internet *see* telecommunications
investment   15, 66–7, 71–2, 127–8, 149
Ireland   6–7, 19, 21, 27–54, 138
Isle of Man   6–8, 297
Isle of Wight   297
Italy   82, 84–5, 305

Jamaica   20, 121, 130, 139, 154, 162, 164,
      166–7, 171, 173, 239
Jersey *see* Channel Islands

Kincaid, Jamaica   19–20, 165, 177–94
kinship   13–15, 35, 56, 71, 74, 96, 186–7, 205,
      237, 244, 260, 262–75, 277–95
Kiribati   4

Lamming, George   19, 162, 166–9, 192
land tenure   29, 59, 63, 77, 126–7, 139–40,
      202–5, 210, 218, 236, 262–3, 279
landscape   182–3, 299, 313
language   17–18, 38–9, 44, 46, 67, 197, 205–9,
      213, 216–31, 240–1, 244, 246, 278, 293,
      310–14
literacy   61
literature   19–21, 38–43, 161–94, 299, 317

Mabo, Eddie   195–212
McKay, Claude   171–2
Madeira   6, 65
Majorca   297, 304–5, 310, 316–17, 318 n.7
Malta   3–4, 11, 17, 19, 297–8, 314, 316
Malthus   79–81, 92
markets   92, 131
marriage   35–6, 68, 99, 100, 139, 143–4, 146,
      246–7, 263–4, 278, 289–90
Marshall Islands   8, 23 n.1
Martinique   7, 19, 121, 162, 165
Mauritius   3, 6, 11
Mediterranean   3–4, 6–8, 297–8
Melanesia   4–5, 195–7, 199, 204–9, 219–20,
      224, 255–76
Micronesia   8, 22
military presence   4–5, 48–9, 57, 172, 217
mining   4–5
missions   197, 208, 219–20, 224, 280
mobility (social)   66–7, 146–9
modernity   97–8, 102, 108, 111, 210, 221
Montserrat   6–7, 10, 12, 15, 117, 137–59, 165,
      298

music   20–1, 39, 132, 152, 157–8, 205–9, 246
Mykonos   13, 315

nationalism   72–3, 201–2
natural disasters   7, 10, 60, 73, 78, 83, 117,
      137–59, 317 n.1
Nauru   4
Netherlands, The   7, 82
Netherlands Antilles   118, 122–4, 127, 130,
      141–2, 145, 147, 165–6, 169
networks   56, 74, 125–7, 132–3, 144, 199, 249,
      274–5, 290–1, 306
Nevis   6, 10, 15, 115–35, 141, 143, 152, 181,
      298
New Brunswick   95–113
New Caledonia   4
New York   177, 180, 182–91
New Zealand   7, 11, 28, 236–7, 241, 281–2,
      284–94
Newfoundland   5
Niue   12, 14, 17, 277, 285
Norfolk Island   18, 21, 202, 213–34
North America   4, 8, 19, 64, 73–4, 298
nostalgia   210–11

Oceania *see* South Pacific
oil   5, 122, 141, 165, 201
Orkneys   5

Panama   121, 141, 164, 180
Papua New Guinea   4–5, 13, 195, 201, 220,
      224
Paracel Islands   5
penal settlement   217–18, 228
Philippines   4, 301
Pitcairn   213, 216, 218–19, 221–3
politics   4, 43–5, 72, 97, 105, 124, 158, 246
Polynesia   6–7, 12, 14, 205, 216, 235–53
Ponam (Papua New Guinea)   13
Portugal   55–79, 82, 310
post–colonialism   178–93
postmodernism   102
poverty   72, 122
Puerto Rico   7, 154

race   139, 158, 182–8
racism   7, 140, 171, 185–6, 191, 246
real estate   5, 8, 21, 150–1, 304–5
religion   96, 151, 209, 218, 229, 246, 293, 310
remittances   11, 13–14, 17, 59, 68, 73, 79,
      82–3, 92, 117, 127–9, 142, 146–7, 149–50,
      181, 186, 237

resettlement   12, 23 n.1, 28, 36, 38, 152–3, 199, 218–19, 228
resistance   95–6, 100–11, 180–2, 207, 225–6, 244, 282
retirement   4, 6, 8, 52, 108–9, 150–1, 201, 297–319
Réunion   6–7
Rockall   5

Saba   6, 12, 169
St Barthélemy   6
St Eustatius   6
St Helena   6, 22
St Kilda   23 n.1, 28, 36
St Kitts   115–26, 130, 152
St Maarten (St Martin)   6, 8, 130
St Pierre and Miquélon   5
St Vincent   121, 165, 171, 298
Samoa   13–14, 17, 19, 239, 244, 247, 277–95
São Tomé   16
secession   72–3, 201–2, 219–21
self–reliance   35, 102, 124
Semple, Ellen C.   2–3, 5
sexuality   246–8
Shetlands   5, 17, 97, 103, 105
slavery   3, 6, 11, 15, 20, 77, 117–20, 138–9, 180
Socotra   3
Solomon Islands   5
South Africa   28
South Pacific   3, 6–7, 16, 20, 221–95
Spain   305, 310, 313–14
Spitzbergen   4
sport   271, 314
Spratly Islands   5
subsistence   6, 12, 197, 236, 238, 258
sugar   6, 10, 16, 65, 118–21, 128, 138–41, 199

Tahiti   222–3, 227
tax havens   6, 8, 14, 18, 22, 215
technology   22, 111, 140, 235
telecommunications   22, 133, 152, 156–7, 220, 235–53
television   22, 132, 134
Timor   5
Tokelau   12–13, 285

Tonga   13–14, 22, 235–53
Torres Strait   16, 21, 195–212
Tory Island   27, 39, 46
tourism   1, 6–8, 10, 13–14, 18, 22, 38, 46–8, 72–3, 78, 100–1, 106–8, 117, 130–1, 178, 180, 187–8, 215–16, 229–30, 298, 301, 316–17
trade   73, 77, 118, 166
transport   5, 29, 44–5, 50, 73, 92, 98, 106–8, 142, 144, 286, 305
Trinidad   5, 118, 129–30, 139, 162, 165–8
Tristan da Cunha   12
Turks and Caicos   6, 8
Tuvalu   4

unemployment *see* employment
unions   141–2
United Kingdom   1, 7, 10, 12, 115, 117, 124–33, 138–58, 168–93, 236, 281, 302–17
United Nations   77, 202, 220, 260, 282
United States of America   7–9, 13, 28, 40, 57, 62–9, 72, 81, 121, 124, 126–7, 130–4, 141, 145, 147, 150–1, 155, 164, 167, 171–3, 182–3, 236–7, 241, 247–8, 281
urbanization   64, 164–5, 200, 207–8, 236–7, 257, 259, 266, 274, 277–8

values   91, 111
Vancouver Island   5
Vanuatu   204
Venzuela   165–6
Virgin Islands   6, 12, 18, 124, 130, 133, 152, 165–6, 298

Wallis and Futuna   13, 22
Western Samoa *see* Samoa
West Indies *see* Caribbean
whaling   61, 81, 92
women   40–1, 66, 68, 82, 84–91, 93, 99, 109–10, 121–2, 127–8, 139, 143, 145–6, 165, 178–94, 228, 239, 241–2, 245–7, 259, 283
see also gender